Maintenance and Function of Biodiversity in Forests

Maintenance and Function of Biodiversity in Forests

Guest Editors

Xugao Wang
Zuoqiang Yuan
Chengjin Chu

MDPI

Basel • Beijing • Wuhan • Barcelona • Belgrade • Novi Sad • Cluj • Manchester

Guest Editors

Xugao Wang
Institute of Applied Ecology
Chinese Academy of Sciences
Shenyang
China

Zuoqiang Yuan
School of Ecology and Environment
Northwestern Polytechnical University
Xi'an
China

Chengjin Chu
School of Ecology
Sun Yat-Sen University
Guangzhou
China

Editorial Office
MDPI AG
Grosspeteranlage 5
4052 Basel, Switzerland

This is a reprint of the Special Issue, published open access by the journal *Plants* (ISSN 2223-7747), freely accessible at: www.mdpi.com/journal/plants/special_issues/D5RFD7G3O2.

For citation purposes, cite each article independently as indicated on the article page online and using the guide below:

Lastname, A.A.; Lastname, B.B. Article Title. *Journal Name* **Year**, *Volume Number*, Page Range.

ISBN 978-3-7258-3692-5 (Hbk)
ISBN 978-3-7258-3691-8 (PDF)
https://doi.org/10.3390/books978-3-7258-3691-8

Cover image courtesy of Zuoqiang Yuan

Contents

plants

Article

Response of Plant Leaf Traits to Environmental Factors in Climax Communities at Varying Latitudes in Karst Regions

Gang Xie [1], Yang Wang [2,*], Zhifeng Chen [2], Yangyang Ji [2], Yao Lu [2], Yi Liang [2], Ruiyu Zhou [2] and Juan Tao [2]

1 Institute of Guizhou Mountainous Resources, Guiyang 550001, China; 13628512665@163.com
2 College of Ecological Engineering, Guizhou University of Engineering Science, Bijie 551700, China; 19192682759@163.com (Z.C.); 15125497829@163.com (Y.J.); 16608803117@163.com (Y.L.); 13595773249@163.com (Y.L.); 18085091257@163.com (R.Z.); 18484643869@163.com (J.T.)
* Correspondence: wangyang20218910@126.com

Abstract: Exploring the changes in plant functional traits and their relationship with the environment in karst climax communities across different latitudes can enhance our understanding of how these communities respond to environmental gradients. In this study, we focus on climax karst climax plant communities in Guizhou Province, China. We selected three sample sites located at varying latitudes and analyzed the variations in functional traits of the plant communities at these latitudes. Additionally, we examined the relationship between functional traits and environmental factors, integrating species characteristics and community structure into our analysis. The results indicated that (1) there were significant differences in both the community leaf aspect ratio and the community-specific leaf area. (2) Soil organic carbon content exhibited significant variations across different latitudes, while soil nitrogen content was notably higher in mid-latitude and low-latitude regions compared to high-latitude areas. The distribution of soil factors was more concentrated in high and mid-latitude regions, whereas low-latitude areas displayed more pronounced variability. (3) The primary environmental factors influencing the climax community in the karst study area included soil water content (SPMC), soil bulk density (BD), soil organic carbon content (SOC), soil nitrogen content (SNC), and soil phosphorus content (SPC). Our findings suggest that karst plant communities exhibit specific combinations of functional traits at distinct latitudes. With increasing latitude, the community demonstrated a gradual shift in ecological strategy from conservative to more opportunistic. Most environmental factors imposed limiting effects on plant functional traits, with plants primarily constrained by BD during growth. Among the responses of plant functional traits to environmental factors, community-weighted leaf area and community-weighted chlorophyll content were the most sensitive to soil conditions.

Keywords: latitudinal gradient; climax communities; functional traits; environmental factors; karst

Academic Editors: Oleg Chertov and Valeria Spagnuolo

Received: 15 November 2024
Revised: 30 December 2024
Accepted: 9 January 2025
Published: 10 January 2025

Citation: Xie, G.; Wang, Y.; Chen, Z.; Ji, Y.; Lu, Y.; Liang, Y.; Zhou, R.; Tao, J. Response of Plant Leaf Traits to Environmental Factors in Climax Communities at Varying Latitudes in Karst Regions. *Plants* **2025**, *14*, 183. https://doi.org/10.3390/plants14020183

1. Introduction

Plant functional traits represent the expression of plant function and morphology under varying environmental conditions. The karst landscape is formed through the geological processes of water interacting with soluble rocks, such as carbonate rocks, gypsum, and rock salt. This interaction primarily involves chemical dissolution and erosion, complemented by mechanical actions, including water erosion, subduction, and collapse, as well as the phenomena resulting from these processes. Erosion is enhanced by mechanical processes such as flowing water, submerged erosion, and collapse, along with the phenomena resulting from these actions [1]. These traits reflect the ecological strategies of species and are frequently utilized to explain and predict community and ecosystem functions [2]. Leaves are essential functional organs in plant photosynthesis and are also the structures that interact most frequently with the surrounding environment. Investigating changes in leaf traits to understand plant adaptation to environmental conditions has become a significant method for studying the structure of plant communities [3]. The climax community refers to the community structure that emerges at the conclusion of vegetation restoration, and its characteristics significantly influence the maintenance and stability of forest ecosystems. Among the factors affecting plant functional traits, soil characteristics are a crucial component [4]. In studies of plant communities, both topographic and soil factors can determine the distribution of functional traits [5], and different soil characteristics can exert varying effects on these traits [6]. Within specific habitats, distinct ecological strategies may converge or diverge among different communities, and functional traits will respond accordingly to environmental conditions [7].

Karst areas are highly heterogeneous and characterized by fragility and complexity. The study of the genesis of plant communities in these regions is significant for understanding the adaptation of plant traits to habitat heterogeneity. Research has confirmed that the leaf functional traits of plants exhibit a pronounced gradient pattern along various environmental gradients, such as climate and soil. Additionally, there are notable differences in the functional traits of different plant species under identical environmental conditions [8]. As a gradient effect encompassing multiple environmental factors, examining the latitudinal patterns of species diversity is essential for exploring the dynamic responses of species to their environment, understanding the processes of community aggregation in different climatic zones, and predicting the future impacts of climate change on biodiversity [9]. Currently, most studies conducted by ecologists on the mechanisms of species coexistence in karst habitats are independent pieces of research that are scattered across various latitudinal belts.

The research conducted by Cheng et al. [10] on the genealogical and functional diversity of tropical cloud forest communities in Bawangling, Hainan Island, demonstrated that habitat filtering is a key factor driving community structure. Kemble et al. [11] found that the genealogical structure of Panama's tropical rainforest plant communities gradually clusters with increasing spatial scales. Additionally, Li et al. [12], in their study of the species within evergreen-deciduous-broad-leaved mixed forests of the Central Subtropical Karst, discovered a strong correlation among different plant functional traits, various functional diversity indicators, and multiple functional redundancy indicators, which existed to varying degrees. In a study examining the phylogeny and functional trait structure of plant communities across different latitudes, Miao et al. [13] found that environmental filtering and competitive exclusion predominantly influenced species aggregation in most communities. Environmental factors such as soil water content, soil acidity and alkalinity, and average annual air temperature significantly impacted the pattern of plant traits. While the findings of these studies provide valuable insights into the community dynamics of their respective zones, the conclusions require further verification and expansion through subsequent research on communities at different latitudes, due to the lack of continuity in the scale of the studies.

This paper focuses on karst climax communities at various latitudes, examining how environmental changes along different latitudinal gradients contribute to regional variations in vegetation and environmental conditions. The study aims to investigate the following questions based on species, traits, and environmental factors. (1) How do the functional traits of karst climax communities vary along the latitudinal gradient? (2) What are the patterns of adaptation regarding functional traits and environmental factors in climax communities across different latitudinal gradients? By addressing these questions, we can evaluate the ecological and evolutionary processes that influence the ecological strategies of plants in karst top communities across varying latitudes.

2. Study Area and Research Methodology

2.1. Overview of the Study Area

Typical karst climax communities in Guizhou Province, China, were selected as the study sample (Figure 1). Field investigations revealed that the soil-forming rocks at the three sample sites were predominantly dolomite and limestone, with minimal soil coverage, more exposed rock surfaces, and shallow soil layers present only in depressions and valley bottoms. In the plant distribution area, the soil-forming rocks include sandstone, sandy shale, siliceous rock, and carbonate rock. The soil is primarily composed of siliceous-aluminous yellow–brown loam, followed by siliceous-aluminous yellow loam, yellow loam, and yellow–brown loam, with the soil layer at the top of the ridge measuring 20 to 30 cm in thickness. We established two criteria for selecting the apex community: first, our research team had previously studied these plots, and second, the biomass of the apex community was the most significant in the entire forest, exhibiting greater species richness, and the plants were larger and less individually differentiated. The external characteristics of the vegetation indicated that trees were the dominant species, while shrubs and herbaceous plants were relatively scarce. After consulting with the staff of the nature reserve, we identified the climax communities in different reserves and used them as sample sites for this study. The climate of the Dashahe area is characterized by high humidity and year-round fog and an annual average temperature of 17.7 °C, with annual precipitation ranging from 1200 to 1360 mm, high relative humidity, and a humid monsoon climate typical of the northern subtropics. The Yuntai Mountain area has an annual average temperature of 14.1 °C, with an extremely high temperature of 39.4 °C and a frost-free period of 209 days; the annual precipitation is 943.3 mm. The Maolan area experiences a middle subtropical monsoon temperate climate, with an average annual temperature of 15.3 °C, an average annual temperature variation of 18.3 °C, and a growing season lasting 237 days. The annual precipitation in this area is 1752.5 mm, and the average annual relative humidity is notably high.

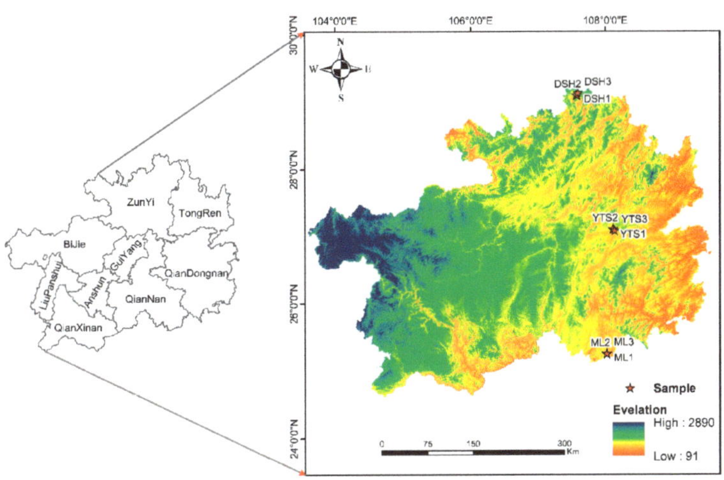

Figure 1. Schematic diagram of the distribution of sample sites in the study area. DSH represents the sample site in the high-latitude region of Dashahe, YTS reaches the sample site in the mid-latitude region of Yuntai Mountain, and ML represents the sample site in the low-latitude region of Maolan.

The minimum sample area was determined to be 900 m^2 (30 m × 30 m) using the 'species-area curve' method. To ensure ecological significance, three sample plots (replicates) were established for each climax stage, resulting in a total of nine sample plots covering an area of 8100 m^2. Initially, we set up the sample plots and assessed the diversity of the plant communities in the study area. During the peak of the growing season, we collected plant leaves and soil samples from the designated plots.

2.2. Sample Setup

The community composition survey referenced the methods of Fang et al. [14]. In each sample plot, nine small sample squares were established, and each square was surveyed for trees, shrubs, and herbs. Specifically, nine tree samples were designated within each plot, with each tree sample covering an area of 10 m × 10 m. Four shrub samples were arranged in each plot along the diagonal, with each shrub sample occupying an area of 5 m × 5 m. Additionally, one herb sample was established within each shrub sample, covering an area of 2 m × 2 m. Thus, in one sample plot, we collected nine tree samples, each measuring 10 m × 10 m, four shrub samples measuring 5 m × 5 m, and four grass samples measuring 2 m × 2 m. We surveyed and recorded all species within the sample plots and measured their respective traits.

The basic information for each sample plot at the terminal stage is presented in Table 1.

Table 1. Basic situation of plots.

Sample Site	Dominant Species	Cover (%)	Elevation (m)	Longitude (°)	Latitude (°)
DSH	*Fagus longipetiolata, Cunninghamia lanceolata, Woonyoungia septentrionalis, Synedrella nodiflora*	85	645.58	107.576389 E	29.102778 N
YTS	*Liquidambar formosana, Rhus chinensis, Styrax confusus, Pteridium aquilinum*	80	667.83	108.116208 E	27.110187 N
ML	*Cornus wilsoniana, Lindera communis, Nandina domestica, Selaginella tamariscina*	90	744.35	108.022222 E	25.258333 N

2.3. Sample Collection and Processing

(1) Plant leaf sampling and processing

Leaf collection is divided into two parts. The first part involves collecting plant leaves from all species present at the sampling site for trait determination. The second part focuses on collecting leaves from the dominant species at the site for nutrient analysis. Prior to this, the dominant species (or those with significant importance values) in each stratum of the community were identified by integrating preliminary plant surveys with species diversity data analysis. Subsequently, samples were collected. In each 30 m × 30 m sample plot containing tree species in the arboreal layer, three to five healthy plants were selected as sample plants. Leaves from the four cardinal directions of the crowns of these sample plants were collected using high-pruning shears, resulting in a total of 27 arboreal sample leaves. The arborvitae leaves were divided into two portions and bagged. A portion of the samples was used to measure blade thickness using vernier calipers. To minimize measurement errors, leaf thickness was recorded on the same day the samples were collected. Additionally, 10 to 20 leaves were prepared for determining leaf area, leaf length, specific leaf area, and other relevant parameters. The second portion was placed in a sealed bag, transported to the laboratory, and dried in an oven at 75 °C until a constant weight was achieved. Subsequently, the dried leaves were crushed in a pulverizer, ground, sieved, labeled, and stored in bags for the analysis of soil organic carbon content (SOC), soil nitrogen content (SNC), and soil phosphorus content (SPC).

(2) Soil Collection

Selection of soils corresponding to the vegetation types of plant communities in the study area. A diagonal sampling method was employed to identify five sampling points within each sample plot, focusing on the soil layer of 0–20 cm. Due to the thin soil layer characteristic of karst areas, the actual depth may be less than 20 cm; thus, the specific depth during sampling shall prevail. The soil samples were mixed in equal volumes to create a composite sample, which was then placed into a sealed bag for elemental analyses, including soil organic carbon content (SOC), soil nitrogen content (SNC), and soil phosphorus content (SPC). Concurrently, a ring knife was used to collect samples for assessing soil water content and soil bulk density. The retrieved soil samples were transported to the laboratory, where gravel and debris were removed. The samples were then air-dried, ground, sieved, bagged, and labeled. The ring knife was dried at 100 °C until a constant weight was achieved, after which the dry weight was determined.

2.4. Determination of Functional Properties and Analysis of Samples

(1) Elemental Determination

The organic carbon content of both plant and soil samples was determined using the potassium dichromate oxidation-external heating method [15]. Plant samples were digested using the H_2SO_4-H_2O_2 method, while total nitrogen was measured by the indophenol blue colorimetric method (NY/T2017-2011) [15]. Total phosphorus was assessed using the molybdenum-antimony colorimetric method (NY/T2017-2011) [16]. For soil samples, total nitrogen was measured using the Kjeldahl method (LY/T1228-2015) [17], and total phosphorus was determined by the NaOH melting-molybdenum-antimony colorimetric method (LY/T1232-2015) [18]. Additionally, the potassium content in both plant and soil samples was analyzed using flame spectrophotometry.

(2) Determination of Functional Traits

Based on the principles of plant growth characteristics, resource acquisition, nutrient content and distribution, ease of trait determination, and reliability, six traits were selected for the assessment of functional traits (Table 2). The determination methods were referenced from the new manual for standardized measurement of functional traits in plants worldwide [19]. Leaf thickness was measured using electronic vernier calipers (Deli, DL91150, Shanghai, China), while chlorophyll content was assessed with a chlorophyll meter (caaKEr, MLCK-A, Beijing, China). Leaf fresh weight was recorded using a 1 in 10,000 electronic analytical balance, and leaf dry weight was determined after drying the samples in an oven and weighing them again with the same analytical balance (Leqi, YT1004, Shanghai, China). Leaf length, width, and area were scanned and calculated using a scanner in conjunction with Photoshop software 13.0 (HP, HPScanJet N92120, Shanghai, China). Additionally, the carbon (C), nitrogen (N), and phosphorus (P) contents, along with their ratios, were obtained through direct measurements.

Table 2. Functional trait indicator selection.

Functional Feature Type	Data Type	Attribute
Leaf thickness (LT)	numerical value	Leaf blade thickness (mm)
Leaf area (LA)	numerical value	Mean leaf blade area of species (cm^2)
Specific leaf area (SLA)	numerical value	Fresh leaf area/leaf dry mass (mm^2/mg)
Chlorophyll (CHL)	numerical value	Leaf blade chlorophyll content (%)
Leaf aspect ratio (LWR)	numerical value	Leaf blade length to width ratio (%)
Leaf tissue density (LTN)	numerical value	Leaf blade dry weight/leaf volume (g/cm^3)

2.5. Data Processing

The community-weighted mean (CWM) of functional traits in karst plant communities was derived from the weighted average of species' functional trait values and their relative abundance [20]. The formula for its calculation is as follows:

$$\mathrm{CWM} = \sum_{\mathrm{i=1 P_i \times V_i}}^{\mathrm{s}}$$

where "S" represents the number of species in the community, "P_i" denotes the relative abundance of species "i", and "V_i" indicates the value of a functional trait characteristic of species "i". The importance value of the arbor layer was calculated as (relative abundance + relative frequency + relative dominance based on diameter at breast height)/3 and shrub and grass layers as (relative abundance + relative frequency + relative cover)/3.

The data were initially collated using Microsoft Excel 2019 and tested for normality and variance using the Kolmogorov–Smirnov normality test before analysis. Data were analyzed using SPSS version 25.0 [21], employing one-way ANOVA and Tukey's HSD for multiple comparisons to assessing variations in leaf functional traits, soil characteristics, and other factors across latitudinal gradients in plant communities. Pearson correlation analysis was conducted to elucidate the relationships between these metrics, with the data presented as mean $\pm$ standard deviation. To further investigate the variation in plant community leaf functional traits and soil factors, we utilized the 'ggcor', 'vegan', and 'ggplot2' software packages in R version 4.3.2 for correlation analyses. We analyzed violin plots using the 'ggpubr' and 'cowplot' packages. Additionally, we conducted Redundancy Analysis (RDA) of environmental factors and plant leaf functional traits across latitudinal gradients using the Vegan software package. Furthermore, we employed 'WGCNA' and 'igraph' for network analyses and 'reshape' for bubble plot analyses. All statistical analyses and visualizations were performed using R version 4.3.2 [22].

3. Results and Analysis

3.1. Changes in Leaf Functional Traits of the Climax Community at Varying Latitudes

As shown in Figure 2, the chlorophyll content of the karst climax community exhibited significant variations across latitudes, with low latitude (36.42) > high latitude (34.97) > mid-latitude (30.29). Leaf thickness was significantly greater at high latitudes (0.25) compared to the other latitudes (both at 0.17). The community leaf area demonstrated a significant difference between mid-latitude and low latitude, with low latitude (22.07) > mid-latitude (20.15). The leaf aspect ratio was significantly different, showing mid-latitude (4.30) > low latitude (3.57) > high latitude (2.79). Significant differences were also observed in the specific leaf area, with low latitude (346.24) > mid-latitude (307.81) > high latitude (199.47). Community leaf tissue density was significantly different between mid-latitude and low latitude, with mid-latitude (2.65) > low latitude (1.61). The distribution of chlorophyll was more dispersed between low and mid-latitudes, while it was more concentrated at low latitudes. The distribution of leaf thickness was more concentrated in the high-latitude cluster and more dispersed in the mid-

and low-latitudes. Chlorophyll distribution was more widely dispersed in the same regional assay, showing distinct dispersion values in high and mid-latitude communities, while it was more dispersed in low-latitude areas. In contrast, the distribution of leaf aspect ratios was more concentrated, displaying distinct discrete values for high- and low-latitude communities, whereas it was more dispersed in mid-latitude regions.

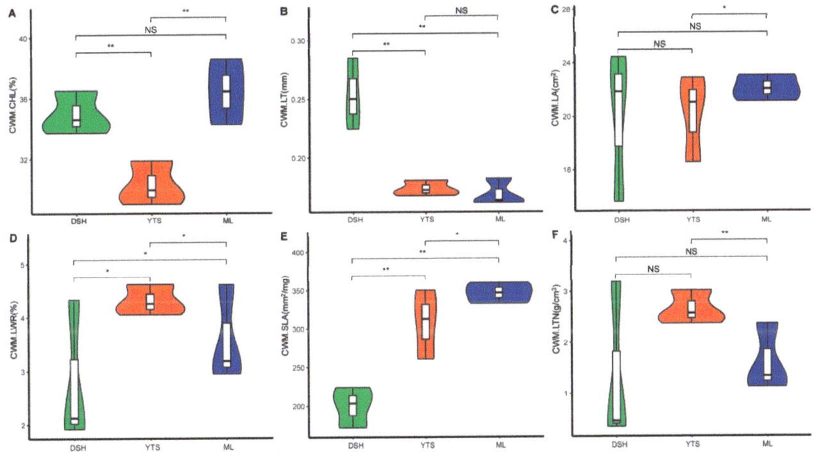

Figure 2. Plot of variation in leaf functional traits in the climax community at different latitudes. In the statistical analysis, a one-way ANOVA was utilized for comparative assessments. The symbol * indicates significant differences ($p < 0.05$), ** indicates highly significant differences ($p < 0.01$), and NS denotes non-significant differences. Three individuals were measured per replicate, with three replicates for each category and three replicates for each plant community. CWM.CHL (**A**)—community-weighted chlorophyll content, CWM.LT (**B**)—community-weighted leaf thickness, CWM.LA (**C**)—community-weighted leaf area content, CWM.LWR (**D**)—community-weighted leaf aspect ratio. CWM.SLA (**E**)—cluster-weighted specific leaf area content, CWM.LTN (**F**)—cluster-weighted leaf tissue density.

The distribution of specific leaf area was more concentrated, exhibiting distinct discrete values for the high-latitude and mid-latitude clusters, while it was more dispersed in the low-latitude region.

3.2. Correlations Between Leaf Traits in the Climax Community

Leaf area exhibited a highly significant positive correlation with specific leaf area and chlorophyll content (Figure 3). In contrast, specific leaf areas demonstrated a highly significant negative correlation with leaf tissue density, chlorophyll, leaf thickness, and leaf aspect ratio. Additionally, leaf tissue density showed a highly significant negative correlation with leaf thickness. Chlorophyll content was found to have a highly significant positive correlation with leaf thickness. Finally, leaf thickness exhibited a significant positive correlation with leaf aspect ratio.

3.3. Changes in Environmental Factors of the Climax Community at Different Latitudes

As illustrated in Figure 4, the soil capacity of the karst climax community exhibited significant variations across different latitudes, with mid-latitude (1.30) > high latitude (1.17) > low latitude (0.68). Soil water content also demonstrated notable differences among latitudes, as indicated by low latitude (31.72) > high latitude (17.38) > mid-latitude (15.75). Furthermore, soil organic carbon content varied significantly across latitudes, with low latitude (22.37) > mid-latitude (11.28) > high latitude (4.31). Soil nitrogen content was significantly higher in mid-latitude (6.66) and low latitude (6.20) compared to high latitude (2.50). Soil phosphorus content showed significant differences across latitudes, with low latitude (1.60) > high latitude

(0.74) > mid-latitude (0.57). The SCN was significantly higher at low latitude (3.62) than at mid-latitude (1.69) and high latitude (1.71). The SCP was significantly greater in low and mid-latitude (19.73) than in high latitudes (5.73). Additionally, the SNP was significantly higher in mid-latitude (11.68) compared to high latitude (3.33) and low latitude (4.47). The distribution of soil factors was more concentrated in the high and mid-latitude regions, while the low-latitude region exhibited more significant discrete values.

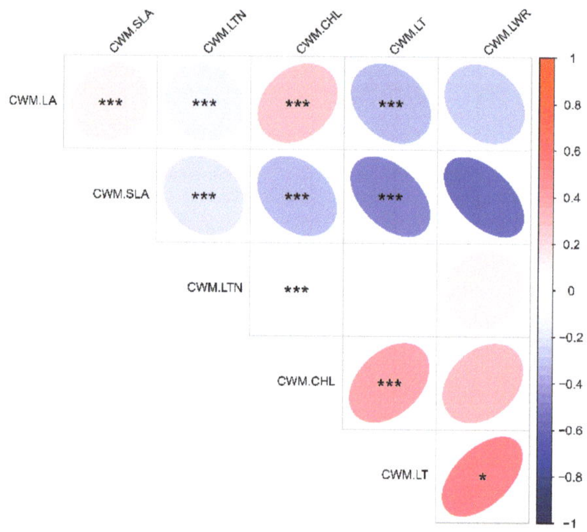

Figure 3. Correlation plots of leaf functional traits in climax communities at different latitudes. * Indicates significant difference ($p < 0.05$), *** indicates highly significant difference ($p < 0.001$). CWM.CHL—community-weighted chlorophyll content, CWM.LT—community-weighted leaf thickness, CWM.LA—community-weighted leaf area content, CWM.LWR—community-weighted leaf aspect ratio, CWM. SLA-cluster-weighted specific leaf area content, CWM.LTN—cluster-weighted leaf tissue density.

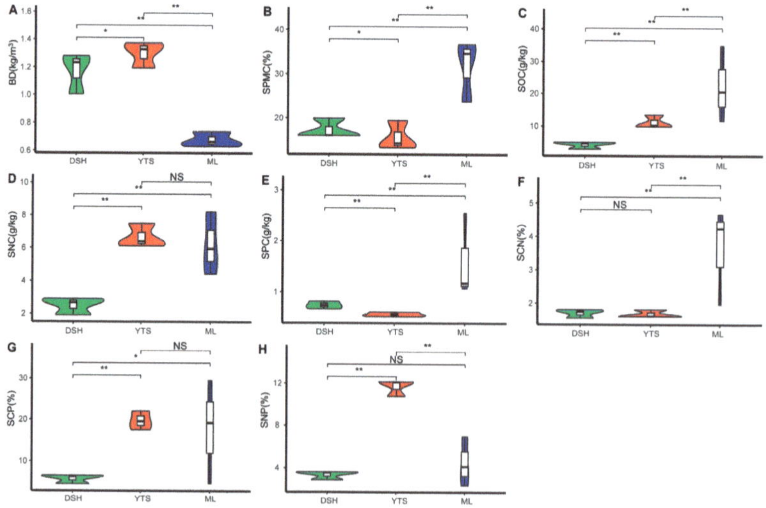

Figure 4. Variation in soil factors in the climax communities at different latitudes. In the statistical analysis, a one-way ANOVA was utilized for comparative assessments. The symbol * indicates significant differences ($p < 0.05$), ** indicates highly significant differences ($p < 0.01$), and NS denotes

non-significant differences. Three individuals were measured per replicate, with three replicates for each category. BD (**A**)—soil bulk density, SPMC (**B**)—soil water content, SOC (**C**)—soil organic carbon content, SNC (**D**)—soil total nitrogen content, SPC (**E**)—soil total phosphorus content, SCN (**F**)—soil carbon to nitrogen ratio, SCP (**G**)—soil carbon to phosphorus ratio, SNP (**H**)—soil nitrogen to phosphorus ratio. Soil nitrogen–phosphorus ratio.

3.4. Relationships Between Environmental Factors and Leaf Traits in the Climax Community

As demonstrated by the Redundancy Analysis (Figure 5), the first axis of environmental factors explained 77.27% of the variance in plant functional traits, while the first two axes of the Redundancy Analysis (RDA) accounted for a total of 88.75% of the explained variance. The primary environmental factors influencing the climax community in the karst study area included soil water content (SPMC), soil bulk density (BD), soil organic carbon content (SOC), soil nitrogen content (SNC), and soil phosphorus content (SPC). Their relative influence, in descending order, was as follows: BD > SPMC > SPC > SNC > SOC. From the distribution of the sample sites, it is evident that in the low-latitude region, the specific leaf area of the community exhibited a positive correlation with soil phosphorus content, soil water content, and soil organic carbon. Additionally, soil phosphorus content and soil organic carbon were positively correlated. The first two axes of the Redundancy Analysis (RDA) explained 88.75% of the total variance, indicating that soil phosphorus content and soil moisture content were the primary environmental factors influencing the area. In the mid-latitude region, community-weighted leaf area and leaf tissue density exhibited a positive correlation with soil bulk density, which was identified as the predominant environmental factor. In high-latitude areas, community leaf thickness, leaf aspect ratio, and leaf dry matter mass exhibited strong correlations and were more concentrated. Furthermore, chlorophyll emerged as a prominent community trait in high-latitude regions, displaying a negative correlation with soil nitrogen content.

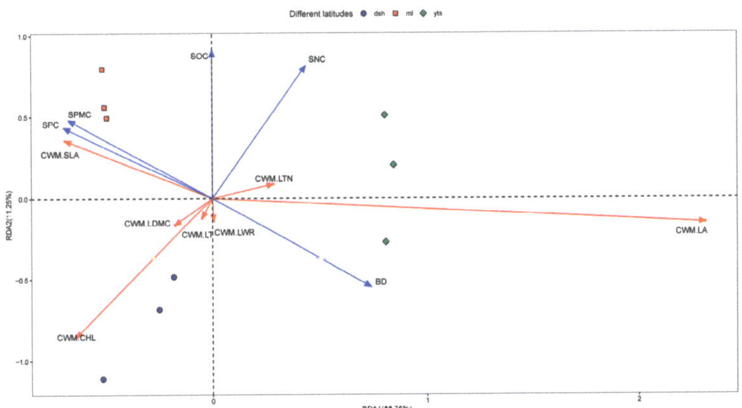

Figure 5. RDA ordination of plant community functional traits with soil factors. The red line is the plant functional trait and the blue line is the environmental factor.

The network analysis of correlations between community traits and leaf-soil nutrients is illustrated in Figure 6A. Measurements with a p-value of less than 0.05 are connected by correlation lines, with red lines indicating positive correlations and blue lines indicating negative correlations. These measurements are categorized into three groups: community traits, leaf nutrients, and soil nutrients. The segments of the correlation lines were counted, resulting in the creation of Figure 6B. As shown in Figure 6B, the size of the circular bubbles represents the number of correlation lines among the three categories, with specific values labeled in the figure. A total of 50 correlation lines were identified, of which 4 (8%) were between community traits and leaf nutrients. Additionally, there were 5 (10%) correlations between community traits and soil nutrients, 13 (26%) between leaf nutrients and soil nutrients, and 28 (56%) correlations among similar measurements (calculated as 4 + 7 + 17).

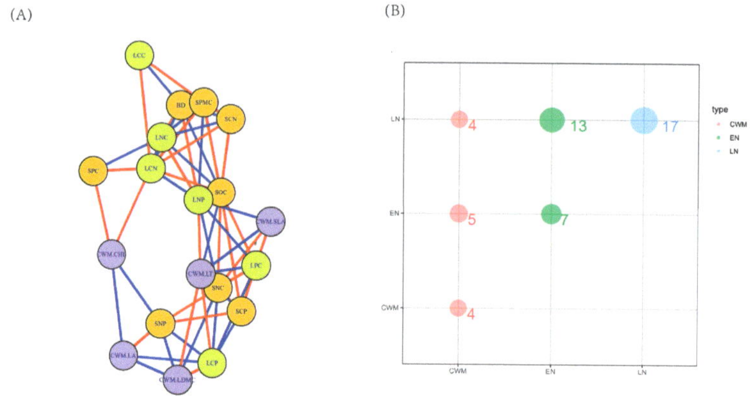

Figure 6. Network relationships of community functional traits with leaf nutrients and soil factors. LCC—leaf organic carbon content, LNC—leaf total nitrogen content, LPC—leaf total phosphorus content, LCN—leaf C:N, LCP—leaf C:P, LNP—leaf N:P, meanings of the rest of the letters refer to Figures 2 and 4. In (**A**), the red line indicates a positive correlation, and the blue line indicates a negative correlation. cwm stands for community functional traits, LN—leaf nutrient content, EN—soil factor. In (**B**), the numbers surrounding the bubbles indicate the total count of correlation links between the variables.

4. Discussion

4.1. Changes in the Laws of Plant Functional Traits in Karst Vertex Communities Across Different Latitudes

The results of this study indicated that the community-weighted mean specific leaf area decreased significantly with increasing latitude, which aligns with the latitudinal distribution patterns of leaf traits identified by Reich [23] and Wright [24]. A lower specific leaf area (SLA) suggests that the plant community is more adept at utilizing environmental resources [25]. The study area is characterized by a typical karst plateau rocky desertification landscape, which has relatively limited soil and water resources. As plant traits change, leaves enhance water use efficiency by minimizing water loss due to transpiration. Additionally, plant communities improve their adaptability to the environment by optimizing nutrient utilization and conserving water. These findings are consistent with the conclusions drawn by Liu et al. [26], Zhang et al. [27], and Li et al. [28]. As latitude increases, hydrothermal conditions diminish, leading to increased environmental stress.

High-latitude polar communities are predominantly characterized by coniferous forests, where coniferous leaves are adapted to minimize water loss by increasing leaf thickness. In this study, the community-weighted mean leaf thickness (CWM.LT) decreased with decreasing latitude, while the community-weighted mean leaf area (CWM.LA) increased with decreasing latitude (Figure 2C). This suggests that plants enhance their leaf surface area by reducing leaf thickness (LT) and increasing leaf area (LA) to effectively capture light resources [29]. Additionally, communities characterized by thinner leaves were observed at lower latitudes, a condition attributed to plant adaptations aimed at minimizing aerobic respiration and conserving nutrients. Furthermore, the results indicated that leaf tissue density was significantly higher in mid-latitudes compared to other regions. This phenomenon is likely due to the pronounced impact of human activities on areas outside the mid-latitude apex community [30]. Species within these stands adapted to their environment under stressful conditions primarily by developing mechanical resistance, reducing nutrient cycling, and employing other defensive strategies against high-input pressures. These strategies included decreasing leaf area (LA), enhancing photosynthetic rates, and increasing leaf thickness density (LTD) to cope with the resource-poor peak scrub landscape characteristic of the region [31]. Species within these stands have adapted to environmental stressors primarily by enhancing their mechanical resistance, reducing nutrient cycling, and employing various defense strategies against high-input pressures (e.g., decreasing leaf area, increasing photosynthetic rates, and enhancing leaf tissue density). These adaptations are essential for survival in the resource-limited peak scrub landscape of the region [32]. Our results indicate that CWM.CHL, CWM.LT, CWM.LA, and CWM.LTN exhibit insignificant changes in trait characteristics across latitudes, demonstrating trait convergence. Trait convergence typically arises from environmental filtering and competitive exclusion. Specifically, environmental pressures can limit the range of viable species, leading to similar traits among different species that adapt to the environment. This phenomenon occurs due to environmental filtering, resulting in trait convergence at the community level [33]. Additionally, trait convergence may also arise when competitively advantaged species with similar trait values exclude competitively disadvantaged species with differing trait values [34].

4.2. Changes in Soil Factors in Karst Vertex Communities at Varying Latitudes

Karst landscapes represent a fragile ecological environment characterized by high habitat heterogeneity and complex geomorphology. The distribution and changes in soil nutrients significantly influence plant functional traits, as well as their growth and development [35,36]. The nutrient content of soil fractions serves as a crucial indicator for characterizing the composition and quality of soil organic matter [37]. The SCN can be utilized to assess the rate of decomposition of soil organic matter, while the SCP indicates the availability of effective phosphorus in the soil. Additionally, the SNP is the most effective indicator for predicting nutrient limitations in forest ecosystems.

In this study, the SCN, SCP, and SNP of the climax community soils were found to be 2.33, 14.41, and 6.50, respectively. These values are relatively low compared to global forest soils [38] (14.5, 211, and 14.6) and national terrestrial surface soils [39] (14.4, 136, and 9.3). This finding aligns with the study by Liu et al. [40], which suggests that the soil phosphorus levels are high while nitrogen levels are deficient, indicating an N-limited type of soil. When comparing soils across different latitude climax communities, it was observed that low-latitude areas exhibit higher soil water content, soil organic carbon, and soil phosphorus content, with significantly higher C:N ratios than other regions. The harsh living environment of the Maolan Karst forests, which grow on dolomite and limestone, results in very little soil presence and a high ratio of bare rock. Soil is primarily found in depressions and at the bottoms of valleys, where a shallow layer of soil exists. Additionally, scattered humus soil can be found in some stone gullies and crevices [41]. The soil surface is often covered with dead leaves, contributing to a soft, moist, and organic matter-rich environment. The distribution of soil factors is more concentrated in high and mid-latitude areas, while low-latitude areas exhibit more pronounced discrete values [42].

4.3. Response of Plant Functional Traits in the Climax Community to Environmental Factors at Different Latitudes

The environmental factors that decisively influence the distribution of plant functional traits are usually different at different scales, and the distribution of functional traits at a given site is often the result of cascading filtration from large to small scales as well as the combined effects of multiple factors [43]. The distribution patterns of community functional traits along spatial environmental gradients at large scales often stem from differences in functional traits within communities [7,24].

In this study, Redundancy Analysis (RDA) demonstrated a strong correlation between soil factors and leaf functional traits, consistent with the findings of Wang et al. [44]. Key environmental factors influencing plant functional traits in the climax community of the karst study area included SNC, BD, SPC, SPMC, and SOC (Figure 5). In low-latitude areas, these environmental factors were influenced by a greater number of additional environmental variables, leading to a stronger dependence of plant nutrient content on soil nutrients (Figure 6). Some of these nutrients were utilized for plant growth and organic matter accumulation, while others helped plants resist nutrient deprivation. The impact of total soil nitrogen was particularly pronounced in low-latitude regions, where communities dominated by glossy balsam and balsam trees exhibited a conservative strategy characterized by a combination of low specific leaf area and leaf thickness [32]. The results of this study also indicated that the functional traits of plant communities at low latitudes may be influenced by soil water content, soil phosphorus content, and soil organic carbon, particularly affecting changes in community-specific leaf area. This finding aligns with the research conducted by Zhou et al. [45]. Specific leaf area serves as a functional trait indicator that characterizes the interactions between plants and their environments; it is generally smaller in barren or harsh conditions and larger in resource- and nutrient-rich environments. Furthermore, both intraspecific and interspecific specific leaf areas tend to increase with rising temperatures but decrease with increased solar radiation, indicating that plant leaves are thinner in warmer environments and thicker in well-lit conditions [46]. Our study revealed a significant decrease in specific leaf areas with increasing latitude, suggesting that habitat barrenness intensifies in karst climax communities as latitude increases. Under similar conditions, temperatures in the karst climax community decline with increasing latitude. The functional traits of plant communities in mid-latitudes may be influenced by soil nitrogen content and soil bulk density, particularly affecting community leaf area and leaf tissue density. The size of the leaf area directly impacts photosynthetic efficiency, which, to some extent, reflects the level of plant production [41]. Mid-latitude regions are more susceptible to human activities, and species within these stands primarily adapt to environmental stressors by developing mechanical resistance, reducing nutrient cycling, and employing other defensive strategies against high input pressures (e.g., reducing leaf area, enhancing photosynthesis rates, increasing leaf thickness density, etc.) [47].

The functional traits of plant communities in high-latitude regions may be influenced by soil nitrogen content and soil organic carbon levels. However, changes in the functional traits of these plant communities—excluding chlorophyll—are not significantly driven by soil factors. This suggests that the structure and complexity of plant communities result from a combination of multiple soil factors rather than the influence of a single soil factor [48].

In this study, the primary soil factors were ranked as follows: BD > SPMC > SPC > SNC > SOC. This ranking may be attributed to the shallow and thin soil layers characteristic of karst areas, which exhibit limited water and fertilizer retention capacity. Additionally, soil erosion can deplete both water and soil organic carbon, resulting in reduced water content and organic carbon levels in the study area. The explanation rate of environmental factors on functional traits, as illustrated in Figure 6, supports this observation. In karst regions, environmental heterogeneity significantly influences plant leaf traits, which can vary considerably both between and within species. Generally, these traits exhibit a combination that results in lower specific leaf area and leaf area. This trait combination suggests that plants are likely to develop a set of drought-resistant traits to adapt to the physiological drought induced by habitat characteristics, such as shallow karst soils and soil water leakage.

5. Conclusions

(1) Karst plant communities display distinct functional trait combinations at varying latitudes. As latitude increases, the resource utilization traits shift to a combination characterized by high leaf thickness (LT) and low specific leaf area (SLA). In terms of overall changes, the functional traits of the apex communities exhibited some convergence effects, which may be attributed to the limited variation in environmental pressures with latitude in the karst region;

(2) Bulk density (BD), soil microbial carbon (SPMC), soil nitrogen carbon (SNC), soil phosphorus content (SPC), and soil organic carbon (SOC) play a significant role in the changes in functional traits during the process of plant restoration. Plants are primarily limited by bulk density (BD) during growth, and most environmental factors exert a limiting effect on plant functional traits. Furthermore, many plant functional traits can only fulfill their roles under specific environmental conditions. Among the various responses of plant functional traits to environmental factors, community-weighted mean leaf area (CWM.LA) and community-weighted mean chlorophyll content (CWM.CHL) were found to be the most sensitive to soil factors.

Author Contributions: Conceptualization, G.X.; methodology, G.X. and Y.W.; software, G.X. and Y.W.; validation, Z.C., Y.J., Y.L. (Yao Lu) and Y.L. (Yi Liang); formal analysis, G.X.; investigation, G.X., R.Z. and J.T.; resources, Y.W.; data curation, G.X.; writing—original draft preparation, G.X.; writing—review and editing, G.X., Y.W. and Z.C.; visualization, Y.W.; supervision, Z.C.; project administration, Y.W.; funding acquisition, G.X. All authors have read and agreed to the published version of the manuscript.

Funding: This research was funded by the Science and Technology Support Project of Guizhou Provincial Science and Technology Department (Qiankehe Support [2023] General 048). Guizhou Provincial Science and Technology Program (Grant number: Qiankehejichu-ZK [2024] General 601).

Data Availability Statement: The original contributions presented in the study are included in the article; further inquiries can be directed to the corresponding author.

Conflicts of Interest: The authors declare no conflicts of interest.

References

1. Wang, J.; Wang, Y.; He, N.; Ye, Z.; Chen, C.; Zang, R.; Feng, Y.; Lu, Q.; Li, J. Plant functional traits regulate soil bacterial diversity across temperate deserts. *Sci. Total Environ.* **2020**, *715*, 136976. [CrossRef] [PubMed]
2. He, N.; Li, Y.; Liu, C.; Xu, L.; Li, M.; Zhang, J.; He, J.; Tang, Z.; Han, X.; Ye, Q.; et al. Plant Trait Networks: Improved resolution of the dimensionality of adaptation. *Trends Ecol. Evol.* **2020**, *35*, 908–918. [CrossRef]
3. Li, T.; Deng, Q.; Yuan, Z.Y.; Jiao, F. Latitudinal changes in plant stoichiometric and soil C,N,P stoichiometry in loess Plateau. *Chin. J. Environ. Sci.* **2015**, *36*, 2988–2996.
4. Zhang, L.; Tan, X.; Dong, Z.; Zheng, J.; Yuan, Z.; Li, C. The relationship between plant functional traits and soil physicochemical properties in the riparian zones of downtown Chongqing. *Acta Ecol. Sin.* **2023**, *43*, 1892–1902, (In Chinese with English Abstract).
5. Liu, M.; Li, L.; Che, Y.; Jiao, J. Functional traits of plant leaves at different succession stages in alpine meadow. *Bull. Bot. Res.* **2019**, *39*, 760–769, (In Chinese with English Abstract).
6. Zhou, X.; Xin, J.; Huang, X.; Li, H.; Li, F.; Song, W. Linking leaf functional traits with soil and climate factors in forest ecosystems in China. *Plants* **2022**, *11*, 3545. [CrossRef]
7. He, N.; Yan, P.; Liu, C.; Xu, L.; Li, M.; Van Meerbeek, K.; Zhou, G.; Zhou, G.; Liu, S.; Zhou, X.; et al. Predicting ecosystem productivity based on plant community traits. *Trends Plant Sci.* **2023**, *28*, 43–53. [CrossRef]
8. Ackerly, D.; Cornwell, W. A trait-based approach to community assembly: Partitioning of species trait values into within- and among- community components. *Ecol. Lett.* **2007**, *10*, 135–145. [CrossRef]
9. Corlett, R.; Westcott, D. Will plant movements keep up with climate change? *Trends Ecol. Evol.* **2013**, *28*, 482–488. [CrossRef]

10. Chen, Y.; Zhang, H.; Wang, X.; Long, W.; Li, C.; Fang, Y.; Fu, M.; Zhu, K. Effects of functional diversity and phylogenetic diversity on the tropical cloud forest community assembly. *Chin. J. Plant Ecol.* **2019**, *43*, 217–226, (In Chinese with English Abstract). [CrossRef]
11. Kembel, S.; Hubbell, S. The phylogenetic structure of a neotropical forest tree community. *Ecology* **2006**, *87*, S86–S99. [CrossRef] [PubMed]
12. Li, Y.; Liang, S.; Zhou, L.; Liang, S.; Ning, J.; Mo, J.; Li, D. Maintenance of species diversity in evergreen deciduous broad-leaved mixed forest of karst hills in central subtropical region based on functional traits. *Acta Ecol. Sin.* **2024**, *44*, 4400–4411, (In Chinese with English Abstract).
13. Miao, L.; Liu, M.; Xiao, Y.; Yang, C.; Wang, Q.; Wang, M. Studies on phylogeny and functional traits structure of plant communities at different latitudes. *Acta Ecol. Sin.* **2024**, *44*, 317–329.
14. Fang, J.; Wang, X.; Shen, Z.; Tang, Z.; He, J.; Yu, D.; Jiang, Y.; Wang, Z.; Zheng, C.; Zhu, J.; et al. Methods and protocols for plant community inventory. *Biodivers. Sci.* **2009**, *17*, 533–548, (In Chinese with English Abstract).
15. Bao, S.D. *Soil Agrochemical Analysis*, 3rd ed.; Agricultural Press: Beijing, China, 2005; pp. 45–52.
16. *NY/T2017-2011*; Determination of Nitrogen, Phosphorus and Potassium in Plants. Ministry of Agriculture: Beijing, China, 2011.
17. *LY/T1228-2015*; Determination of Nitrogen in Forest Soil. Ministry of Agriculture: Beijing, China, 2015.
18. *LY/T1232-2015*; Phosphorus Determination Methods of Forest Soils. Ministry of Agriculture: Beijing, China, 2015.
19. Pérezharguindeguy, N.; Díaz, S.; Garnier, E. New handbook for standardised measurement of plant functional traits worldwide. *Aust. J. Bot.* **2013**, *61*, 167–234. [CrossRef]
20. Lavorel, S.; Grigulis, K.; McIntyre, S.; Williams, N.; Garden, D.; Dorrough, J.; Berman, S.; Quétier, F.; Thébault, A.; Bonis, A. Assessing functional diversity in the field-methodology matters. *Fun. Ecol.* **2008**, *22*, 134–147. [CrossRef]
21. The SPSSAU Project. SPSSAU, Version 25.0. Online Application Software. 2022. Available online: https://www.spssau.com (accessed on 12 February 2024).
22. R Core Team. R: A Language and Environment for Statistical Computing. R Foundation for Statistical Computing. 2020. Available online: https://www.R-project.org/ (accessed on 12 February 2024).
23. Reich, P. The world-wide "fast-slow" plant economics spectrum: A traits manifesto. *J. Ecol.* **2014**, *102*, 275–301. [CrossRef]
24. Wright, I.; Reich, P.; Westoby, M.; Ackerly, D.; Baruch, Z.; Bongers, F.; Cavender-Bares, J.; Chapin, T.; Cornelissen, J.; Diemer, M.; et al. The worldwide leaf economics spectrum. *Nature* **2004**, *428*, 821–827. [CrossRef]
25. Liu, R.H.; Bai, J.L.; Bao, H.; Zhao, J.J.; Jiang, Y.; Liang, S.C.; Li, Y.J. Variation and correlation in functional traits of main woody plants in the cyclobalanopsis glauca community in the karst hills of Guilin, southwest China. *Chin. J. Plant Ecol.* **2020**, *44*, 828–841, (In Chinese with English Abstract). [CrossRef]
26. Liu, M.X.; Nan, X.N.; Zhang, G.J.; Li, B.W.; Xu, L.; Mu, R.L.; Li, L.; Yu, R.X. Relationship between species diversity and functional diversity of plant communities on different slopes in alpine meadow. *Acta Ecol. Sin.* **2021**, *41*, 5398–5407, (In Chinese with English Abstract).
27. Zhang, Z.K.; Zheng, X.X.; Lin, H.Z.; Lin, X.; Hang, L.J. Summary of changes in plant functional traits and environmental factors in different successional stages of island plants. *Acta Ecol. Sin.* **2019**, *39*, 3749–3758, (In Chinese with English Abstract).
28. Li, Y.J.; Zheng, J.M.; Wang, G.Z.; Zhou, J.X.; Liu, Y.G.; Ha, W.X. A study of functional traits of natural secondary forests and their influencing factors in different succession stages in Karst areas: A case study of Dahei mountain, Yunnan province. *Acta Geosci. Sin.* **2021**, *42*, 397–406, (In Chinese with English Abstract).
29. Wu, X. Effects of plants functional traits and diversity on forest biomass and productivity along latitudinal gradient and during successional stages. *Chin. J. Plant Ecol.* **2018**, *42*, 562–572, (In Chinese with English Abstract).
30. Zhang, J.Y. Characteristics of plant communities across the natural tropical coniferous forest -broadleaved forest ecotones in Hainan Island, China. *Trends Ecol. Evol.* **2020**, *35*, 908–918, (In Chinese with English Abstract).
31. Lu, M.M.; Huang, X.C.; Ci, X.Q.; Yang, G.P.; Li, J. Phylogenetic community structure of subtropical forests along elevational gradients in Ailao Mountains of southwest China. *Chin. Biodivers.* **2014**, *22*, 438–448, (In Chinese with English Abstract).
32. Wang, Y.; Zhang, L.; Chen, J.; Feng, L.; Li, F.; Yu, L. Functional diversity of plant communities in relationship to leaf and soil stoichiometry in karst areas of southwest China. *Forests* **2022**, *13*, 864. [CrossRef]
33. Keddy, E. Assembly and response rules: Two goals for predictive community ecology. *J. Veg. Sci.* **1992**, *3*, 157–164. [CrossRef]
34. Mayfield, M.; Levine, J. Opposing effects of competitive exclusion on the phylogenetic structure of communities. *Ecol. Lett.* **2010**, *13*, 1085–1093. [CrossRef]
35. Liu, M.X. Studies on physiological and leaf morphological traits for photosynthesis on different slopes in a subalpine meadow. *Acta Ecol. Sin.* **2017**, *37*, 8526–8536, (In Chinese with English Abstract).
36. Kang, L.L. Effects of Karst Habitat Heterogeneity on Plant Leaf Functional Traits and Soil Nutrients. Master's Thesis, Guizhou University, Guiyang, China, 2021. (In Chinese with English Abstract).
37. Wang, S.Q.; Yu, G.R. Ecological stoichiometry characteristics of ecosystem carbon, nitrogen and phosphorus elements. *Acta Ecol. Sin.* **2008**, *28*, 3937–3947, (In Chinese with English Abstract).

38. Cleveland, C.C.; Liptzin, D. C: N: P Stoichiometry in soil: Is there a "redfield ratio" for the microbial biomass? *Biogeochemistry* **2007**, *85*, 235–246. [CrossRef]
39. Tian, H.; Chen, G.; Zhang, C.; Jerry, M.; Chars, A. Pattern and variation of C: N: P ratios in China's soils: A synthesis of observational data. *Biogeochemistry* **2010**, *98*, 139–151. [CrossRef]
40. Liu, N.; Yu, L.F.; Zhao, Q.; Wu, Y.N.; Yan, L.B. C: N: P stoichiometry of leaf-litter-soil continuum in secondary forests of the rocky desertification regions of the karst plateau. *Chin. J. Appl. Environ. Biol.* **2020**, *26*, 681–688, (In Chinese with English Abstract).
41. Liu, Y. Effects of Soil Depth on Plant Community Structure and Biomass in Karst Region. Ph.D. Thesis, Southwest University, Chongqing, China, 2020. (In Chinese with English Abstract).
42. Wang, Q.; Rong, L.; Wang, M.J.; Ye, T.M.; Li, T.T.; Yang, W.S.; Li, X. Response of leaf functional traits of woody plants to degraded karst soil in central Guizhou. *Earth Environ.* **2022**, *50*, 639–647, (In Chinese with English Abstract).
43. Liu, X.J.; Ma, K.P. Plant functional traits-concepts, applications and future directions. *Sci. Sin. Vitae* **2015**, *45*, 325–339, (In Chinese with English Abstract). [CrossRef]
44. Wang, Y.; Zhang, L.; Chen, J.; Feng, L.; Li, F.; Yu, L. Study on the relationship between functional characteristics and environmental factors in karst plant communities. *Ecol. Evol.* **2022**, *12*, e9335. [CrossRef]
45. Zhou, T.; Cui, Y.C.; Ye, Y.Y.; Zhao, W.J.; Hou, Y.J.; Wu, P.; Ding, F.J. Leaf functional traits of typical karst forest plants under different niches. *J. Cent. South Univ. For. Technol.* **2022**, *42*, 129–140, (In Chinese with English Abstract).
46. Shui, W.; Guo, P.P.; Zhu, S.F.; Feng, J.; Sun, X.; Li, H. Variation of plant functional traits and adaptive strategy of woody species in degraded karst tiankeng of Yunnan province. *Sci. Geogr. Sin.* **2022**, *42*, 1295–1306, (In Chinese with English Abstract).
47. Huang, L.J.; Yu, Y.M.; An, X.F.; Yu, L.L.; Xue, Y.G. Leaf functional traits, species diversity and functional diversity of plant community in Tiankeng forests. *Acta Ecol. Sin.* **2022**, *42*, 10264–10275, (In Chinese with English Abstract).
48. Wang, X.R.; Xing, Y.J. Research progress on the effects of environmental factors and community dynamics on plant species diversity. *Int. J. Ecol.* **2021**, *10*, 608–617, (In Chinese with English Abstract). [CrossRef]

plants

Article

Impact of Microtopography and Neighborhood Effects on Individual Survival Across Life History Stages

Haonan Zhang [1,2,*,†], Xiao Zheng [2,†], Yi Wu [2], Baokun Xu [1,2], Peng Cui [2], Xu Zhou [2], Yanming Fang [3], Lei Xie [3] and Hui Ding [2,*]

[1] Innovative Research Team for Forest Restoration Mechanisms, Chishui National Ecological Quality Comprehensive Monitoring Stations, Nanjing Institute of Environmental Sciences, Ministry of Ecology and Environment (MEE), Nanjing 210042, China

[2] Research Center for Nature Conservation and Biodiversity, State Environmental Protection Scientific Observation and Research Station for Ecology and Environment of Wuyi Mountains, State Environmental Protection Key Laboratory on Biosafety, Nanjing Institute of Environmental Sciences, Ministry of Ecology and Environment (MEE), Nanjing 210042, China

[3] Co-Innovation Center for Sustainable Forestry in Southern China, College of Life Sciences, Key Laboratory of State Forestry and Grassland Administration on Subtropical Forest Biodiversity Conservation, Nanjing Forestry University, Nanjing 210037, China

* Correspondence: zhn@nies.org (H.Z.); nldinghui@sina.com (H.D.); Tel.: +86-18614055380 (H.Z.)

† These authors contributed equally to this work.

Abstract: Understanding drivers of plant community assembly and individual survival in forest ecosystems is crucial for effective conservation and management. While macro-scale factors influencing vegetation patterns are well documented, the combined impact of microtopographic variations and neighborhood effects at neighborhood scales, particularly in subtropical forests, requires further study. To contribute to this area of research, we established a 9.6 ha dynamic plot in a subtropical evergreen broad-leaved forest to examine the interplay between microtopographic factors and neighborhood effects on individual plant survival across different life stages. We conducted a comprehensive analysis of microtopographic variables and neighborhood effects, with individual plant survival censused through repeated surveys at 5-year intervals. Mixed-effects models were employed to assess the combined influence of these factors across life stages. Our results reveal that both microtopographic factors and neighborhood effects significantly influence plant survival, with their impacts varying across life stages. Water availability, represented by flow direction, emerged as a consistently critical factor throughout all life stages. Elevation and the topographic position index showed significant positive effects on survival, particularly in later life stages, possibly reflecting adaptations to light acquisition and water drainage. The influence of topographic factors intensified with succession, while the impact of neighborhood effects, particularly asymmetric competition and conspecific negative density dependence, changed as plants matured. This study enhances our understanding of forest community assembly, emphasizing the importance of considering abiotic and biotic factors across multiple scales for effective forest conservation and management. It provides insights into mechanisms driving spatial variation in community composition, crucial for preserving biodiversity in heterogeneous forest landscapes.

Keywords: microtopographic variables; neighborhood effects; plant survival; subtropical forest

Citation: Zhang, H.; Zheng, X.; Wu, Y.; Xu, B.; Cui, P.; Zhou, X.; Fang, Y.; Xie, L.; Ding, H. Impact of Microtopography and Neighborhood Effects on Individual Survival Across Life History Stages. *Plants* **2024**, *13*, 3216. https://doi.org/10.3390/plants13223216

Academic Editors: Xugao Wang, Zuoqiang Yuan and Chengjin Chu

Received: 11 October 2024
Revised: 8 November 2024
Accepted: 14 November 2024
Published: 15 November 2024

1. Introduction

Plants, as autotrophic and sessile organisms, exhibit life-long immobility following establishment, leading to a strong correlation between the spatial distribution of plant species and local environmental factors. At both global and regional scales, climate, topography, and geomorphology play pivotal roles in determining the types and distribution patterns of vegetation [1,2]. Locally, although climatic variables such as precipitation and temperature tend to be more homogeneous, microtopographic variations become pronounced, particularly in tropical and subtropical mountainous forests [3–5]. These variations in terrain create diverse microhabitats and microclimates within short distances, significantly influencing vegetation patterns and ecological processes [3–9]. Investigating such variations is crucial for understanding the drivers of spatial variation in community species composition, especially for the conservation of rare and endangered plants and the management of forests [10–12]. Insights from these studies can significantly inform strategies for forest conservation and restoration following environmental disturbances [13].

Topography emerges as a critical factor in shaping forest landscape heterogeneity [3]. Factors like elevation, slope, aspect, and terrain curvature substantially influence environmental conditions such as soil moisture, nutrient availability, sunlight, and microclimate [14–17]. These factors, in turn, affect plant growth, reproduction, and survival, consequently determining forest structure and composition. Micro-scale topographic variations promote spatial heterogeneity, essential for plant distribution at finer scales [18,19]. Topography acts as a key driver and filter in forest regeneration, influencing species distribution based on specific habitat requirements [20–22]. This influence manifests in distinct patterns, with gullies supporting higher biodiversity, ridges favoring drought-tolerant species, and valleys accommodating resource-demanding species [23,24]. Consequently, these topography-induced variations regulate essential resources such as water and light availability, affecting plant survival and growth, which ultimately shapes species distribution and canopy architecture, contributing to the overall forest structure and composition [1,25]. Microtopography can significantly affect the spatial distribution of species by indirectly affecting soil, moisture, and other environmental factors, especially some species with obvious habitat preferences [4,5]. To elucidate the mechanisms underlying these topography-driven patterns, researchers often quantify the influence of environmental regulation on species distribution by examining individual plant survival as explained by environmental variables [4,26–28]. This approach provides a crucial pathway for understanding how microtopography impacts forest species distribution patterns, offering insights into the complex interplay between topographic features and plant community dynamics.

Species interactions, such as resource competition and facilitative mutualism, importantly influence the spatial structure of species at neighborhood scales in addition to environmental factors [9–11,29,30]. In forests, although the distribution of environmental resources like light, water, and nutrients may appear relatively uniform locally, they are typically limited. Variations in individual plant sizes and local densities lead to differences in competitive abilities. Consequently, smaller plants differ significantly in their resource utilization efficiency compared to larger plants, resulting in a size-proportional (asymmetrical), uneven distribution of these limited resources among individuals [31,32]. This competition is significantly influenced by the density and size of neighboring trees surrounding a focal individual, affecting the resources available for its growth and survival [33,34]. The impact of these neighborhood interactions on the survival–mortality trade-offs in plant individuals is often attributed to density-dependent mortality (NDD), typically categorized into conspecific and heterospecific negative density dependence (CNDD and HNDD) [31,35–40]. Furthermore, studies investigating the relationship between biodiversity and individual survival have become more prevalent [41–43]. The observed positive correlation between diversity and survival suggests that diversity in heterospecific assemblies at small spatial scales may enhance microclimatic conditions and improve light interception [30,44–46].

Collectively, abiotic and biotic factors interact to influence individual survival within forest dynamics, underscoring the importance of considering resource competition and plant interactions in microenvironmental studies. Species interactions exhibit significant scale dependency, with richer interspecies interactions like neighborhood diversity and asymmetrical competition observed at smaller scales. Therefore, analyzing microtopography's impact on individual survival at the neighborhood level aids in quantifying the synergistic effects of species interactions, as noted by numerous scientists who partition community structure variation into components explained by environmental conditions and species interactions [4,40,47–49]. This approach further elucidates the roles of ecological processes such as environmental filtering, competitive exclusion, and mutualistic coexistence in forest dynamics [4,12,28,50–53]. Quantifying the interactions among forest community structure, individual spatial distribution, environmental factors, and neighborhood effects is thus essential for understanding their driving mechanisms.

To address these challenges, we established a dynamic plot of 9.6 hectares (240 m × 400 m) within a subtropical evergreen broad-leaved forest [54,55], aiming to systematically analyze the effects of microtopography and neighborhood species interactions on tree survival. Based on the preceding literature review and theoretical insights, we propose several scientific hypotheses to guide our research: (H1) Microtopographic factors and neighborhood effects jointly influence the survival of individuals. (H2) The asymmetrical competition resulting from individual size is significantly correlated with plant survival, with larger individuals having a higher probability of survival. (H3) The influence of microtopographic factors and neighborhood effects on survival varies across different life history stages. (H4) Different species exhibit distinct survival probabilities in response to microtopographic conditions and neighborhood effects. Specifically, the 32 dominant species demonstrate significant interspecific differences. This integrated perspective is crucial for advancing conservation and restoration strategies, particularly in heterogeneous forest landscapes.

2. Results

2.1. Microtopographic and Neighborhood Effects on Tree Survival

Microtopography and neighborhood effects were analyzed for their influence on individual tree survival across different spatial scales. Elevation, topographic position index (TPI), and flow direction consistently exhibited positive effects on survival across all scales, with relationships generally significant (Figure 1a,d,f). This suggests that topographic factors regulating light availability (elevation and TPI) and water availability (flow direction) strongly influence individual survival. Flow direction, in particular, not only significantly promoted survival at all scales but also demonstrated the largest effect size, indirectly indicating that water availability is a crucial determinant of survival. Aspect also positively affected survival probability, but only at smaller scales (Figure 1b). In this study, higher aspect values corresponded to south-facing slopes, while lower values represented north-facing (shaded) slopes. The positive correlation suggests higher survival probabilities for individuals on shaded slopes. While TRI and slope exhibited positive and negative influences, respectively, these effects were not statistically significant (Figure 1c,e).

Regarding neighborhood effects, asymmetric competition based on individual diameter at breast height (DBH) significantly impacted survival. A strong positive correlation was observed between individual DBH and survival rate, indicating higher survival for larger trees (Figure 1g). Conspecific negative density dependence (CNDD) was also detected, demonstrating a significant negative impact on individual survival (Figure 1i). Interestingly, a positive effect of diversity on individual survival was also observed (Figure 1h).

In summary, as hypothesized (H1 and H2), both microtopographic factors and neighborhood effects play a role in influencing individual tree survival. Furthermore, asymmetric competition related to individual size significantly correlates with survival, with larger individuals exhibiting higher survival rates.

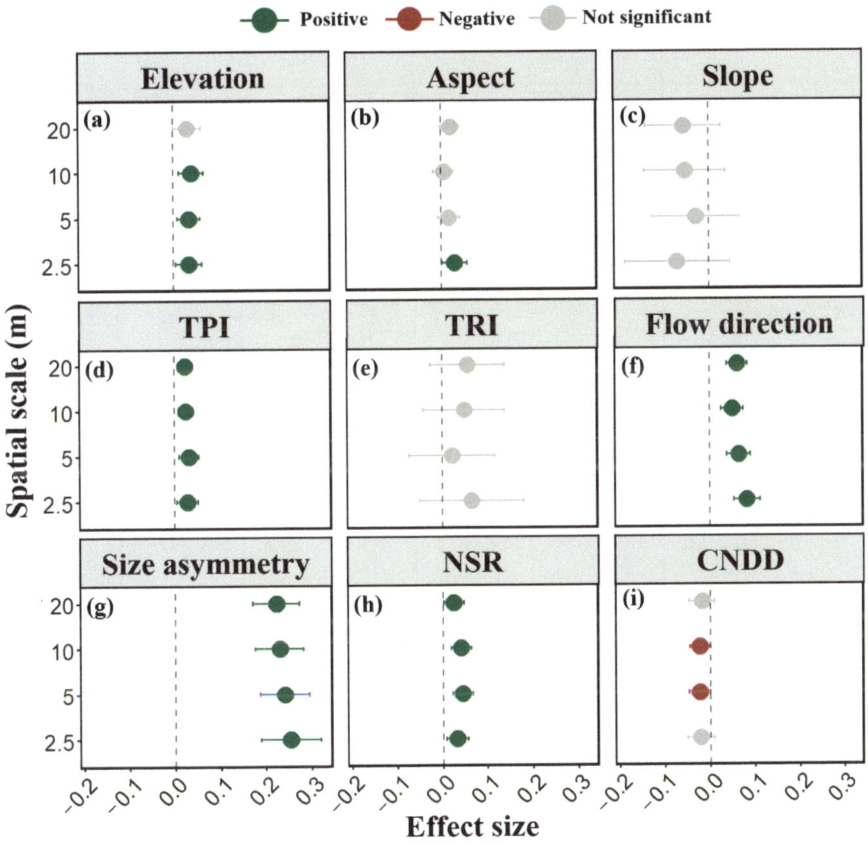

Figure 1. Parameter estimates of microtopographic (**a–f**) and neighborhood effects (**g–i**) on individual tree survival at neighborhood scales. Dots represent estimated parameter effects, with error bars indicating standard errors. A semi-transparent gray dashed line indicates a null effect (parameter estimate of zero) in each subplot.

2.2. Microtopographic and Neighborhood Effects Across Life Stages

The influence of microtopography and neighborhood effects, including individual size, on individual tree survival was analyzed across different life history stages. The positive effects of elevation, aspect, TPI, TRI, and flow direction on survival progressively increased from early to late life stages (Figure 2a,b,d–f), while the negative effect of slope intensified with successional stage (Figure 2c). Although some parameter estimates were non-significant at certain stages, this overall trend suggests that as individuals grow and develop, their survival becomes increasingly influenced by these relatively stable environmental factors. Adult trees, in particular, exhibit a stronger dependence on these factors.

In contrast to the shifting influence of environmental factors, the positive effect of asymmetric competition on survival decreased significantly with successional stage and individual growth (Figure 2g). This indicates that asymmetric competition among adult trees has a considerably weaker impact on survival compared to earlier life stages, and the effect is non-significant for adult trees. The analysis of conspecific negative density dependence (CNDD) revealed a significant negative effect on survival, which weakened with life stage progression (Figure 2h). This suggests that adult trees (i.e., larger trees) exhibit greater tolerance to CNDD, although the magnitude of this change is relatively small ($R^2_{CNDD} = 0.13$). Neighborhood diversity consistently demonstrated a significant positive effect on survival across life stages, but similarly to CNDD (Figure 2i), the magnitude of change with succession was minimal (R^2_{NSR} = 0.16).

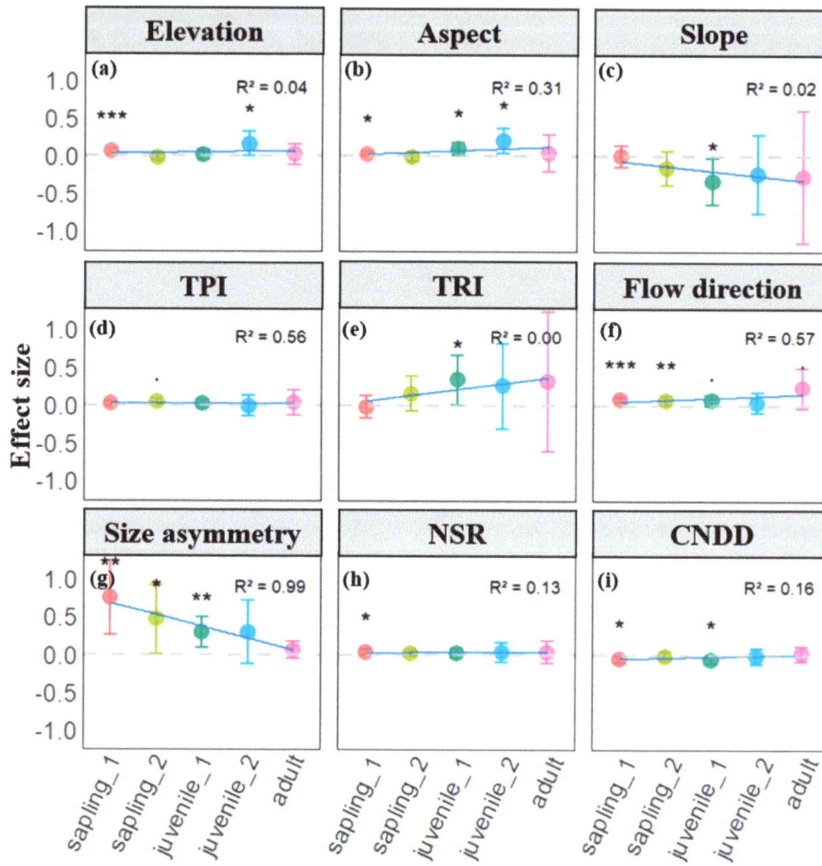

Figure 2. Parameter estimates of microtopographic (**a–f**) and neighborhood effects (**g–i**) on individual tree survival across different life history stage. Dots represent estimated coefficients with error bars depicting standard errors. Positive coefficients indicate positive effects, while negative coefficients indicate negative effects. Significance levels: • $p < 0.1$; * $p < 0.05$; ** $p < 0.01$; *** $p < 0.001$. In the inset figure, R-squared values represent the regression coefficient for changes across life stages, and light blue lines illustrate the trend of neighborhood effects across life stages.

Overall, these findings support Hypothesis 3, demonstrating that the influence of microtopography and neighborhood effects on survival varies across life history stages. Specifically, the impact of microtopographic factors on individual survival increases from early to late stages, while the influence of neighborhood effects decreases or remains relatively constant.

2.3. *Interspecific Variability*

As hypothesized (H4), different species exhibited varying survival probabilities in response to microtopography and neighborhood effects. Survival probabilities were predicted for 32 dominant tree species using mixed-effects models, revealing significant interspecific variation (Figures 3 and 4). Among microtopographic factors, elevation, aspect, TPI, and flow direction generally exerted positive effects on survival during early life stages, particularly the Sapling_1 stage (Figure 3(a1,a2,a4,a6)). However, this consistency diminished across life stages, suggesting that habitat preferences diverge as individuals grow. Notably, the influence of flow direction remained relatively stable throughout the life cycle, indicating the persistent importance of water availability for survival within the observed timeframe.

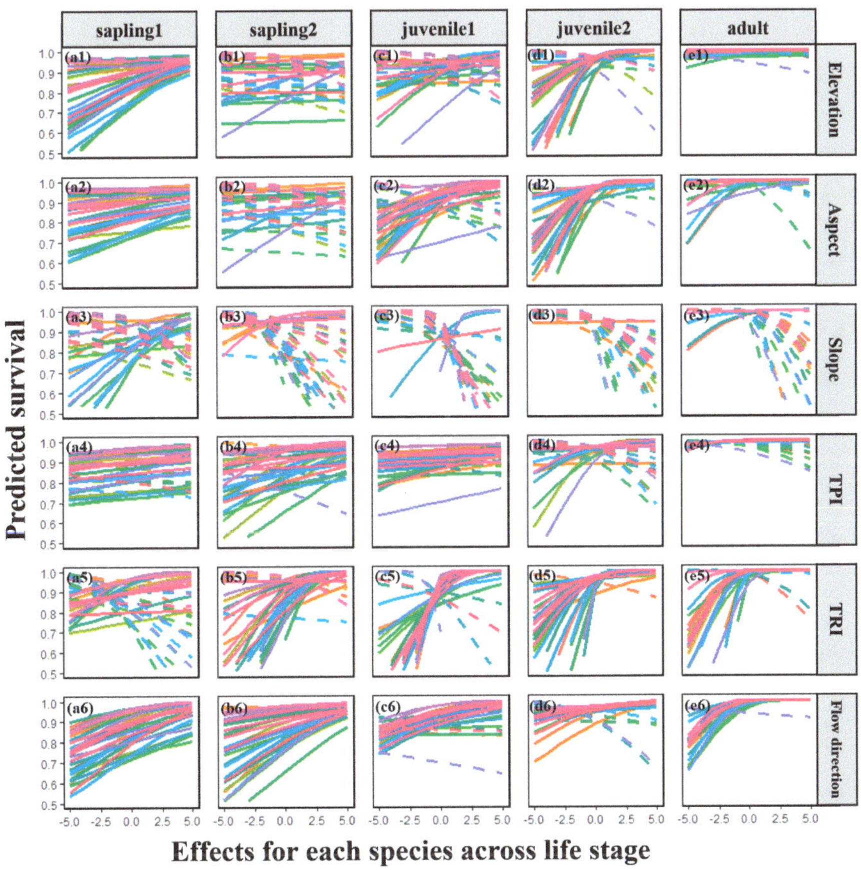

Figure 3. Relationship between six microtopographic factors (elevation: (**a1–e1**), aspect: (**a2–e2**), slope: (**a3–e3**), TPI: (**a4–e4**), TRI: (**a5–e5**) and flow direction: (**a6–e6**)) and survival probability among 32 species across life stages at a scale of 2.5 m. The figure displays predicted survival probability for each of the 32 observed species across life stages at a scale of 2.5 m, and other test scales are 5 m, 10 m, and 20 m; see Figure S3–S5 in the Supplementary Files. Different species are represented by different colors in the lines (see Figure S5 in the Supplementary Files): the solid line represents a positive effect and the dotted line represents a negative effect. Predicted survival probability is back-transformed from the generalized linear mixed model as described in the text, and all neighborhood effects were Z-score-transformed at quantification.

Regarding neighborhood effects, the positive influence of asymmetric competition on survival was relatively consistent across species but gradually weakened (decreasing slope) throughout the life cycle. This decline was particularly pronounced in the later adult stage (Figure 4(e1)), which explains the non-significant overall effect observed in Figure 2g. Furthermore, neighborhood species richness (NSR) and CNDD exhibited relatively consistent negative and positive effects, respectively, during the seedling stage

(Figure 4(a2,a3)). However, these effects became increasingly variable in later life stages (Figure 4(d2,e2,d3,e3)).

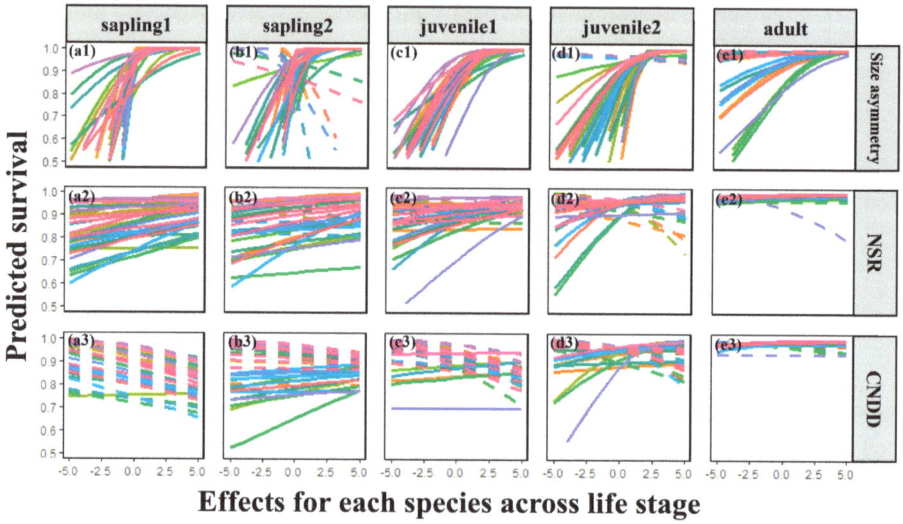

Figure 4. Relationship between three neighborhood effects (size asymmetry: (**a1–e1**), NSR: (**a2–e2**), and CNDD: (**a3–e3**)) and survival probability among 32 species across life stages at a scale of 2.5 m. The figure displays predicted survival probability for each of the 32 observed species across life stages at a scale of 2.5 m, and other test scales are 5 m, 10 m, and 20 m; see Figure S6 in the Supplementary Files. Different species are represented by different colors in the lines (see Figure S3 in the Supplementary Files): the solid line represents a positive effect and the dotted line represents a negative effect. Predicted survival probability is back-transformed from the generalized linear mixed model as described in the text, and all neighborhood effects were Z-score-transformed at quantification.

3. Discussion

Topography acts as both a driver and filter in forest natural regeneration, significantly influencing species distribution based on specific habitat requirements [20–22]. This filtering effect is particularly evident in its impact on individual plant survival during natural regeneration processes [3–5]. Characteristically, valleys and lowlands, with their deeper, fertile soils and ample moisture, support species with higher resource demands [3,14,24]. Our study provides empirical evidence for these topographical influences. At scales between 2.5 m and 5 m, individual survival rates were notably higher on shaded aspect (Figure 2b). Furthermore, flow direction significantly promoted individual survival across all tested scales (Figure 2f), supporting the notion that water availability is a crucial factor in individual survival. This finding aligns with previous research indicating that flow direction influences soil moisture distribution, which in turn affects seedling establishment and growth [5]. Areas aligned with optimal flow directions tend to retain more moisture, thereby reducing drought stress and enhancing nutrient uptake efficiency. Consistent water availability can also buffer plants against extreme weather conditions, contributing to higher survival rates [5,56,57]. Interestingly, we also detected significant positive effects of elevation and TPI on individual survival (Figure 2a,d), possibly reflecting the adaptation of light-demanding or drought-tolerant species to ridges with shallower, well-drained soils [15–17]. Although we did not directly assess soil characteristics, the existing literature suggests that higher elevations and specific topographic positions often correlate with distinct soil moisture regimes and nutrient availability [56–59]. For instance, ridges may offer better drainage, which benefits species adapted to lower moisture conditions, while valleys may retain more nutrients and moisture [5]. Future studies incorporating soil analyses would provide a more comprehensive understanding of how elevation and TPI influence plant survival through soil-mediated mechanisms.

Beyond topographical factors, neighborhood effects also significantly influenced individual survival. Notably, DBH showed a strong positive correlation with survival probability, indicating that initial plant size is crucial in determining resource acquisition capabilities within our 5-year monitoring period [5,13,51,52]. Consistent with classic studies, conspecific negative density dependence (CNDD) significantly reduced species survival [31,35–37,40]. Conversely, aligning with recent research, diversity promoted individual plant survival [30,44–46]. In summary, our study reveals that small-scale topographic factors exhibit strong spatial variation, with their impact on individual survival often rivaling the degree of variation seen across broad environmental or biogeographic gradients [1,2].

The process of plant community assembly is dynamic and unfolds over time, with distinct assembly mechanisms emerging at various stages of plant growth [5,9,27,30,38–40,49]. This mechanism is profoundly influenced by external environmental factors and interspecific interactions among individual plants [3–5,18,40,49]. Our study quantifies the impact of these influential factors on individual survival across different life history stages, decomposing them into two crucial aspects: microtopography and neighborhood effects. Interestingly, we observed that as succession progresses, plant growth significantly alters the response patterns of individual survival to microtopographic factors and neighborhood effects. Specifically, topographic factors that promote individual survival exhibit stronger positive effects in later life stages, while those with negative impacts show more pronounced inhibitory effects. In particular, the positive effects of elevation, TPI, TRI, flow direction, and aspect on survival intensify with succession, potentially reflecting the increased demand for resources such as light and water as plants grow and their size increases [49]. Conversely, the negative impact of slope on survival strengthens over time, indicating the influence of environmental filtering on individual survival [49]. Notably, despite considerable variation in species' responses to different topographic factors (Figure 4), flow direction maintains a relatively stable influence throughout the life cycle (Figure 3(a6–e6)). This suggests that water availability remains one of the most critical factors affecting individual survival throughout the observed life history stages.

Furthermore, by examining the effects of asymmetric competition, diversity, and conspecific negative density dependence (CNDD) on individual survival, we observed that plant growth also increases tolerance to competition. We attribute this to varying regulatory mechanisms of asymmetric competition across different life history stages. Initially, we observed a significant positive correlation between individual size and survival probability in early life stages. However, this correlation diminishes as life stages progress, with the promoting effect of individual size on species survival decreasing and becoming non-significant in the mature tree stage (Figure 2g). This indicates the presence of significant asymmetric competition within the plot, where larger individuals possess stronger competitive abilities [5,51–53]. As growth and development continue, the disparity in individual sizes gradually decreases, further reinforcing the notion that asymmetric competition among large trees has a reduced impact on individual survival [51]. Notably, the relationships between CNDD, diversity, and survival become less stable in early life stages as life stages progress (Figure 4(d2,e2,d3,e3)). Specifically, plants' tolerance to CNDD gradually increases with individual growth. This result supports classic conclusions from CNDD research, namely, that negative density-dependent mortality is most pronounced in early life stages of plants [39]. It also aligns with the aforementioned perspective on asymmetric competition, suggesting that competition among mature trees gradually becomes more symmetric, potentially leading to a decrease in density-dependent mortality caused by CNDD [51,53]

While this study provides valuable insights into the effects of topography and neighborhood dynamics on individual plant survival, it is important to acknowledge certain limitations. One notable limitation is the absence of an analysis of soil nutrient availability across the study area. In a landscape covering nearly 10 hectares, soil properties can exhibit significant spatial variability, leading to differences in nutrient availability that may influence plant growth and survival. Soil nutrients play a critical role in sustaining plant health and resilience, and variations in nutrient levels could potentially affect the outcomes observed in our study. Consequently, the omission of soil nutrient analysis may limit the comprehensiveness of our findings. Future research should incorporate detailed assessments of soil nutrient profiles to better elucidate their impact on forest natural regeneration and to provide a more holistic understanding of the factors driving individual plant survival.

4. Materials and Methods

4.1. Study Area and Plot Establishment

The research was carried out in Wuyishan National Park, situated in the northwest region of Fujian Province, China. This locale experiences an average annual temperature of 19.2 °C and receives roughly 1600 mm of precipitation each year. In this region, red soil is the most extensively distributed zonal soil type, ranging from riverbeds at an elevation of 160 m to mountainous areas up to 1100 m. The predominant vegetation in this area consists of subtropical evergreen broad-leaved forests (Wu, 1980), although past commercial logging has predominantly transformed these original forests into secondary growth [49,54]. The site underwent selective logging of the Chinese fir plantations originally used for timber production, with efforts made to preserve native forest species during the thinning process.

Within this secondary subtropical evergreen broad-leaved forest, a 9.6-hectare dynamic plot was established in 2013, located at 27°35′24.23′′ N, 117°45′55.43′′ E, and measuring 400 m by 240 m (Figure 5). The plot features moderate topographic variability, with elevations ranging from 450 to 580 m. Formerly part of the Sixin Forestry plantation, the area was subject to logging in the 1960s and has since experienced six decades of natural regeneration.

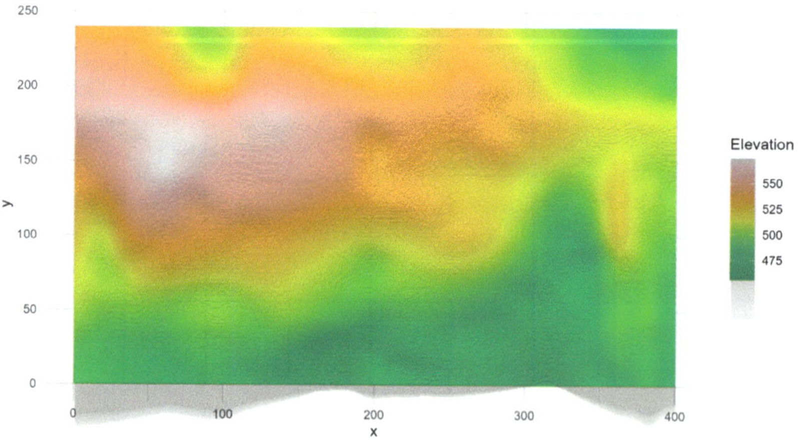

Figure 5. Microtopography in dynamic plot. This map only shows the combined terrain factors and the basic conditions of elevation; specific factors such as aspect and slope are shown in Figure S1 in the Supplementary Files. The shadow effect illustrates terrain undulations in orthographic view.

4.2. Repeated Censuses and Individual Tree Survival

In accordance with the CTFS (Center for Tropical Forest Science) survey protocols, the entire plot was divided into 240 large quadrats (20 m × 20 m), and each large quadrat was further subdivided into 16 smaller plots (5 m × 5 m), totaling 3840 small plots. These smaller quadrats were used as work units to measure the relative position, DBH (diameter at breast height), and other individual attributes of all trees. Two censuses were conducted in the dynamic plot between 2013 and 2018. During these censuses, species identity, location, DBH, height, and crown base height were recorded for all trees with DBH $\geq$ 1 cm. The first census showed a total of 68,336 tree individuals (including branches and sprouts) with DBH $\geq$ 1 cm, belonging to 173 species, 88 genera, and 48 families. The co-dominant families included Fagaceae, Ericaceae, and Elaeocarpaceae, with co-dominant species including *Castanopsis carlesii*, *Castanopsis fordii*, *Castanopsis eyrei*, and *Schima superba*. No single species was overwhelmingly dominant (Table S1 in the Supplementary Files), and the stand structure indicated that the forest community in our study was still in the early stage of secondary succession because most tree individuals were saplings [49,54]

Tree survival was determined using binary code for each individual (1 for survival, 0 for death), focusing solely on trees alive at the start of the interval to compute survival probability. We selected 32 co-dominant tree species based on their importance values, abundance, and average DBH (diameter at breast height) from the plot for this study (Tables S1 and S2 in the Supplementary Files). Importance values (IVs) were calculated by combining the percentages of relative abundance, relative dominance (DBH), and relative frequency for these 32 co-dominant species, reflecting their overall ecological significance within the community [54,55].

4.3. Microtopographic Factors and Neighborhood Effects

Within our dynamic forest plot, fundamental microtopographic variables, including mean elevation, aspect, and slope, were assessed for each subplot (Figure 5 and Figure S1 in the Supplementary Files). Following Center for Tropical Forest Science (CTFS) protocols, the plot was divided into 240 quadrats (20 m $\times$ 20 m), each further subdivided into 16 subplots (5 m $\times$ 5 m). Elevation, aspect, and slope were measured in each subplot using a total station. These measurements were then used to generate a digital elevation model (DEM) of the plot, from which microtopographic variables were derived [4,49]. Higher aspect values in the DEM correspond to south-facing slopes, while lower values represent north-facing (shaded) slopes. To explore the influence of microtopography on ecological processes within plant communities, we also derived and quantified more complex microtopographic factors from the DEM data, such as the terrain position index (TPI), terrain ruggedness index (TRI), and flow direction [56,57]. The TPI evaluates terrain position by calculating the mean elevation difference between each pixel and all neighboring pixels in a DEM. The TRI quantifies terrain roughness by computing the root mean square of the elevation differences between each pixel and its eight adjacent pixels in a DEM. Flow direction, derived from the DEM, represents the direction of water flow based on elevation differences between a central pixel and its eight neighbors. Higher flow direction values indicate areas of greater water accumulation, suggesting increased water availability for plants in those locations [56,57]. Microtopographic variables for focal tree species at different neighborhood scales were calculated and visualized using R Studio (based on R version 4.2.3), utilizing the packages "spatstat (version 3.0-3)" and "raster (version 3.6-2)".

Size-asymmetric competition was assessed by examining the relationship between individual tree size (diameter at breast height, DBH) and survival [40,49,51,53]. DBH was measured for each tree, and its location within the plot was recorded (Figure S2a in the Supplementary Files). Repeated censuses over a five-year period tracked individual survival, allowing us to test Hypothesis 2 (H2), which posited that larger individuals (greater DBH) exhibit higher survival rates. DBH was chosen as a proxy for individual size due to its reliability and ease of measurement in large-scale field surveys (with over 60,000 individuals censused per survey). Neighborhood species richness (NS; Figure S2b in the Supplementary Files) assesses biodiversity by counting distinct tree species within a specified radius around each focal tree. For any given focal tree i, NSR is precisely calculated as the total number of immediate heterospecific neighbor species, mathematically represented as $NSR_i = \sum_{j \neq i} N_j$, where N denotes the recorded number of species for each neighboring tree j. This measure allows for an in-depth analysis of how the survival of a focal tree is influenced by the species diversity of its immediate surroundings. The NSR around each individual was calculated for four radii (2.5, 5, 10, or 20 m). The conspecific neighborhood competition indices (CNDD; Figure S2c in the Supplementary Files) were calculated by evaluating the DBH area of neighboring trees of the same species [30,38,39]. The indices were formulated as $CNDD_i = \sum_{j \neq i} \frac{\pi D_j^2}{4}$, where Dj represents the diameter at breast height (DBH) of neighboring trees [30,39]. We computed the microtopography factor and neighborhood effect indices for neighborhoods of different radii: 2.5, 5, 10, and 20 m [5,30,39]. Neighborhood effects for focal tree species at different neighborhood scales were calculated and visualized using R Studio (based on R version 4.2.3), utilizing the package "spatstat (version 3.0-3)".

4.4. Microtopographic and Neighborhood Effects on Tree Survival

Generalized linear mixed-effects models (GLMMs) with binomial error structures were used to analyze the influence of microtopographic factors and neighborhood effects on tree survival across life history stages. Microtopographic factors (elevation, slope, aspect, TPI, TRI, and flow direction) and neighborhood effects (size-asymmetric competition, NSR, and CNDD) were included as predictors [5,30,39,40]. To facilitate comparison and improve model stability and interpretability, given the disparate scales of predictors like competition indices and species richness, all predictors were Z-score-transformed. For all individuals, a basic model was fitted at each of life stages, from the focal individual.

$$\begin{aligned}\ln\left(p_{ij}/\left(1-p_{ij}\right)\right) \\ &= \beta_{0j} + \beta_1 Elevation_{ijp} + \beta_2 Aspect_{ijp} + \beta_3 Slope_{ijp} \\ &+\beta_4 TPI_{ijp} + \beta_5 TRI_{ijp} + \beta_6 Flow\ direction_{ijp} \\ &+\beta_7 \text{Size asymmetric}_{ijp} + \beta_8 NSR_{ijp} \\ &+\beta_9 CNDD_{ijp} + \varepsilon_j + p_q\end{aligned}$$

where p_{ij} is the predicted survival probability of each focal tree i of species j growing in quadrats q. We included all microtopographic factor and neighborhood effect variables as fixed effects in our model. The coefficients β_1 to β_9 represent the effects of these variables on survival probability. The random effect structure incorporates crossed random effects for species identity and plot identity to account for the variability in survival probabilities across different species and quadrats (small plots). This includes ε_j, random intercepts and slopes for species j. This term accounts for baseline survival probability differences across species and allows the effects of microtopographic factor and neighborhood interaction to vary among species. It also includes p_q, random intercepts for quadrats, accounting for potential differences in baseline survival probabilities across different quadrats [30,39,40].

Furthermore, the inclusion of random intercepts and slopes for species *j* in the models allowed us to predict species-specific responses (i.e., survival probabilities) to microtopographic factors and neighborhood interactions. This approach facilitated testing Hypothesis 4, which stated that different species exhibit distinct survival probabilities in response to these factors. We examined the survival responses of 32 dominant tree species to nine predictors. We used the "lme4 1.1-32" package in R to run the basic model for each life stage (based on R version 4.2.3).

5. Conclusions

Our study provides crucial insights into the complex dynamics of plant community assembly in forest ecosystems, focusing on the understudied impact of microtopographic variations and neighborhood effects at local scales. We demonstrate that both microtopographic factors and neighborhood effects significantly influence individual plant survival across different life stages in a subtropical evergreen broad-leaved forest. This research reveals that topographic factors, particularly those related to water availability (flow direction) and light acquisition (elevation and TPI), consistently impact survival throughout plant life cycles. Notably, the effects of these factors intensify as succession progresses, reflecting changing resource demands and environmental filtering. At fine scales (2.5 m to 5 m), survival rates were higher on shaded aspects. Elevation and TPI also showed significant positive effects on survival, though the mechanisms behind these relationships require further investigation. We observed that the influence of both topographic factors and neighborhood effects (size-asymmetric competition, CNDD, and NSR) varies across life history stages. As plants mature, their tolerance to competition increases, with the impact of individual size (DBH) on survival probability decreasing in later life stages. Conspecific negative density dependence (CNDD) shows a decreasing influence on mortality in mature plants, supporting classic theories of density-dependent effects being most pronounced in early life stages. These findings underscore the need for multiscale, temporally explicit approaches in forest ecology research and management. By elucidating the interplay between abiotic and biotic factors across plant life stages, our study contributes to a more nuanced understanding of forest community assembly processes. This comprehensive view has important implications for conservation and restoration strategies, particularly in heterogeneous forest landscapes.

One notable limitation of our study is the absence of an analysis of soil nutrient availability across the study area. In a landscape covering nearly 10 hectares, soil properties can exhibit significant spatial variability, leading to differences in nutrient availability that may influence plant growth and survival. We observed that the survival rates of different species are influenced by topographic factors and biotic interactions, reflecting species-specific traits. However, further investigation is needed to understand the underlying causes of these phenomena. Future research could integrate functional trait data to distinguish the resource acquisition strategies of different taxa, such as comparing pioneer species with resource-conservative species in their responses to topographic factors and species interactions. Long-term studies across diverse forest ecosystems are necessary to unravel these complex ecological interactions and inform effective forest management practices in the face of global environmental changes. Understanding these interactions thoroughly will be foundational in formulating strategies that ensure the resilience and sustainability of forest ecosystems.

Supplementary Materials: The following Supplementary Files can be downloaded at: https://www.mdpi.com/article/10.3390/plants13223216/s1, Figure S1. Spatial variation in microtopographic predictors at the neighborhood scale: elevation (a), aspect (b), slope (c), terrain position index (d), terrain ruggedness index (e), and flow direction (f). The maps were generated using an Epanechnikov kernel with a bandwidth of 5, and the intensity values range from blue (low) to purple (high). Figure S2. Spatial variation in neighborhood effect predictors at the neighborhood scale: DBH size asymmetry (a), neighborhood species richness (b), and CNDD (c). The maps were generated using an Epanechnikov kernel with a bandwidth of 5, and the intensity values range from blue (low) to purple (high). Figure S3. Relationship between microtopographic (elevation, aspect, slope, TPI, TRI, and flow direction) and individual survival across life stages at a scale of 5 m. The inter-census predicted survival probability for each of the 32 co-dominant tree species is represented by lines of different colors, with solid lines indicating a positive relationship and dashed lines indicating a negative one. The predicted individual survival was obtained by back-transforming from the general linear mixed models, with all diversity effects quantified by Z-score transformation. Figure S4. Relationship between microtopographic (elevation, aspect, slope, TPI, TRI, and flow direction) and individual survival across life stages at a scale of 10 m. The inter-census predicted survival probability for each of the 32 co-dominant tree species is represented by lines of different colors, with solid lines indicating a positive relationship and dashed lines indicating a negative one. The predicted individual survival was obtained by back-transforming from the general linear mixed models, with all diversity effects quantified by Z-score transformation. Figure S5. Relationship between micro topographic (elevation, aspect, slope, TPI, TRI, and flow direction) and individual survival across life stages at a scale of 20 m. The inter-census predicted survival probability for each of the 32 co-dominant tree species is represented by lines of different colors, with solid lines indicating a positive relationship and dashed lines indicating a negative one. The predicted individual survival was obtained by back-transforming from the general linear mixed models, with all diversity effects quantified by Z-score transformation.

Figure S6. Relationship between neighborhood effect (size asymmetry, CNDD, and NSR) and individual survival across life stage at a scale of 5 m, 10 m, and 20m. The inter-census predicted survival probability for each of the 32 co-dominant tree species is represented by lines of different colors, with solid lines indicating a positive relationship and dashed lines indicating a negative one. The predicted individual survival was obtained by back-transforming from the general linear mixed models, with all diversity effects quantified by Z-score transformation. Table S1. Basic characteristics of the 32 co-dominant tree species in a dynamic forest plot in the Wuyi Mountains, China. Note: Species names and family assignments in the table are based on the Flora of China and conform to the accepted names in Plants of the World Online (PoWO; https://powo.science.kew.org/). Table S2. Forest dynamics of the 32 co-dominant tree species in the subtropical evergreen broad-leaved forest plot in the Wuyi Mountains, China, for the years 2013 and 2018.

Author Contributions: Conceptualization, H.Z.; funding acquisition, H.D. and H.Z.; investigation, H.Z., X.Z. (Xiao Zheng), Y.W., B.X., X.Z. (Xu Zhou) and L.X.; writing—original draft, H.Z.; writing—review and editing, H.Z.; project administration, H.Z., H.D., P.C. and Y.F. All authors have read and agreed to the published version of the manuscript.

Funding: This work was funded by the Innovative Team Project of the Nanjing Institute of Environmental Sciences MEE (ZX2023QT022); the Central Public Welfare Scientific Institution Basal Research Fund, MEE (GYZX240104, GYZX210302); the Sino-German Environmental Partnership Phase III: Strengthening Chinese environmental and nature conservation policy through dialog (SGEP2022-2027); the Biodiversity Survey, Observation and Assessment Program of the Ministry of Ecology and Environment of China (BSOA2019-2024); the National Environmental Protection Public Welfare Industry Targeted Research Fund: Research and demonstration of key technologies for dynamic supervision of nature reserves (201509042); the Special Foundation for National Science and Technology Basic Resources Investigation of China (2019FY202300); and the Jiangsu Postdoctoral Research Funding Program (2021K038A).

Data Availability Statement: The forest census data are available from the corresponding author on reasonable request.

Acknowledgments: We are grateful to the people who helped to collect field survey data.

Conflicts of Interest: The authors declare no conflicts of interest.

References

1. Quesada, C.A.; Phillips, O.L.; Schwarz, M.; Czimczik, C.I.; Baker, T.R.; Patiño, S.; Fyllas, N.M.; Hodnett, M.G.; Herrera, R.; Almeida, S.; et al. Basin-wide variations in Amazon forest structure and function are mediated by both soils and climate. *Biogeosciences* **2012**, *9*, 2203–2246. [CrossRef]
2. Sullivan, M.J.; Talbot, J.; Lewis, S.L.; Phillips, O.L.; Qie, L.; Begne, S.K.; Chave, J.; Cuni-Sanchez, A.; Hubau, W.; Lopez-Gonzalez, G.; et al. Diversity and carbon storage across the tropical forest biome. *Sci. Rep.* **2017**, *7*, 39102. [CrossRef] [PubMed]
3. Werner, F.A.; Homeier, J. Is tropical montane forest heterogeneity promoted by a resource-driven feedback cycle? Evidence from nutrient relations, herbivory and litter decomposition along a topographical gradient. *Funct. Ecol.* **2015**, *29*, 430–440. [CrossRef]
4. Punchi-Manage, R.; Getzin, S.; Wiegand, T.; Kanagaraj, R.; Savitri Gunatilleke, C.V.; Nimal Gunatilleke, I.A.U.; Wiegand, K.; Huth, A. Effects of topography on structuring local species assemblages in a Sri Lankan mixed dipterocarp forest. *J. Ecol.* **2013**, *101*, 149–160. [CrossRef]
5. Zhang, H.; Zhang, X.; Lv, Y.; Ni, Y.; Xu, B.; Han, X.; Cao, X.; Yang, Q.; Xu, W.; Qian, Z. How Topography and Neighbor Shape the Fate of Trees in Subtropical Forest Restoration: Environmental Filtering and Resource Competition Drive Natural Regeneration. *For. Ecosyst.* **2024**, *11*, 100169. [CrossRef]
6. Givnish, T.J. On the causes of gradients in tropical tree diversity. *J. Ecol.* **1999**, *87*, 193–210. [CrossRef]
7. John, R.; Dalling, J.W.; Harms, K.E.; Yavitt, J.B.; Stallard, R.F.; Mirabello, M.; Hubbell, S.P.; Valencia, R.; Navarrete, H.; Vallejo, M.; et al. Soil nutrients influence spatial distributions of tropical tree species. *Proc. Natl. Acad. Sci. USA* **2007**, *104*, 864–869. [CrossRef]
8. Russo, S.E.; Brown, P.; Tan, S.; Davies, S.J. Interspecific demographic trade-offs and soil-related habitat associations of tree species along resource gradients. *J. Ecol.* **2008**, *96*, 192–203. [CrossRef]
9. Punchi-Manage, R.; Wiegand, T.; Wiegand, K.; Getzin, S.; Huth, A.; Gunatilleke, C.S.; Gunatilleke, I.N. Neighborhood diversity of large trees shows independent species patterns in a mixed dipterocarp forest in Sri Lanka. *Ecology* **2015**, *96*, 1823–1834. [CrossRef]
10. Hui, G.Y.; Gadow, K.V. *Quantitative Analysis of Forest Spatial Structure*; China Science & Technology Press: Beijing, China, 2003. (In Chinese with English Abstract)
11. Wiegand, T.; Gunatilleke, C.V.S.; Gunatilleke, I.A.U.N.; Huth, A. How individual species structure diversity in tropical forests. *Proc. Natl. Acad. Sci. USA* **2007**, *104*, 19029–19033. [CrossRef]

12. Page, N.V.; Shanker, K. Environment and dispersal influence changes in species composition at different scales in woody plants of the Western Ghats, India. *J. Veg. Sci.* **2018**, *29*, 74–83. [CrossRef]
13. Seidl, R.; Turner, M.G. Post-disturbance reorganization of forest ecosystems in a changing world. *Proc. Natl. Acad. Sci. USA* **2022**, *119*, e2202190119. [CrossRef] [PubMed]
14. Banin, L.; Feldpausch, T.R.; Phillips, O.L.; Baker, T.R.; Lloyd, J.; Affum-Baffoe, K.; Arets, E.J.; Berry, N.J.; Bradford, M.; Brienen, R.J.; et al. What controls tropical forest architecture? Testing environmental, structural and floristic drivers. *Glob. Ecol. Biogeogr.* **2012**, *21*, 1179–1190. [CrossRef]
15. Paoli, G.D. Divergent leaf traits among congeneric tropical trees with contrasting habitat associations on Borneo. *J. Trop. Ecol.* **2006**, *22*, 397–408. [CrossRef]
16. Heineman, K.D.; Jensen, E.; Shapland, A.; Bogenrief, B.; Tan, S.; Rebarber, R.; Russo, S.E. The effects of belowground resources on aboveground allometric growth in Bornean tree species. *For. Ecol. Manag.* **2011**, *261*, 1820–1832. [CrossRef]
17. Holdaway, R.J.; Richardson, S.J.; Dickie, I.A.; Peltzer, D.A.; Coomes, D.A. Species- and community-level patterns in fine root traits along a 120000-year soil chronosequence in temperate rain forest. *J. Ecol.* **2011**, *99*, 954–963. [CrossRef]
18. Tiessen, H.; Chacon, P.; Cuevas, E. Phosphorus and nitrogen status in soils and vegetation along a toposequence of dystrophic rainforests on the upper Rio Negro. *Oecologia* **1994**, *99*, 145–150. [CrossRef]
19. Xia, S.W.; Chen, J.; Schaefer, D.; Goodale, U.M. Effect of topography and litterfall input on fine-scale patch consistency of soil chemical properties in a tropical rainforest. *Plant Soil* **2016**, *404*, 385–398. [CrossRef]
20. Dodson, E.K.; Root, H.T. Conifer regeneration following stand-replacing wildfire varies along an elevation gradient in a ponderosa pine forest, Oregon, USA. *For. Ecol. Manag.* **2013**, *302*, 163–170. [CrossRef]
21. Francis, D.; Ex, S.; Hoffman, C. Stand composition and aspect are related to conifer regeneration densities following hazardous fuels treatments in Colorado, USA. *For. Ecol. Manag.* **2018**, *409*, 417–424. [CrossRef]
22. Redmond, M.D.; Kelsey, K.C. Topography and overstory mortality interact to control tree regeneration in spruce-fir forests of the southern Rocky Mountains. *For. Ecol. Manag.* **2018**, *427*, 106–113. [CrossRef]
23. Gunatilleke, C.V.S.; Gunatilleke, I.A.U.N.; Esufali, S.; Harms, K.E.; Ashton, P.M.S.; Burslem, D.F.; Ashton, P.S. Species habitat associations in a Sri Lankan dipterocarp forest. *J. Trop. Ecol.* **2006**, *22*, 371–384. [CrossRef]
24. Paoli, G.D.; Curran, L.M.; Slik, J.W.F. Soil nutrients affect spatial patterns of aboveground biomass and emergent tree density in southwestern Borneo. *Oecologia* **2008**, *155*, 287–299. [CrossRef] [PubMed]
25. Aiba, S.I.; Takyu, M.; Kitayama, K. Dynamics, productivity and species richness of tropical rainforests along elevational and edaphic gradients on Mount Kinabalu, Borneo. *Ecol. Res.* **2005**, *20*, 279–286. [CrossRef]
26. Whittaker, R.H. Vegetation of the Great Smoky Mountains. *Ecol. Monogr.* **1956**, *26*, 1–80. [CrossRef]
27. Legendre, P.; Borcard, D.; Peres-Neto, P.R. Analyzing beta diversity: Partitioning the spatial variation of community composition data. *Ecol. Monogr.* **2005**, *75*, 435–450. [CrossRef]
28. Chang, Y.M.; Baddeley, A.; Wallace, J.; Canci, M. Spatial statistical analysis of tree deaths using airborne digital imagery. *Int. J. Appl. Earth Obs. Geoinf.* **2013**, *21*, 418–426. [CrossRef]
29. Zhang, C.; Jin, W.; Gao, L.; Zhao, X. Scale dependent structuring of spatial diversity in two temperate forest communities. *For. Ecol. Manage.* **2014**, *316*, 110–116. [CrossRef]
30. Fichtner, A.; Härdtle, W.; Li, Y.; Bruelheide, H.; Kunz, M.; von Oheimb, G. From competition to facilitation: How tree species respond to neighbourhood diversity. *Ecol. Lett.* **2017**, *20*, 892–900. [CrossRef]
31. Zhang, H.; Xu, J.; Meng, W.; Li, Z.; Ni, Y.; Li, W.; Chen, H.; Zhang, X.; Yuan, H.; Wang, Z. Positive interactions in shaping neighborhood diversity during secondary forests recovery: Revisiting the classical paradigm. *For. Ecol. Manag.* **2024**, *552*, 121586. [CrossRef]
32. DeMalach, N.; Zaady, E.; Weiner, J.; Kadmon, R. Size asymmetry of resource competition and the structure of plant communities. *J. Ecol.* **2016**, *104*, 899–910. [CrossRef]
33. Stoll, P.; Newbery, D.M. Evidence of species-specific neighborhood effects in the Dipterocarpaceae of a Bornean rain forest. *Ecology* **2005**, *86*, 3048–3062. [CrossRef]
34. Matsushita, M.; Takata, K.; Hitsuma, G.; Yagihashi, T.; Noguchi, M.; Shibata, M.; Masaki, T. A novel growth model evaluating age–size effect on long-term trends in tree growth. *Funct. Ecol.* **2015**, *29*, 1250–1259. [CrossRef]
35. Janzen, D.H. Herbivores and the number of tree species in tropical forests. *Am. Nat.* **1970**, *104*, 501–528. [CrossRef]
36. Connell, J.H. On the role of natural enemies in preventing competitive exclusion in some marine animals and in rain forest trees. In *Dynamics of Populations*; den Boer, P.J., Gradwell, G.R., Eds.; Centre for Agricultural Publishing and Documentation: Wageningen, The Netherlands, 1971; pp. 298–312.
37. Comita, L.S.; Muller-Landau, H.C.; Aguilar, S.; Hubbell, S.P. Asymmetric density dependence shapes species abundances in a tropical tree community. *Science* **2010**, *329*, 330–332. [CrossRef] [PubMed]
38. Zhu, Y.; Mi, X.; Ren, H.; Ma, K. Density dependence is prevalent in a heterogeneous subtropical forest. *Oikos* **2010**, *119*, 109–119. [CrossRef]
39. Zhu, Y.; Comita, L.S.; Hubbell, S.P.; Ma, K. Conspecific and phylogenetic density-dependent survival differs across life stages in a tropical forest. *J. Ecol.* **2015**, *103*, 957–966. [CrossRef]

40. Zhang, H.N.; Chen, S.; Xia, X.; Ge, X.; Zhou, D.; Wang, Z. The competitive mechanism between post-abandonment Chinese fir plantations and rehabilitated natural secondary forest species under an in situ conservation policy. *For. Ecol. Manag.* **2021**, *502*, 119725. [CrossRef]
41. Yang, J.; Cao, M.; Swenson, N.G. Why functional traits do not predict tree demographic rates. *Trends Ecol. Evol.* **2013**, *28*, 561–563. [CrossRef]
42. Liu, X.; Swenson, N.G.; Lin, D.; Mi, X.; Umaña, M.N.; Schmid, B.; Ma, K. Linking individual-level functional traits to tree growth in a subtropical forest. *Ecology* **2022**, *103*, e3426. [CrossRef]
43. Hülsmann, L.; Chisholm, R.A.; Comita, L.; Visser, M.D.; de Souza Leite, M.; Aguilar, S.; Anderson-Teixeira, K.J.; Bourg, N.A.; Brockelman, W.Y.; Bunyavejchewin, S.; et al. Latitudinal patterns in stabilizing density dependence of forest communities. *Nature* **2024**, *627*, 564–571. [CrossRef] [PubMed]
44. Sapijanskas, J.; Paquette, A.; Potvin, C.; Kunert, N.; Loreau, M. Tropical tree diversity enhances light capture through crown plasticity and spatial and temporal niche differences. *Ecology* **2014**, *95*, 2479–2492. [CrossRef]
45. Wright, A.; Schnitzer, S.A.; Reich, P.B. Living close to your neighbors: The importance of both competition and facilitation in plant communities. *Ecology* **2014**, *95*, 2213–2223. [CrossRef] [PubMed]
46. Williams, L.; Paquette, A.; Cavender-Bares, J.; Messier, C.; Reich, P. Spatial complementarity in tree crowns explains overyielding in species mixtures. *Nat. Ecol. Evol.* **2017**, *1*, 63. [CrossRef]
47. Chase, J.M.; Myers, J.A. Disentangling the importance of ecological niches from stochastic processes across scales. *Philos. Trans. R. Soc. B* **2011**, *366*, 2351–2363. [CrossRef]
48. Zhang, H.N.; Yang, Q.; Zhou, D.; Xu, W.; Gao, J.; Wang, Z. How evergreen and deciduous trees coexist during secondary forest succession: Insights into forest restoration mechanisms in Chinese subtropical forest. *Glob. Ecol. Conserv.* **2021**, *25*, e01418. [CrossRef]
49. Zhang, H.; Chen, S.; Zheng, X.; Ge, X.; Li, Y.; Fang, Y.; Cui, P.; Ding, H. Neighborhood diversity structure and neighborhood species richness effects differ across life stages in a subtropical natural secondary forest. *For. Ecosyst.* **2022**, *9*, 100075. [CrossRef]
50. Gilbert, B.; Lechowicz, M.J. Neutrality, niches, and dispersal in a temperate forest understory. *Proc. Natl. Acad. Sci. USA* **2004**, *101*, 7651–7656. [CrossRef]
51. Pommerening, A.; Särkkä, A. What mark variograms tell about spatial plant interactions. *Ecol. Model.* **2013**, *251*, 64–72. [CrossRef]
52. Tatsumi, S.; Loreau, M. Partitioning the biodiversity effects on productivity into density and size components. *Ecol. Lett.* **2023**, *26*, 1963–1973. [CrossRef]
53. Zhang, H.N.; Xue, J.H. Spatial pattern and competitive relationships of moso bamboo in a native subtropical rainforest community. *Forests* **2018**, *9*, 774. [CrossRef]
54. Ding, H.; Chen, S.F.; Xu, H.; Luo, X.Q.; Li, Y.; Xu, B.K.; Wu, Y.; Wu, Y.Q.; Yong, F.; Zhou, W.L.; et al. Dynamics of arbor layer in the Subtropical Evergreen Broad-leaved Forest in the Wuyi Mountains, Fujian Province, southeastern China in 2013–2018. *Acta Ecol. Sin.* **2022**, *42*, 3458–3469, (In Chinese with English Abstract). [CrossRef]
55. Wu, Z.Y. *Vegetation of China*; Science Press: Beijing, China, 1980; pp. 823–888, (In Chinese with English Abstract).
56. Amatulli, G.; McInerney, D.; Sethi, T.; Strobl, P.; Domisch, S. Geomorpho90m, empirical evaluation and accuracy assessment of global high-resolution geomorphometric layers. *Sci. Data* **2020**, *7*, 162. [CrossRef] [PubMed]
57. Du, H.; Ning, B.; Jiao, J.; Gasanova, Z.U.; Stepanova, N.Y.; Konyushkova, M.V. Temporal change in plant communities and its relationship to soil salinity and microtopography on the Caspian Sea coast. *Sci. Rep.* **2020**, *12*, 18082. [CrossRef]
58. Fortunel, C.; Lasky, J.R.; Uriarte, M.; Valencia, R.; Wright, S.J.; Garwood, N.C.; Kraft, N.J. Topography and neighborhood crowding can interact to shape species growth and distribution in a diverse Amazonian forest. *Ecology* **2018**, *99*, 2272–2283. [CrossRef] [PubMed]
59. Johnson, D.J.; Condit, R.; Hubbell, S.P.; Comita, L.S. Abiotic niche partitioning and negative density dependence drive tree seedling survival in a tropical forest. *Proc. R. Soc. B* **2017**, *284*, 20172210. [CrossRef]

plants

MDPI

Article

Influence of Functional Traits of Dominant Species of Different Life Forms and Plant Communities on Ecological Stoichiometric Traits in Karst Landscapes

Yang Wang [1,2,*], Limin Zhang [3], Ling Feng [4], Zhuhong Fan [1,2], Ying Deng [1,2] and Tu Feng [1,2]

1 School of Ecological Engineering, Guizhou University of Engineering Science, Bijie 551700, China; fanzh961226@163.com (Z.F.); dy3594142@163.com (Y.D.); fengtu@gues.edu.cn (T.F.)
2 Guizhou Province Key Laboratory of Ecological Protection and Restoration of Typical Plateau Wetlands, Bijie 551700, China
3 Institute of Mountain Resources of Guizhou Academy of Sciences, Guiyang 550025, China; zhanglimin563406@163.com
4 College of Life Sciences, Guizhou University, Guiyang 550025, China; 15870149772@163.com
* Correspondence: wangyang20218910@126.com

Abstract: Assessing the functional traits and ecological stoichiometric characteristics of dominant species across different life forms within plant communities in karst environments and investigating the inherent connection between them can provide insights into how species adjust their functional attributes in response to habitat heterogeneity. This approach offers a more comprehensive understanding of ecosystem processes and functions in contrast to examination of the taxonomic diversity of species. This study examines the relationship between the functional characteristics of dominant species in plant communities of various life forms in karst environments, focusing on deciduous leaf–soil ecological stoichiometry. The investigation relies on community science surveys, as well as the determination and calculation of plant functional traits and ecological stoichiometries, in plant communities of various life forms in Guizhou (a province of China). The findings of our study revealed considerable variability in the functional trait characteristics of dominant species across different plant-community life forms. Specifically, strong positive correlations were observed among plant height (PLH), leaf area (LA), leaf dry matter content (LDMC), and specific leaf area (SLA) in the dominant species. Additionally, our results indicated no significant differences in leaf ecological stoichiometry among different life forms. However, we did observe significant differences and strong positive correlations between soil N:P, withered material C:N, and apomictic C:P. Furthermore, our study found that plant height (PLH), leaf area (LA), and specific leaf area (SLA) were particularly sensitive to the ecological stoichiometry of soil and apomixis. The results of our study suggest that the functional traits of diverse plant-community life forms in karst regions are capable of adapting to environmental changes through various expressions and survival strategies. The development of various plant-community life forms in karst areas is particularly vulnerable to phosphorus limitation, and the potential for litter decomposition and soil nutrient mineralization is comparatively weaker. The functional traits of various plant-community life forms in karst regions exhibit greater sensitivity to both the soil's C:N ratio and the C:N ratio of apomictic material. Habitat variations may influence the ecological stoichiometric characteristics of the plant leaf–apomictic soil continuum.

Keywords: life forms; plant community; functional traits; ecological stoichiometry; karst

check for updates

Citation: Wang, Y.; Zhang, L.; Feng, L.; Fan, Z.; Deng, Y.; Feng, T. Influence of Functional Traits of Dominant Species of Different Life Forms and Plant Communities on Ecological Stoichiometric Traits in Karst Landscapes. *Plants* **2024**, *13*, 2407. https://doi.org/10.3390/plants13172407

Academic Editor: Ladislav Paule

Received: 23 July 2024
Revised: 26 August 2024
Accepted: 27 August 2024
Published: 28 August 2024

1. Introduction

As ecology continues to evolve, plant functional traits have proven to be an important means of exploring various ecological frontiers [1]. Plant functional traits serve as a crucial link between plants and their environment [2] and also reflect the variability of species in terms of growth, survival, and reproduction. This statement contributes to a more comprehensive understanding of the distribution of plant species, the process of community formation, and ecosystem function [3]. Presently, there is a substantial body of research focused on plant functional traits, which includes the examination of leaf, branch, and stem traits, as well as root traits, in order to elucidate the relationships among these traits [4]. Moreover, it is essential to investigate the correlation between plant characteristics and the environment in order to comprehend the mechanisms of coexistence at different spatial and temporal levels and to analyze the dynamics of species in reaction to environmental fluctuations. Secondly, it is essential to investigate the correlation between plant characteristics and the environment in order to comprehend the coexistence mechanisms at different spatial and temporal levels and to analyze the dynamic attributes of species in reaction to environmental fluctuations [5]. The functional traits of plant leaves have been shown to be closely associated with plants' capacity to utilize resources and, to some extent, can indicate their survival strategies for adapting to environmental changes [6]. The examination of functional traits in plant leaves can enhance our comprehension of how leaf physiology reacts to environmental fluctuations [7]. Various indicators of plant functional traits, such as specific leaf area, leaf nitrogen content, leaf thickness, leaf dry matter content, chlorophyll content, leaf relative water content, stem tissue density, leaf carbon content, stem nitrogen content, stem phosphorus content, and other traits, are commonly selected [8–10].

Ecological stoichiometry pertains to the proportion of vital chemical elements implicated in ecological interactions and processes [11]. The interconnected organic systems of plants and soils necessitate the analysis of their carbon (C), nitrogen (N), and phosphorus (P) contents and ratios, which are essential for the energy cycle and stability of ecosystems [12]. The southwestern karst region is considered to be one of the most ecologically vulnerable areas in China, and the ecological and environmental challenges have emerged as a bottleneck constraining the economic and social development of the region [13]. Ecological chemometrics combines the basic principles of biology, chemistry, and physics, including those of ecology and chemometrics, taking into account the first law of thermodynamics, the principle of natural selection for biological evolution, and the theory of the central law of molecular biology [14]. The rapid advancement of ecological chemometrics has attracted growing attention from scholars, who are exploring various aspects of this technique. Previous studies have documented the ecological and chemometric properties of soils in diverse karst ecosystems [15,16]. Recently, a study was conducted to analyze the eco-chemical quantification of leaves from dominant species in secondary karst forests. The C:N:P stoichiometry of the leaf–deciduous leaf–soil continuum in secondary forests was calculated [17]. Carbon (C), nitrogen (N), and phosphorus (P) are essential elements for life, with nitrogen and phosphorus serving as the primary limiting factors in natural terrestrial ecosystems [18]. The contents of these elements have the potential to influence the species composition and productivity of plant communities. The biogeochemical processes of systems are responsible for regulating the cycling of nutrients, as indicated by reference [19]. Hence, carbon (C), nitrogen (N), and phosphorus (P) are crucial in ecosystem processes and serve as connections between community species diversity, plant functional traits, and ecological stoichiometry [20]. Utilizing this framework to investigate the inherent connection of ecosystem processes and their feedback effects can lead to a deeper understanding of ecosystem nutrient cycling principles and system stabilization mechanisms. Variations in leaf traits among different life forms lead to differences in the requirement for and utilization of light, precipitation, temperature, nutrients, and other resources [21]. These variations also result in a wide range of adaptations to different topographic conditions [22]. Hence, it is essential to examine the relationship between functional traits and ecological stoichiometry in various plant life forms in order to comprehend the mechanisms underlying biodiversity maintenance and the survival tactics of plants.

Plant functional traits play a crucial role in illustrating the connection between plants and their environment, revealing the adaptive strategies and regulatory mechanisms that plants employ in particular environments [23]. The capacity of plants to obtain and utilize resources is also pertinent. The stoichiometric characteristics of elements such as carbon (C), nitrogen (N), and phosphorus (P) in organisms are closely linked to their stability, community species composition, and biogeochemical cycling processes. These characteristics serve as potent instruments for examining nutrient restriction and cycling [24]. The functional traits and ecological stoichiometric characteristics of forest plants are subject to influence by environmental factors, which in turn constrain ecosystem processes and functions. Consequently, an examination of their fundamental characteristics and inherent relationships can lay the groundwork for the sustainable management of forest ecosystems.

Based on this premise, this paper aims to uncover the functional traits and ecological stoichiometric characteristics of various life forms within plant communities in karst regions and to finally clarify the interaction between soil and vegetation. This will be achieved through community surveys and the measurement or calculation of plant functional traits and ecological stoichiometric indexes. The ultimate goal is to offer insights for research on the mechanisms involved in the establishment of forest communities and the preservation of biodiversity in karst environments.

2. Research Methods

2.1. Study Area

The Huaxi District is located in the southern region of Guiyang City, Guizhou Province, in China (106°27′–106°52′ E, 26°11′–26°34′ N), within the watershed area that separates the Yangtze River system and the Zhujiang River system. The total land area encompasses 953.83 km^2, with 94% of the territory exhibiting a karstic landscape. This includes a mountainous and hilly topography, delicate habitats, and a low land-carrying capacity. Due to its elevated plateau, the region experiences a highland monsoon humid climate, characterized by an average annual rainfall of 1178.3 mm and an average annual temperature of 14.9 °C (Figure 1).

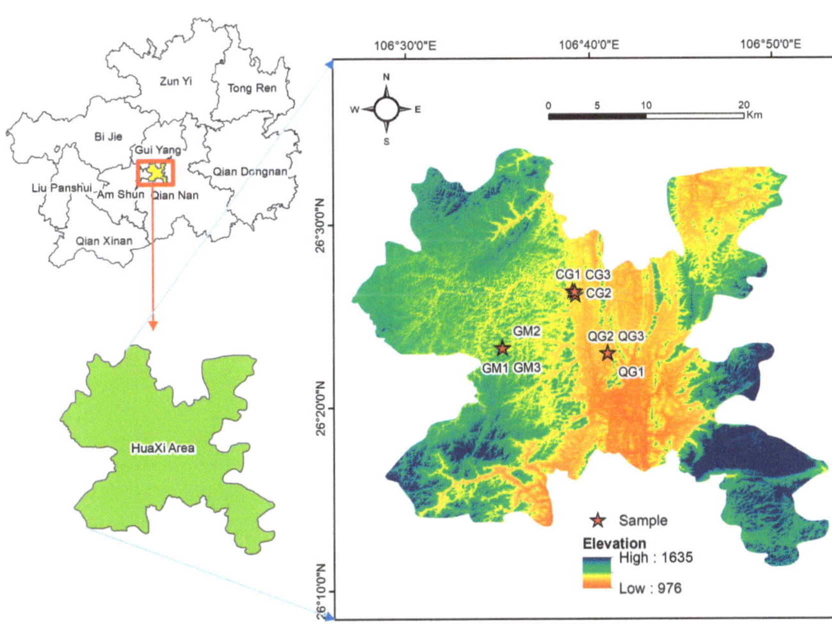

Figure 1. Schematic diagram of the sample area. CG—grass–shrub community, GM—shrub community, QG—tree–shrub community. (The map is original).

2.1.1. Soil Environment

The study area is situated in the hilly regions of Guizhou and Zhongshan on the Yunnan–Guizhou Plateau. The terrain is characterized by higher elevations in the southwest and lower elevations in the northeast. Denuded hills are interspersed with basins, valleys, and depressions, with a relative height difference of 100–200 m. The geological structure is primarily marked by north–south-trending structures, with rock layers in Huaxi District occurring in a monoclinal form. Large folds or faults are absent from the surface of the study area, indicating a relatively simple structural complexity. The main exposed strata in Huaxi District consist of the Quaternary and Middle Triassic Yangliujing Formation, with uniform rock strata and no structural passage. However, due to regional structural influences, the rock mass is relatively fractured. The predominant soil type in the study area is yellow soil, which is the most prevalent zonal soil type. Yellow soil is formed under warm and humid subtropical monsoon bioclimate conditions, with predominantly acidic pH levels, followed by neutral and less alkaline characteristics [25].

2.1.2. Description of Karst Vegetation

Three distinct plant communities in the region, namely, the grass–shrub plant community (CG), the shrub plant community (GM), and the tree–shrub plant community (QG), were selected for the study based on the findings of previous research [26]. Two equations were utilized to ascertain the predominant species of trees, shrubs, and herbs. The final dominant species were identified as follows: dominant tree species include *Lindera communis, Itea yunnanensis* Franch., *Quercus fabri*, and chestnut *Castanea mollissima* Blume, among others, while the dominant shrub species mainly consist of *Rhamnus leptophylla, Rosa multiflora, Glochidion puberum, Pyracantha fortuneana, Coriaria nepalensis, Sarcococca ruscifolia, Rosa cymosa, Rhamnus heterophylla, Viburnum foetidum* var *ceanothoides*, and *Myrsine africana*. The dominant herbaceous plant species mainly include *Imperata cylindrica, Senecio asperifolius, Erigeron canadensis, Carex capilliformis*, and *Ficus tikoua*.

2.2. Sample Setup

Three sample plots were chosen as replicates for each plant-community life type in this study, resulting in a total of nine standard sample plots.

The research team conducted a survey and sampled plant communities between April and August 2021. In accordance with the principles of representativeness, typicality, and consistency, the research sample plots were chosen based on areas with similar slope positions, slope directions, and other land conditions. The sample plots were established at a size of 10 m × 10 m for the herbaceous stage and 30 m × 30 m for the other stages. Small sampling squares were designated within the sample plots to aid in conducting plant surveys.

Nine small sample plots, each measuring 10 m × 10 m, were established within the arborvitae community. A single 4 m × 4 m shrub-layer sample plot was established within each tree plot. A 1 m × 1 m herb-layer sub-sample square was positioned within each shrub-layer sub-sample square. Consequently, there were nine sample plots designated for the tree layer, nine for the shrub layer, and nine for the herb layer within the tree–shrub plant community. Given the presence of three replicate sample plots, the total count of sample plots within the arboreal–shrub community amounted to 81.

Nine small sample squares, each measuring 4 m × 4 m in area, were established in both the grass–shrub plant community and the shrub plant community to represent the shrub layer. A 1 m × 1 m herbaceous-layer sample plot was established within each shrub-layer sample plot. Consequently, there were nine sub-samples of shrubs and nine sub-samples of herbaceous plants in both the grass–shrub plant community and the shrub plant community. Therefore, there were a total of 36 small sample squares within these two plant communities. The two plant communities comprised a total of 108 small sample squares, with each community having 3 replicate sample plots.

In total, three plant communities with varying life forms were investigated, encompassing 81 herbaceous sample plots, 81 shrub sample plots, and 9 tree sample plots. The study ultimately documented the nomenclature, height, diameter at breast height, and crown width of both tree and shrub species. The study also recorded the names, quantities, average heights, and ground coverage of herbaceous plants to assist in the assessment of the functional traits of dominant species.

2.3. Sample Collection and Processing

The measurement of the plant sample was divided into two parts. The initial step entailed the selection of all plants within the sampling area, followed by the use of high pruning shears to cut the branches of the sampled plants in the four directions of the crown (south, east, north, west). Approximately 20 healthy and disease-free leaves were then collected from each branch to form a mixed sample. The second step involved selecting the top three significant values (dominant species) within the sampling area as the sampled plants and combining each dominant species from each stage into a single sample. The importance value of the tree stratum was determined by aggregating the relative multiplicity, the relative frequency, and the relative significance, then dividing the sum by 3. Similarly, the significance of the shrub and herb stratum was calculated by combining the relative abundance, the relative frequency, and the relative cover, then dividing by 3.

A total of 27 collection nets, each measuring 1 m × 1 m, were deployed across three plant communities, with 9 nets in each community. The samples were collected in September 2021 and retrieved in March 2022. The "S" sampling method [27] was utilized to collect soil samples from the 0–20 cm depth range in each plant community. The samples were obtained from a limited sampling area (less than 20 cm, based on the actual depth) and were subsequently combined in equal volumes to form a composite sample. Several leaf samples were collected from the initial section, and 27 leaf samples, 27 deadwood samples, and 27 soil samples were collected from the subsequent section. The plant and litter samples were subjected to initial heating at 105 °C for 2 h, followed by drying at 75 °C until a constant weight was achieved. The soil samples were dried at room temperature in an indoor environment. All samples were finely ground, sieved through a 60-mesh sieve, and stored in a well-ventilated area for nutrient analysis.

2.4. Determination of Functional Properties and Analysis of Samples

Six metrics were selected to measure plant functional traits, comprising plant height (PLH), leaf thickness (LT), chlorophyll content (CHL), leaf dry matter content (LDMC), leaf area (LA), and specific leaf area (SLA). The approach for determination was established according to the recently developed manual for standardized measurement of global plant functional traits [28]. The fresh weight of the leaves from each plant sample was determined. The samples underwent baking at 60 °C for 72 h until they achieved a consistent weight. The measurement of the leaves' dry weight was conducted. LDMC is determined by dividing the dry weight of leaves by the fresh weight of leaves. Leaf thickness was assessed with electronic vernier calipers (Deli, DL91150, Qingdao, China), whereas leaf length and area were determined through scanning and subsequent calculations using an HP scanner in conjunction with Photoshop software 2023 (HPScanJetN92120, Wuhan, China). The specific leaf area (SLA) was determined by dividing the leaf area by the leaf dry weight. The chlorophyll content of the leaves was assessed using a chlorophyll meter (Linde, LD-YD, Jining, China). Plant leaf area (LA) was determined through direct field measurements, while leaf dry matter content (LDMC) was calculated as the ratio of the leaf's weight after drying to its fresh weight. The organic carbon content (Soil_C) was quantified using the potassium dichromate oxidation–external heating method [29]. Plant and litter samples underwent digestion using the H_2SO_4-H_2O_2 method. The total nitrogen content in litter and leaves was assessed through the indophenol blue colorimetric method (Litter_C, Leaf_C) (NY/T2017-2011) [30], whereas the total phosphorus content in litter was determined using the molybdenum antimony colorimetric method (Litter_P, Leaf_P) (NY/T2017-2011) [30]. Soil samples underwent testing for total nitrogen utilizing the Kjeldahl nitrogen determination method (Soil_N), as per the LY/T1228-2015 standard [31], and for total phosphorus using the NaOH melting–molybdenum antimony colorimetric method (Soil_P), in accordance with LY/T1232-2015 [32].

2.5. Data Processing

The data were initially organized using Microsoft Excel 2019. Prior to analysis, the data underwent normality testing and were then subjected to ANOVA. The data were analyzed utilizing SPSS 25.0 statistical software [33]. One-way analysis of variance (ANOVA) and Tukey's honestly significant difference multiple-comparisons test were utilized to evaluate the differences in functional trait values, leaf-litter–soil nutrient content, and ecological stoichiometry across different life forms within plant communities. Pearson correlation analysis was employed to elucidate the association between these metrics, and the data were reported as means ± standard deviations. To further examine the patterns of variation in functional traits and leaf litter–soil stoichiometry, multivariate analysis was conducted using the "ggcor", "vegan", "dplyr", and "ggplot2" packages in R version 4.3.2. Software packages in the R 4.3.2 programming language were used for psychological analysis [34].

3. Results and Analysis

3.1. Characterizing Changes in Functional Traits of Dominant Species in Plant Communities with Different Life Forms

As illustrated in Figure 2, there was notable variation in the functional traits of dominant species across different plant-community life forms. The plant leaf area index (PLH) exhibited significant variation among different life forms within karst plant communities, with tree–shrub communities demonstrating the highest values, followed by shrub communities and grass–shrub communities. Meanwhile, there was no significant difference in leaf thickness (LT) between grass–shrub and shrub communities. However, leaf thickness was significantly greater in shrub communities compared to tree–shrub communities. Moreover, the leaf area (LA) exhibited a statistically significant increase in tree–shrub communities compared to grass–shrub communities. No significant differences were observed in LDMC, SLA, or CHL among plant communities with different life forms. The correlation analysis of plant functional traits indicated a strong positive correlation between leaf area (LA) and plant height (PLH), as well as between leaf area (LA) and leaf dry matter content (LDMC). The study found a significant positive correlation between PLH and LDMC, as well as a significant positive correlation between specific leaf area (SLA) and LA. Furthermore, there was a notable inverse relationship between SLA and chlorophyll (CHL) and LDMC (Figure 3).

3.2. C, N, and P Contents of Leaf Litter and Soil of Dominant Species in Plant Communities with Different Life Forms

The carbon (C), nitrogen (N), and phosphorus (P) contents of leaf litter and soil for dominant species in plant communities with different life forms indicate that there were no significant differences in Leaf_C or Leaf_P among the various plant communities. However, Leaf_N exhibited notable variations across the various life forms, with arborvitae demonstrating the highest values, followed by shrubs and then grass–shrubs (Figure 4). Moreover, the Litter_C content was notably higher in the shrub areas in comparison to the grassy shrub areas. Significant variations in Litter_N and Litter_P were observed among different life forms within arboreal communities, with the ranking being grass–shrub > shrub > arboreal communities. Significant differences were observed in Soil_C and Soil_N among various life forms of plant communities, with the order being shrub > grass–shrub > arboreal communities. Furthermore, Soil_P displayed notable variations among various life forms within plant communities, with the ranking being grass–shrub community > shrub community > tree–shrub community.

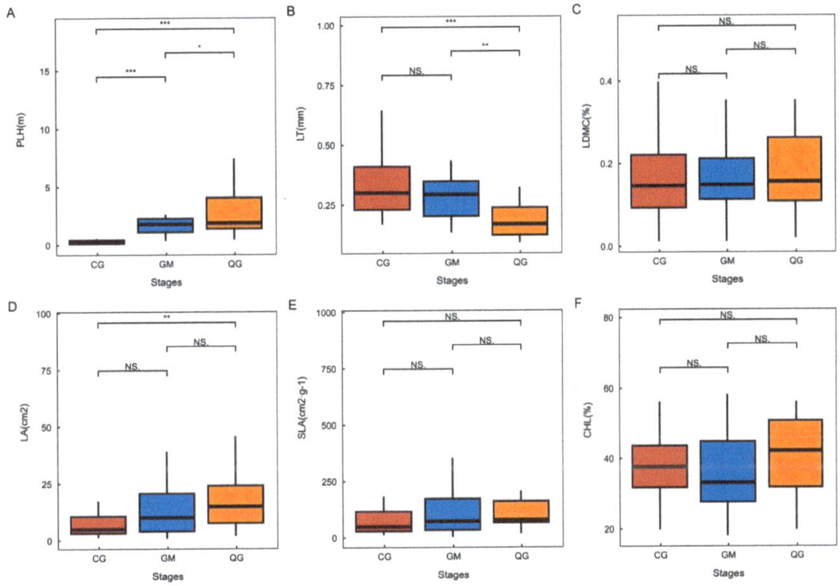

Figure 2. Variations in functional traits of dominant species within plant communities of diverse life forms. In the field of statistics, one-way ANOVA is employed for conducting comparative analysis. * Indicates significant difference ($p < 0.05$), ** indicates highly significant difference ($p < 0.01$) *** indicates highly significant difference ($p < 0.001$). NS signifies no significant difference and lack of ecological statistical significance. Three individuals were measured for each replication, and three replications were conducted for each category of life. Stages are indicative of plant communities characterized by varying life forms. PLH—plant height, LT—leaf thickness, LDMC—leaf dry matter content, LA—leaf area, SLA—specific leaf area, CHL—chlorophyll. The term "CG" denotes a plant community consisting of grass and shrubs, "GM" refers to a plant community dominated by shrubs, and "QG" represents a plant community characterized by a combination of trees and shrubs. The changes in functional traits of plant communities under different life forms are characterized by PLH (**A**), LT (**B**), LDMC (**C**), LA (**D**), SLA (**E**), and CHL (**F**).

3.3. Characterization of Changes in Leaf-Litter–Soil Ecological Stoichiometry of Dominant Species in Plant Communities with Different Life Forms

The investigation into the variations in the leaf-litter–soil ecological stoichiometry of dominant species in plant communities with diverse life forms indicated a significant difference in Leaf_C.N between grass–shrub (58.35) and tree–shrub (27.22) communities. However, the leaf ecological stoichiometry showed no significant differences among dominant species in plant communities with different life forms. Additionally, Leaf_N.P (CG—16.27, GM—17.33, QG—17.81) and Leaf_C.P (CG—625.68, GM—602.56, QG—643.59) did not display significant variations across plant communities with different life forms. Conversely, Litter_C.N (CG—16.68, GM—24.25, QG—19.92) and Litter_C.P (CG—362.34, GM—478.86, QG—436.22) consistently exhibited notable variances across various plant-community types. Specifically, the shrub community displayed significantly elevated levels compared to the tree–shrub community, while the tree–shrub community exhibited significantly higher levels than the grass–shrub community. Litter biomass, however, showed no significant variations across various types of plant communities. Moreover, Soil_N.P demonstrated notable variations across various plant-community types, with the shrub community (6.92) showing superior performance compared to the tree–shrub (6.02) community and the latter outperforming the grass–shrub (4.83) community. The value of N exhibited significant variation across various plant communities, with the highest levels found in shrub communities, followed by arboreal–shrub communities and then grass–shrub communities. Furthermore, Soil_C.P was notably elevated in the shrub (84.75) and tree–shrub (88.64) communities in comparison to the grass–shrub (34.29) community (Figure 5). Further examination of the ecological stoichiometry of leaf litter and soil for dominant species across various plant-community life forms uncovered notable correlations. Specifically, Leaf_N.P demonstrated a highly significant positive correlation with Leaf_C.P, whereas Litter_C.N displayed a highly significant positive correlation with Litter_C.P. Furthermore, there were highly significant positive correlations observed between Soil_C.N, Soil_C.P, and Litter_C.P, and Soil_N.P showed a highly significant positive correlation with Soil_C.P (Figure 6).

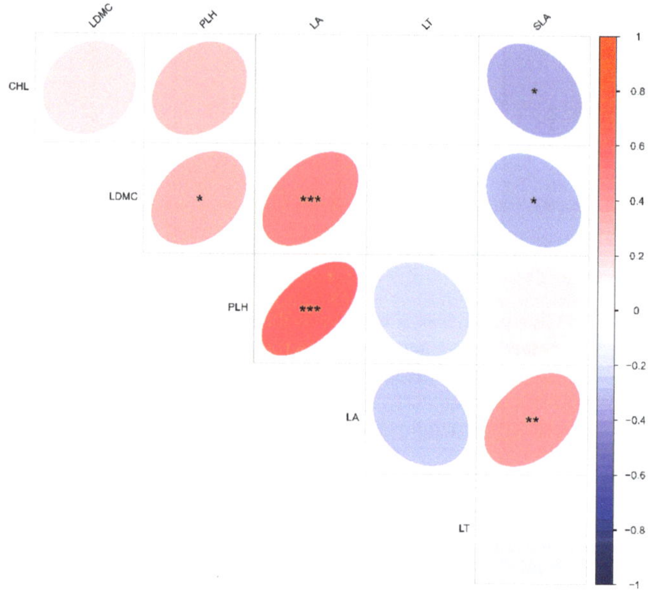

Figure 3. Correlations of functional traits of dominant species in plant communities with varying life forms. PLH—plant height, LT—leaf thickness, LDMC—leaf dry matter content, LA—leaf area, SLA—specific leaf area, CHL—chlorophyll. * Indicates significant difference ($p < 0.05$), ** indicates highly significant difference ($p < 0.01$), *** indicates highly significant difference ($p < 0.001$).

3.4. Relationships between Functional Traits of Dominant Species and Deciduous Leaf-Litter–Soil Ecological Stoichiometry in Plant Communities with Different Life Forms

The relationships between the functional traits of dominant species in various life forms within plant communities and the ecological stoichiometry between deciduous leaf litter and soil were examined. The results revealed that specific leaf area (SLA) was significantly positively correlated with Soil_N.P; plant height (PLH) exhibited a significant positive correlation with Litter_C.N, Litter_C.P, Soil_C.P, and Soil_C.N; leaf area (LA) showed a significant positive correlation with Litter_C.N; and Soil_C.N displayed a significant positive correlation. Additionally, leaf dry matter content (LDMC) was found to have a significant positive correlation with Soil_C.N (Figure 7).

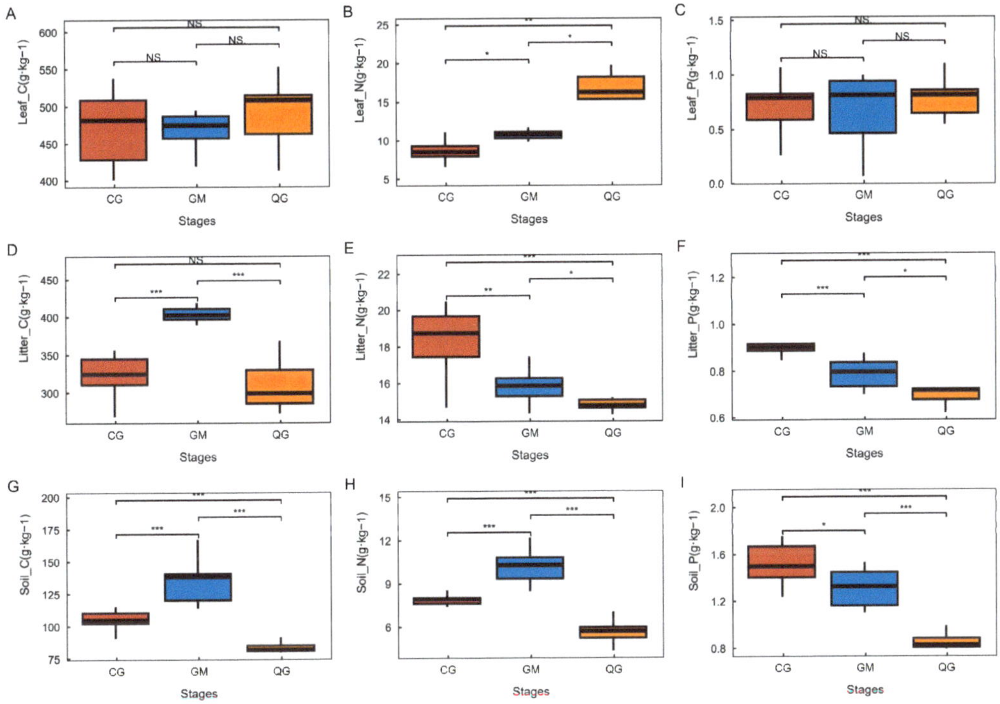

Figure 4. The study examined the carbon, nitrogen, and phosphorus contents of leaf litter and soil for dominant species across various plant-community life forms. In the field of statistics, one-way analysis of variance (ANOVA) is employed for conducting comparative analyses. * Indicates significant difference ($p < 0.05$), ** indicates highly significant difference ($p < 0.01$), *** indicates highly significant difference ($p < 0.001$). Additionally, NS signifies a non-significant difference lacking ecological statistical significance. Each type of life form was subjected to three replications, with three individuals measured in each replication. The stages refer to the plant communities characterized by different life forms, including Leaf_C (leaf carbon content), Leaf_N (leaf nitrogen content), Leaf_P (leaf phosphorus content), Litter_C (litter carbon content), Litter_N (litter nitrogen content), Litter_P (litter phosphorus content), Soil_C (soil carbon content), Soil_N (soil nitrogen content), and Soil_P (soil phosphorus content). These characteristics were observed in CG (grass–shrub plant community), GM (shrub plant community), and QG (tree–shrub plant community). The variables Leaf_C (**A**), Leaf_N (**B**), Leaf_P (**C**), Litter_C (**D**), Litter_N (**E**), Litter_P (**F**), Soil_C (**G**), Soil_N (**H**), and Soil_P (**I**) represent the changes in carbon, nitrogen, and phosphorus contents of leaves, dead litter, and soil in plant communities with different life forms.

To examine the influence of dominant species' functional traits on the ecological stoichiometry of plant communities across various life forms, we categorized the plant functional shapes into two groups, namely, "Fun01" and "Fun02". In this study, the three functional traits, PLH, LT, and LA, were collectively referred to as "Fun01" traits, signifying notable variations in plant communities across different life forms. The three functional traits, LDMC, SLA, and CHL, were also taken into consideration as "Fun02" traits, indicating no significant variations in plant communities across different life forms. The findings indicated that the characteristics of class A had a notable effect on the ratio of carbon to nitrogen in the soil and litter, whereas the traits of class B exerted a substantial influence on the ratio of nitrogen to phosphorus in the soil and the ratio of carbon to phosphorus (Figure 8). Mantel analysis revealed that soil ecological stoichiometry significantly influenced the functional traits of plant communities with diverse life forms in karst environments. Furthermore, the ecological stoichiometry of leaves and litter had a significant impact on the functional traits of plant communities with varying life forms in karst environments. The plant communities in karst areas exhibited diverse responses to functional traits across different life forms, while the ecological stoichiometry of leaves and litter remained unaffected by these responses. Through Mantel analysis, we conducted a comprehensive investigation to explore the relationship between plant functional traits and ecological stoichiometry. The results of our study indicated notable positive associations between plant leaf height (PLH) and soil C:N, as well as litter C:N. Moreover, leaf thickness (LT) displayed significant negative correlations with soil C:N, while leaf area (LA) exhibited significant positive correlations with soil C:N and litter C:N. Additionally, specific leaf area (SLA) showed significant positive correlations with soil N:P and soil C:P (Figure 9). The diagram depicts relationships. The ratio of carbon to nitrogen in the soil is expected to be a significant factor affecting plant leaf height, leaf thickness, and leaf area.

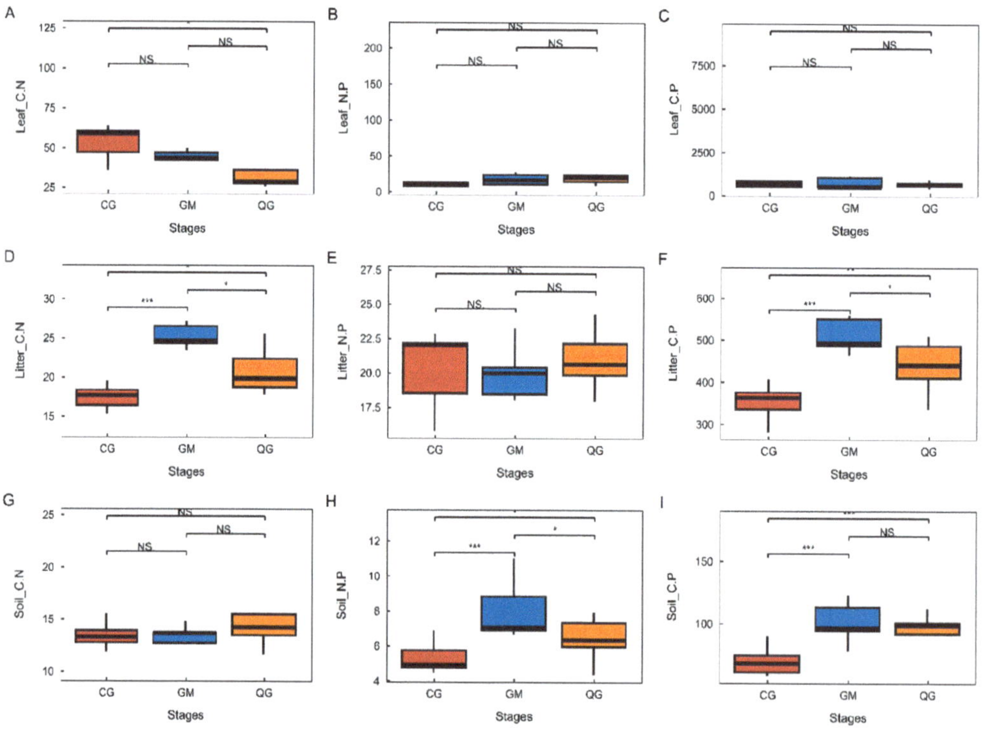

Figure 5. Alterations in the ecological stoichiometry of leaf litter and soil for predominant species within plant communities exhibiting diverse life forms. In statistical analysis, one-way ANOVA is employed to compare groups. * Indicates significant difference ($p < 0.05$), ** indicates highly significant difference ($p < 0.01$), *** indicates highly significant difference ($p < 0.001$), NS signifies a non-significant difference lacking ecological statistical significance. Each life form was subjected to three replicates, with measurements taken for three individuals in each replicate. The study examined the variations in the leaf–litter–soil ecological stoichiometry of plant communities with different life forms, including stages of plant communities and their associated leaf, litter, and soil characteristics. These characteristics include Leaf_C.N (**A**), Leaf_N.P (**B**), Leaf_C.P (**C**), Litter_C.N (**D**), Litter_N.P (**E**), Litter_C.P (**F**), Soil_C.N (**G**), Soil_N.P (**H**), and Soil_C.P (**I**). The plant communities were categorized into CG (grass–shrub), GM (shrub), and QG (tree–shrub) plant communities.

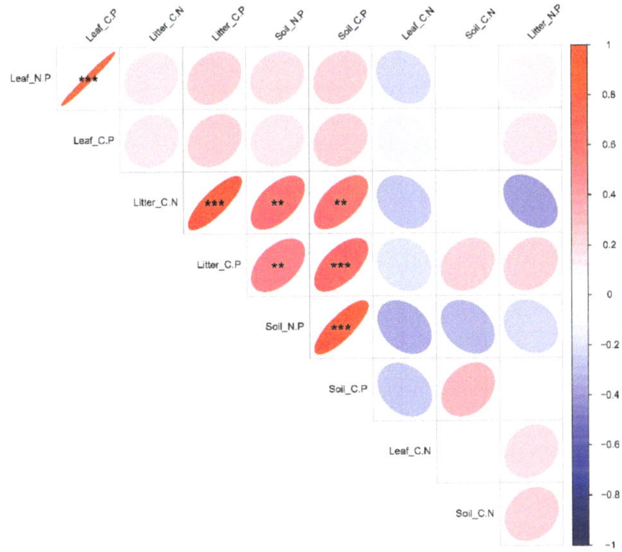

Figure 6. The study examined the correlation between the ecological stoichiometry of leaf, litter, and soil for dominant species within plant communities of various life forms. The variables Leaf_C.N, Leaf_N.P, and Leaf_C.P denote leaf carbon–nitrogen content, leaf nitrogen–phosphorus content, and leaf phosphorus content, respectively. Similarly, Litter_C.N, Litter_N.P, and Litter_C.P denote the contents of carbon–nitrogen, nitrogen–phosphorus, and phosphorus in litter, respectively. Soil_C.N, Soil_N.P, and Soil_C.P denote the levels of carbon–nitrogen, nitrogen–phosphorus, and phosphorus in the soil, respectively. The plant communities under investigation comprised CG (grass–shrub), GM (shrub), and QG (arboreal–shrub) communities. ** and triple *** symbols indicate highly significant differences at $p < 0.01$ and $p < 0.001$, respectively.

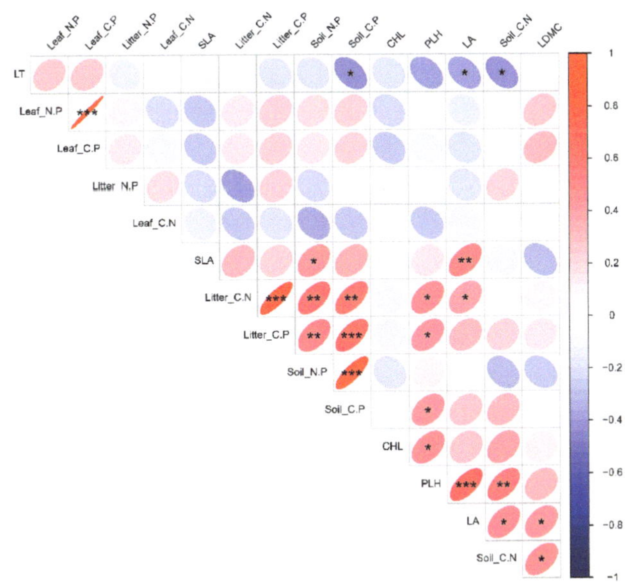

Figure 7. The study examined the relationship between the functional traits of dominant species in plant communities and the ecological stoichiometry of leaf litter and soil. The functional traits

comprised plant height (PLH), leaf thickness (LT), leaf dry matter content (LDMC), leaf area (LA), specific leaf area (SLA), chlorophyll content (CHL), leaf carbon content (Leaf_C.N), leaf nitrogen content (Leaf_N.P), leaf phosphorus content (Leaf_C.P), litter carbon content (Litter_C.N), litter nitrogen content (Litter_N.P), litter phosphorus content (Litter_C.P), soil carbon content (Soil_C.N), soil nitrogen content (Soil_N.P), and soil phosphorus content (Soil_C.P). The plant communities under investigation included CG (grass–shrub), GM (shrub), and QG (tree–shrub) communities. Refer to comments on Figure 3. * Indicates a statistically significant difference at $p < 0.05$; ** and triple *** symbols indicate highly significant differences at $p < 0.01$ and $p < 0.001$, respectively.

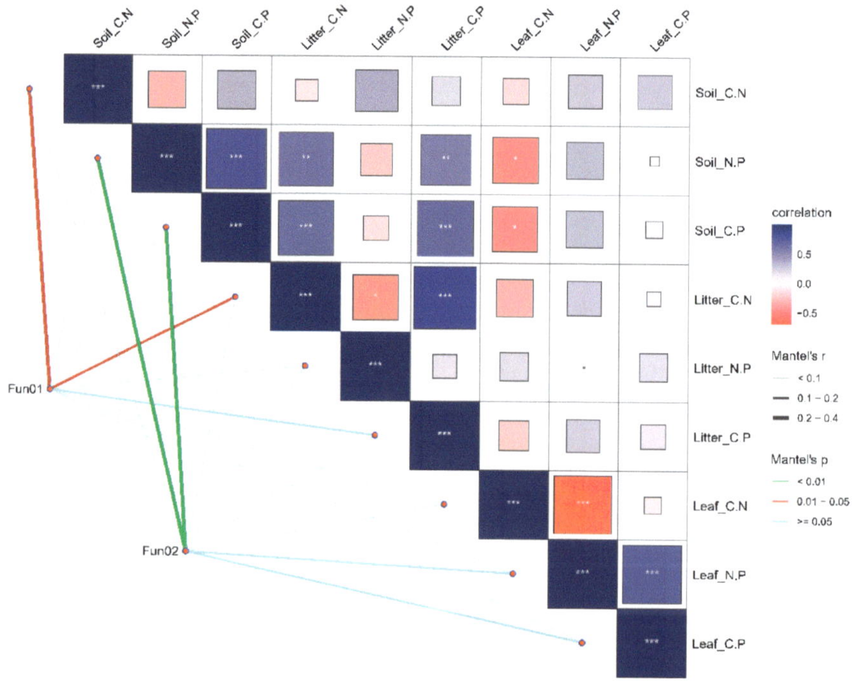

Figure 8. Mantel analysis of the functional shapes of dominant species in plant communities and deciduous leaf–soil ecological stoichiometry. The figure illustrates pairwise comparisons of functional shapes with ecological stoichiometry for various plant life forms, with color gradients representing Spearman's correlation coefficients. The width of the edge corresponds to the Mantel r statistic for distance correlation, while the color of the edge indicates the statistic based on 9999 alignments. * Indicates a statistically significant difference at $p < 0.05$; ** and triple *** symbols indicate highly significant differences at $p < 0.01$ and $p < 0.001$, respectively.

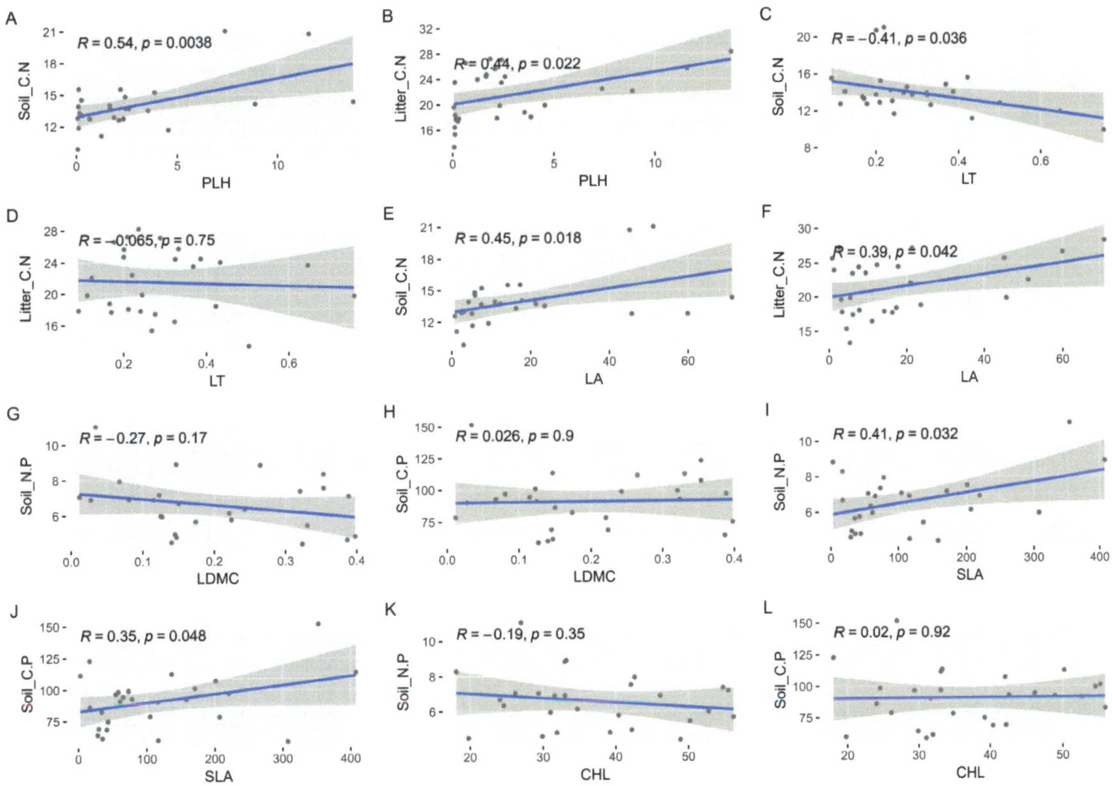

Figure 9. A line plot illustrating the relationship between ecological stoichiometry and functional traits. R represents the magnitude of correlation between the two variables, while *p* represents the level of significance. When *p* is less than or equal to 0.05, it signifies a statistically significant linear correlation between the two variables. The study investigated the relationships between various plant traits and soil properties. The plant traits included plant height (PLH), leaf thickness (LT), leaf dry mass (LDMC), leaf area (LA), specific leaf area (SLA), and chlorophyll content (CHL). The soil properties examined were soil C:N, litter C:N, soil N:P, and soil C:P. The study found linear correlations between PLH and soil C:N (**A**), litter C:N (**B**), LT and soil C:N (**C**), litter C:N (**D**), LA and soil C:N (**E**), litter C:N (**F**), LDMC with soil N:P (**G**), soil C:P (**H**), SLA with soil N:P (**I**), soil C:P (**J**), CHL with soil N:P (**K**), and soil C:P (**L**). The dots represent sample data, blue lines represent correlation trends, and gray areas represent confidence intervals.

4. Discussion

4.1. Changes in Functional Traits of Dominant Species in Plant Communities with Different Life Forms in Karst Areas

Its life form is a characteristic reflection of a plant's physiological, structural, and external morphology with a certain degree of stability after long-term adaptation to an integrated habitat [35]. In this study, we found that LDMC, SLA, and CHL showed no significant differences among different life forms. PLH exhibited significant differences among different life forms. LT showed significant differences between the shrub community and the arboreal–shrub community, as well as between the grass–shrub community and the arboreal–shrub community. LA displayed significant differences between the grass–shrub community and the arboreal–shrub community. These findings align with Yao et al.'s research and indicate that in the same environment, plants with different life forms adopt distinct leaf traits to adapt to their surroundings [36]. In addition, PLH was significantly higher in the arborvitae community than in the shrub and grass–shrub communities (Figure 2A). LT was significantly lower than in the shrub and grass–shrub communities, and LA was significantly higher than in the other life forms (Figure 2B,D). These results are consistent with the findings of Kong et al., suggesting that the competitiveness, productivity, and restoration of vegetation in the arborvitae and grass–shrub communities were stronger compared to those in the shrub and grass–shrub communities. The grass–shrub communities were found to be in more severe environments in terms of life forms, and leaves could better adapt to the environment by increasing their thickness. In the grass–shrub community, the leaf blades could better adapt to the environment and retain soil and water by increasing their thickness, thus enabling them to resist the harsher environment [37]. In addition, the tree–shrub community could improve light and water use efficiency by reducing leaf thickness and increasing leaf area. This adjustment also helps enhance transpiration and promote the rapid growth of plants [38].

4.2. Changes in C, N, and P in the Leaf-Litter–Soil Ecological Stoichiometry of Dominant Species in Plant Communities with Different Life Forms in Karst Areas

Leaf C:N and C:P ratios can characterize a plant's ability to absorb mineral elements for the assimilation of organic matter, reflecting the plant's nutrient use efficiency. In this study, there was no significant difference in the C:N or C:P ratios among different life forms, which contrasts with the results of Liu et al. [17]. This discrepancy may be attributed to the variations in soil nutrient availability and the genetic characteristics of the species studied. Leaf N:P ratio can be used as a diagnostic indicator to determine ecosystems that are subject to nitrogen (N) and phosphorus (P) limitation. Studies have shown that when the ratio of nitrogen to phosphorus (N:P) is less than 14, vegetation growth is limited by nitrogen; when N:P is greater than 16, vegetation growth is limited by phosphorus; and when the N:P ratio falls between 14 and 16, vegetation growth is limited by both nitrogen and phosphorus. In this study, the leaf N:P ratios of various plant communities were all greater than 16 (CG—16.27, GM—17.33, QG—17.81), suggesting that the growth of different plant communities in karst areas is more likely to be limited by phosphorus (P). This aligns with the concept proposed by many scholars that vegetation growth is commonly constrained by phosphorus, especially as vegetation succession progresses [39]. Lower litter C:N ratios are generally considered to indicate high decomposition rates, suggesting faster rates of forest litter decomposition in a region. Among the plant communities with different life types, the grass–shrub community had the lowest C:N ratio (16.68), indicating the fastest decomposition rate of litter and high nutrient cycling efficiency. In contrast, the shrub plant community had the highest C:N ratio (24.25), resulting in the slowest decomposition rate of apomictic litter. The modified results were consistent with the findings of Yu et al. Both studies demonstrated that organic matter decomposition was faster in the grass–shrub and tree–shrub phases. This suggests that in forest communities, maintaining the structural integrity of tree, shrub, and herb hierarchies is essential to enhance microbial utilization of substrates and to increase nutrient turnover and cycling rates [24].

Lower N:P ratios in litter indicate easier decomposition. In this paper, the litter N:P ratio ranged from 19.14 to 33.33, as reported in Pan et al.'s study [40,41]. This suggests that the overall unfavorable decomposition could be attributed to the low nitrogen content and high lignin content in the region and anthropogenic disturbances, as well as the scarcity of soil fauna and microorganisms. These factors contribute to low litter decomposition rates. It can be seen that although karst areas are generally unfavorable to the decomposition of plant communities with different life forms, there are variations among life forms and they are more reliant on soil texture and external disturbances.

Soil ecological stoichiometry is an important indicator for characterizing the composition and quality of soil organic matter [42]. The C:N ratio can be used to assess the rate of decomposition of soil organic matter, while the C:P ratio can characterize the level of effective phosphorus in the soil. Additionally, the N:P ratio is the most effective predictor of nutrient limitation in forests. Compared with global forest soils (14.5, 211, and 14.6) [43] and national terrestrial surface soils (14.4, 136, and 9.3) [44], this study revealed ecological stoichiometries of C:N, C:P, and N:P for forest soils of 14.3, 88.64, and 6.02, respectively. The study found lower C:P and N:P ratios in karst areas with plant communities of various life forms, indicating a high soil P regression level and deficient soil N regression. This suggests that the soil belonged to the N-limited type, further indicating that different plant communities experienced significant habitat pressure [45]. In conclusion, when constructing various living plant communities in karst regions, it is essential to ensure that the layers are complete; enhance the decomposition of apomictic material and the mineralization of soil nutrients; facilitate the accumulation of C, N, P, and other nutrients; and self-regulate their stoichiometric relationships [12]. This will help strengthen their ability to resist fragile habitats.

4.3. Ecological Stoichiometric Correlations between Functional Traits of Dominant Species and Leaf Litter–Soil Interactions in Karst Areas with Plant Communities of Different Life Forms

In this study, we found that plant height (PLH), leaf thickness (LT), specific leaf area (SLA), and leaf area (LA) correlated more strongly with soil and litter ecological stoichiometry and were insensitive to leaf ecological stoichiometry. In this study, we demonstrated that the soil C:N ratio was maintained between 10 and 15, which closely resembled the findings of He et al.'s study. This suggests that mineralization and nitrogen release happen rapidly, aiding nutrient uptake by plants. A lower C:N ratio corresponds to a quicker nitrogen release [46]. Combined with our results, the soil carbon-to-nitrogen ratio in karst areas is more sensitive to plant leaf area, leaf thickness, and leaf turnover. In addition, the most favorable soil C:N ratio is usually considered to be around 25:1. When the soil C:N ratio is too high, microbial decomposition and mineralization are slow, and the available nitrogen in the soil needs to be utilized [42]. On the contrary, the C:N ratio is too small in karst areas (CG—14.21, GM—14.53, QG—14.72), resulting in high soil nitrogen contents. To enhance microbial decomposition, it is necessary to increase soil organic matter appropriately to facilitate apoplastic decomposition. The initial C:N ratio of apoplastic matter responds to the impact of climate change on the carbon-to-nitrogen ratio of plants. For instance, nitrogen deposition causes an increase in nitrogen elements, leading to a lower C:N ratio [47]. In our study, apoplastic C:N showed a significant positive correlation with PLH and LA. This suggests that in karst environments, arboreal and shrub plant communities exhibit higher C:N ratios. Plants with high C:N ratios demonstrate increased nitrogen use efficiency, while apoplastic materials with low C:N ratios exhibit fast decomposition traits. Low apoplastic C:N and a wetter climate favor soil organic carbon (SOC) accumulation. This process can be enhanced by both natural processes and human interventions that reduce apoplastic C:N levels. Additionally, global climate change may increase the wetness index, further promoting SOC accumulation [48,49].

Currently, researchers have directed their attention to the impact of soil quality on plant functional traits in forest stands. The findings suggest that soil fertility influences plant functional traits, demonstrating the screening effect of soil elemental ecological stoichiometry on functional traits during the establishment of plant communities [49]. This implies a close relationship between belowground habitats and aboveground components. The reciprocal regulation between the aboveground and belowground components plays a crucial role in shaping ecosystem development and provides the basis for improving vegetation productivity [50]. In this study, a significant positive correlation was observed between soil N:P and SLA, which aligns with the findings of Yu et al. [23]. This may be attributed to the fact that N and P are primary elements composing plant organisms, and the content of N and P also influences the efficiency of photosynthesis, thereby affecting plant growth and physiological state. The findings indicate that the survival and growth of plants are influenced by the stoichiometric balance of limiting elements, and they adapt to various habitats by balancing trade-offs and synergies of functional traits [50,51]. In the future, there will be a comprehensive exploration of the correlation between nutrient element content, microbial population, and biomass, as well as their stoichiometric relationships in the subsurface space, in relation to the functional traits of aboveground plants. This research aims to provide support for the regulation of ecosystem function.

The findings of this study demonstrated that the association between the functional traits and ecological stoichiometry of plant communities with diverse life forms in karst areas was primarily defined by the interplay between plant functional traits and the ecological stoichiometry of soil and litter. The reasons were analyzed as follows: The high incidence of rocky desertification and habitat heterogeneity in the region, along with the complex karst geological environment, may lead to a greater influence of geological and geographical elements on plant functional traits. These elements include the depth of groundwater, development of fissures, slope, and slope position [8]. However, further exploration is needed to understand the influence of lithology and other geological conditions on plant functional traits. Additionally, factors such as species, succession, and habitat affect the ecological stoichiometry of the leaf–apomictic soil continuum [52]. The genetic characteristics of species play a significant role in determining their selective nutrient absorption and utilization. Additionally, plant communities with varying lifestyles exert regulatory influences on apomictic reserves and stand quality [53]. Consequently, these factors exhibit a strong interdependent relationship with one another. In the future, it is imperative to thoroughly elucidate the mechanism by which habitats influence ecological processes, explore trade-off strategies between different ecological processes, and enhance the self-regulation and self-balancing capacity of ecosystems through moderate disturbance.

5. Conclusions

(1) The functional traits of various life forms within plant communities in karst regions demonstrate adaptation to environmental changes through diverse expressions and survival strategies. The growth of various plant communities in karst areas is more likely to be constrained by phosphorus (P), and the potential for apomictic decomposition and soil nutrient mineralization is comparatively weaker.

(2) The functional traits of various life forms within karst plant communities exhibit greater sensitivity to soil C: N and apomictic C:N. Disparities in habitat may influence the ecological stoichiometry of the plant leaf–apomictic soil continuum.

Author Contributions: Conceptualization, Y.W.; methodology, Y.W. and L.F.; software, Y.W. and L.Z.; validation, Y.D., Z.F. and L.Z.; formal analysis, Y.W.; investigation, Y.W., Z.F. and Y.D.; resources, Y.W.; data curation, Y.W.; writing—original draft preparation, Y.W.; writing—review and editing, Y.W., T.F. and L.Z.; visualization, Y.W.; supervision, Z.F.; project administration, Y.W.; funding acquisition, Y.W. All authors have read and agreed to the published version of the manuscript.

Funding: This research was funded by the Guizhou Provincial Science and Technology Program (grant number: Qiankehejichu—ZK [2024] General 601); the Ministry of Education's Industry–University Cooperation Collaborative Education Program (grant number: 220802313202357); the Science and Technology Project of Bijie City open competition mechanism to select the best candidates (grant number: BKHZDZX[2023]1); the Bijie Talent Team of Biological Protection and Ecological Restoration in the Liuchong River Basin (grant number: 202112); and the Guizhou Province Key Laboratory of Ecological Protection and Restoration of Typical Plateau Wetlands (grant number: Qian Kehe Platform Talents [2020] 2002).

Data Availability Statement: The original contributions presented in the study are included in the article, further inquiries can be directed to the corresponding author.

Conflicts of Interest: The authors declare no conflicts of interest.

References

1. Liu, X.J.; Ma, K.P. Plant functional traits-concepts, applications and future directions. *Sci. Sin. Vitae* **2015**, *45*, 325–339. [CrossRef]
2. Diaz, S.; Lavorel, S.; de Bello, F.; Quetier, F.; Grigulis, K.; Robson, T.M. Incorporating plant functional diversity effects in ecosystem service assessments. *Proc. Natl. Acad. Sci. USA* **2007**, *104*, 20684–20689. [CrossRef]

3. Portela, A.P.; Durance, I.; Vieira, C.; Honrado, J. Response-effect trait overlap and correlation in riparian plant communities suggests sensitivity of ecosystem functioning and services to environmental change. *Sci. Total Environ.* **2023**, *860*, 160549. [CrossRef]
4. Gladstone-Gallagher, R.V.; Pilditch, C.A.; Stephenson, F.; Thrush, S.F. Linking traits across ecological scales determines functional resilience. *Trends Ecol. Evol.* **2019**, *34*, 1080–1091. [CrossRef] [PubMed]
5. Tsakalos, J.L.; Renton, M.; Riviera, F.; Veneklaas, E.J.; Dobrowolski, M.P.; Mucina, L. Trait-based formal definition of plant functional forms and functional communities in the multi-species and multi-traits context. *Ecol. Complex.* **2019**, *40*, 100787. [CrossRef]
6. Wang, Y.; Zhang, L.; Chen, J.; Feng, L.; Li, F.; Yu, L. Study on the relationship between functional characteristics and environmental factors in karst plant communities. *Ecol. Evol.* **2022**, *12*, e9335. [CrossRef]
7. Wang, Y.; Zhang, L.; Chen, J.; Feng, L.; Li, F.; Yu, L. Functional diversity of plant communities in relationship to leaf and soil stoichiometry in karst areas of southwest China. *Forests* **2022**, *13*, 864. [CrossRef]
8. Lu, Z.; Wang, P.; Ou, H.; Wei, S.; Wu, L.; Jiang, Y.; Wang, R.; Liu, X.; Wang, Z.; Chen, L.; et al. Effects of different vegetation restoration on soil nutrients, enzyme activities, and microbial communities in degraded karst landscapes in southwest China. *Forest Ecol. Manag.* **2022**, *508*, 120002. [CrossRef]
9. Liu, C.; Li, Y.; Yan, P.; He, N. How to improve the predictions of plant functional traits on ecosystem functioning? *Front. Plant Sci.* **2021**, *12*, 622260. [CrossRef] [PubMed]
10. Wang, J.; Wang, Y.; He, N.; Ye, Z.; Chen, C.; Zang, R.; Feng, Y.; Lu, Q.; Li, J. Plant functional traits regulate soil bacterial diversity across temperate deserts. *Sci. Total Environ.* **2020**, *715*, 136976. [CrossRef]
11. He, J.S.; Han, X.G. Ecological stoichiometry: Searching for unifying principle from individuals to ecosystems. *Chin. J. Plan Ecol.* **2010**, *34*, 2–6. [CrossRef]
12. Elser, J.J.; Fagan, W.F.; Kerkhoff, A.J.; Swenson, N.G.; Enquist, B.J. Biological stoichiometry of plant production: Metabolism, scaling and ecological response to global change. *New Phytol.* **2010**, *186*, 593–608. [CrossRef]
13. Wang, S.J. An outline of karst geomorphology zoning in the karst areas of southern China. *Bull. Mineral. Petrol. Geochem.* **2003**, *33*, 120–126. [CrossRef]
14. Zeng, D.H.; Chen, G.S. Ecological stoichiometry: A science to explore the complexity of living systems. *Acta Ecol. Sin.* **2005**, *29*, 1007–1019. [CrossRef]
15. Wang, L.J.; Wang, P.; Sheng, M.Y. Stoichiometry characteristics of soil nutrient elements and its influencing factors in typical in karst rocky desertification ecosystems, southwest China. *Acta Ecol. Sin.* **2018**, *38*, 6580–6593. [CrossRef]
16. Yang, D.L.; Yu, Y.H.; Qin, S.Y.; Zhong, X.P. Contents and ecological stoichiometry characteristics of soil nutrients under different land utilization forms in stony desertification area. *Southwest Chin. J. Agric. Sci.* **2018**, *31*, 1875–1881. [CrossRef]
17. Liu, N.; Yu, L.F.; Zhao, Q.; Wu, Y.N.; Yan, L.B. C: N: P stoichiometry of leaf-litter-soil continuum in secondary forests of the rocky desertification regions of the karst plateau. *Chin. J. Appl. Environ. Biol.* **2020**, *26*, 681–688. [CrossRef]
18. Han, W.; Fang, J.; Guo, D.; Zhang, Y. Leaf nitrogen and phosphorus stoichiometry across 753 terrestrial plant species in China. *New Phytol.* **2005**, *168*, 377–385. [CrossRef]
19. Wang, S.Q.; Yu, G.R. Ecological stoichiometry characteristics of ecosystem carbon, nitrogen and phosphorus elements. *Acta Ecol. Sin.* **2008**, *28*, 3937–3947. [CrossRef]
20. Guittar, J.; Goldberg, D.; Klanderud, K.; Telford, R.J.; Vandvik, V. Can trait patterns along gradients predict plant community responses to climate change? *Ecology* **2016**, *97*, 2791–2801. [CrossRef] [PubMed]
21. Li, J.X.; Xu, W.T.; Xiong, G.M.; Wang, Y.; Zhao, C.M.; Lu, Z.J.; Li, Y.L.; Xie, Z.Q. Leaf nitrogen and phosphorus concentration and the empirical regulations in dominant woody plants of shrublands across southern China. *Chin. J. Plant Ecol.* **2017**, *41*, 31–42.
22. Zhong, Q.L.; Liu, L.B.; Xu, X.; Yang, Y.; Guo, Y.M.; Xu, H.Y.; Cai, X.L.; Ni, J. Variations of plant functional traits and adaptive strategy of woody species in a karst forest of central Guizhou province, southwestern China. *Chin. J. Plant Ecol.* **2018**, *42*, 562–572. [CrossRef]
23. He, N.; Li, Y.; Liu, C.; Xu, L.; Li, M.; Zhang, J.; He, J.; Tang, Z.; Han, X.; Ye, Q.; et al. Plant trait networks: Improved resolution of the dimensionality of adaptation. *Trends Ecol. Evol.* **2020**, *35*, 908–918. [CrossRef]
24. Yu, Y.H.; Zhong, X.P.; Zheng, W.; Chen, Z.X.; Wang, J.X. Species diversity, functional traits, stoichiometry and correlation of plant communities in different succession stages of karst forest. *Acta Ecol. Sin.* **2021**, *41*, 2408–2417. [CrossRef]
25. Yan, M.M.; Zhou, Z.; Wang, J.; Gu, X.P.; Xiao, J.Y. Study on the dynamic change of soil moisture in karst area: A case of Huaxi district in Guiyang city. *Carsol. Sin.* **2016**, *35*, 446–452.
26. Wang, Y.; Chen, J.; Zhang, L.; Feng, L.; Yan, L.; Li, F.; Zhao, X.; Yu, L.; Liu, N. Relationship between diversity and stability of a karst plant community. *Ecol. Evol.* **2022**, *12*, e9254. [CrossRef] [PubMed]
27. Sheng, M.Y.; Xiong, K.N.; Cui, G.Y.; Liu, Y. Plant diversity and soil physical-chemical properties in karst rocky desertification ecosystem of Guizhou, China. *Acta Ecol. Sin.* **2015**, *35*, 434–448. [CrossRef]
28. Pérezharguindeguy, N.; Díaz, S.; Garnier, E. New handbook for standardised measurement of plant functional traits worldwide. *Aust. J. Bot.* **2013**, *61*, 167–234. [CrossRef]
29. Bao, S.D. *Soil Agrochemical Analysis*, 3rd ed.; Agricultural Press: Beijing, China, 2005; pp. 45–52.
30. *NY/T2017-2011*; Determination of Nitrogen, Phosphorus and Potassium in Plants. Ministry of Agriculture: Beijing, China, 2011.
31. *LY/T1228-2015*; Determination of Nitrogen in Forest Soil. Ministry of Agriculture: Beijing, China, 2015.

32. *LY/T1232-2015*; Phosphorus Determination Methods of Forest Soils. Ministry of Agriculture: Beijing, China, 2015.
33. The SPSSAU Project. SPSSAU, Version 25.0. 2022. Online Application Software. Available online: https://www.spssau.com (accessed on 12 February 2024).
34. R Core Team. R: A Language and Environment for Statistical Computing. R Foundation for Statistical Computing. 2020. Available online: https://www.R-project.org/ (accessed on 12 February 2024).
35. Wang, J.; Zhu, J.; Ai, X.R.; Yao, L.; Huang, X.; Wu, M.L.; Zhu, Q.; Hong, J.F. Effects of topography on leaf functional traits across plant life forms in Xingdou mountain, Hubei, China. *Chin. J. Plant Ecol.* **2019**, *43*, 447–457. [CrossRef]
36. Yao, X.Y.; Hu, Y.S.; Liu, Y.H. Plant function traits and functional diversities of different communities in broad-leaved Korean pine forests in the Changbai mountain. *Nat. Sci. Ed.* **2014**, *42*, 77–84. [CrossRef]
37. Kong, L.W.; Yu, Y.H.; Xiong, K.N.; Wei, C.S.; Zhang, S.H. Leaf functional traits of *Zanthoxylum planispinum* var. dintanensis and their response to management measures. *J. Forest Environ.* **2022**, *42*, 364–373. [CrossRef]
38. Shui, W.; Guo, P.P.; Zhu, S.F.; Feng, J.; Sun, X.; Li, H. Variation of plant functional traits and adaptive strategy of woody species in degraded karst Tiankeng of Yunnan province. *Sci. Geogra. Sin.* **2022**, *42*, 1295–1306. [CrossRef]
39. Koerselman, W.; Meuleman, A. The vegetation N:P ratio: A new tool to detect the nature of nutrient limitation. *J. Appl. Ecol.* **1996**, *33*, 1441–1450. [CrossRef]
40. Han, W.X.; Wu, Y.; Tang, L.Y.; Chen, Y.H.; Li, L.P.; He, J.S.; Fang, J.Y. Leaf carbon, nitrogen and phosphorus stoichiometry across plant species in Beijing and its periphery. *Acta Scin. Nat. Univ. Pekinensis* **2009**, *45*, 855–860. [CrossRef]
41. Pan, F.J.; Zhang, W.; Wang, K.L.; He, X.Y.; Liang, S.C.; Wei, G.F. Litter C:N:P ecological stoichiometry character of plant communities in typical karst Peak-Cluster depression. *Acta Ecol. Sin.* **2011**, *31*, 335–343.
42. Song, Y.F.; Lu, Y.J.; Liu, T.J.; Liu, H.W.; Yan, Z.Y.; Wang, H.Q. Soil-plant-microbial C, N, P and their stoichiometric characteristics in different rainfall zones of desert steppe. *Acta Ecol. Sin.* **2020**, *40*, 4011–4023.
43. Li, F.L.; Bao, W.K. Responses of the morphological and anatomical structure of the plant leaf to environmental change. *Chin. Bull. Bot.* **2005**, *23*, 118–127.
44. Xu, Y.S.; Huang, H.X.; Shi, Q.R.; Yang, X.D.; Zhou, L.L.; Zhao, Y.T.; Zhang, Q.Q.; Yan, E.R. Response of soil water content to change in plant functional traits in evergreen broadleaved forests in eastern Zhejiang province. *J. Plant Ecol.* **2015**, *39*, 857–866. [CrossRef]
45. Jiang, S.; Zhang, J.; Tang, Y.; Li, Z.; Liu, H.; Wang, L.; Wu, Y.; Liang, C. Plant functional traits and biodiversity can reveal the response of ecosystem functions to grazing. *Sci. Total Environ.* **2023**, *899*, 165636. [CrossRef]
46. He, J.; Tian, Q.; Song, L.L. Community characteristics of herbaceous plants and analysis of C, N and P contents of main plant functional groups in high altitude area on the northern slope of Minshan Mountains. *Ecol. Environ. Sci.* **2020**, *29*, 489–497. [CrossRef]
47. Zhou, G.; Xu, S.; Ciais, P.; Manzoni, S.; Fang, J.; Yu, G.; Tang, X.; Zhou, P.; Wang, W.; Yan, J.; et al. Climate and litter C/N ratio constrain soil organic carbon accumulation. *Natl. Sci. Rev.* **2019**, *6*, 746–757. [CrossRef]
48. Funk, J.L.; Standish, R.J.; Stock, W.D.; Valladares, F. Plant functional traits of dominant native and invasive species in mediterranean-climate ecosystems. *Ecology* **2016**, *97*, 75–83. [CrossRef] [PubMed]
49. Guo, Z.W.; Chen, S.L.; Yang, Q.P.; Li, Y.C. Effects of stand density on *Oligostachyum lubricum* leaf carbon, nitrogen, and phosphorus stoichiometry and nutrient resorption. *Chin. J. Appl. Ecol.* **2013**, *24*, 893–899. [CrossRef]
50. Ma, K.; Guo, Q. Progress and recent trends of vegetation ecology in China. *Sci. Sin. Vitae* **2021**, *51*, 215–218. [CrossRef]
51. Zhang, X.; Li, B.; Peñuelas, J.; Sardans, J.; Cheng, D.; Yu, H.; Zhong, Q. Resource-acquisitive species have greater plasticity in leaf functional traits than resource-conservative species in response to nitrogen addition in subtropical China. *Sci. Total Environ.* **2023**, *903*, 166177. [CrossRef] [PubMed]
52. Chen, L.L.; Deng, Q.; Yuan, Z.Y.; Mu, X.M.; Kallenbach, R.L. Age-related C: N: P stoichiometry in two plantation forests in the loess Plateau of China. *Ecol. Eng.* **2018**, *120*, 857–866. [CrossRef]
53. Zhang, Z.H.; Hu, G.; Zhu, J.D.; Ni, J. Scale-dependent spatial variation of species abundance and richness in two mixed evergreen-deciduous broad-leaved karst forests, southwest China. *Acta Ecol. Sin.* **2012**, *32*, 5663–5672. [CrossRef]

plants

MDPI

Article

Neighborhood Diversity Promotes Tree Growth in a Secondary Forest: The Interplay of Intraspecific Competition, Interspecific Competition, and Spatial Scale

Haonan Zhang [1,2,*], Yuanyun Gao [1], Xiao Zheng [2], Yaping Hu [2], Xu Zhou [2], Yanming Fang [3], Yao Li [3], Lei Xie [3] and Hui Ding [2,*]

1 Innovative Research Team for Forest Restoration Mechanisms, Chishui National Ecological Quality Comprehensive Monitoring Stations, Nanjing Institute of Environmental Sciences, Ministry of Ecology and Environment (MEE), Nanjing 210042, China
2 Research Center for Nature Conservation and Biodiversity, State Environmental Protection Scientific Observation and Research Station for Ecology and Environment of Wuyi Mountains, State Environmental Protection Key Laboratory on Biosafety, Nanjing Institute of Environmental Sciences, Ministry of Ecology and Environment (MEE), Nanjing 210042, China
3 Co-Innovation Center for Sustainable Forestry in Southern China, College of Life Sciences, Key Laboratory of State Forestry and Grassland Administration on Subtropical Forest Biodiversity Conservation, Nanjing Forestry University, Nanjing 210037, China
* Correspondence: zhn@nies.org (H.Z.); nldinghui@sina.com (H.D.); Tel.: +86-18614055380 (H.Z.)

Abstract: Understanding the biodiversity–productivity relationship (BPR) is crucial for biodiversity conservation and ecosystem management. While it is known that diversity enhances forest productivity, the underlying mechanisms at the local neighborhood level remain poorly understood. We established a 9.6 ha dynamic forest plot to study how neighborhood diversity, intraspecific competition, and interspecific competition influence tree growth across spatial scales using linear mixed-effects models. Our analysis reveals a significant positive correlation between neighborhood species richness (NSR) and relative growth rate (RGR). Notably, intraspecific competition, measured by conspecific neighborhood density and resource competition, negatively impacts RGR at finer scales, indicating intense competition among conspecifics for limited resources. In contrast, interspecific competition, measured by heterospecific density and resource competition, has a negligible impact on RGR. The relative importance of diversity and intra/interspecific competition in influencing tree growth varies with scale. At fine scales, intraspecific competition dominates negatively, while at larger scales, the positive effect of NSR on RGR increases, contributing to a positive BPR. These findings highlight the intricate interplay between local interactions and spatial scale in modulating tree growth, emphasizing the importance of considering biotic interactions and spatial variability in studying BPR.

Keywords: tree growth; species diversity; intraspecific competition; interspecific competition; scale-dependent effects; secondary forests

check for updates

Citation: Zhang, H.; Gao, Y.; Zheng, X.; Hu, Y.; Zhou, X.; Fang, Y.; Li, Y.; Xie, L.; Ding, H. Neighborhood Diversity Promotes Tree Growth in a Secondary Forest: The Interplay of Intraspecific Competition, Interspecific Competition, and Spatial Scale. *Plants* **2024**, *13*, 1994. https://doi.org/10.3390/plants13141994

Academic Editor: Xugao Wang

Received: 28 May 2024
Revised: 17 July 2024
Accepted: 19 July 2024
Published: 21 July 2024

1. Introduction

The interplay between species diversity and ecosystem productivity, known as the biodiversity–productivity relationship (BPR), represents a cornerstone of ecological research. Understanding this relationship is crucial for comprehending the consequences of biodiversity loss on ecosystem functionality and services [1–4]. The insights gained from studying the BPR have profound implications for ecosystem function, conservation biology, and the sustainable management of natural resources in the face of global biodiversity decline [4–7].

Central to the BPR is the premise that increased species diversity leads to enhanced ecosystem productivity, a concept supported by extensive research across various ecosystems [1,5].

Empirical evidence supporting a positive BPR spans diverse landscapes, including grasslands, forests, and aquatic environments [6]. For instance, investigations in grassland ecosystems have shown that plots with increased species richness exhibit higher biomass production, directly indicating enhanced ecosystem productivity [2,4,6–9]. Similarly, forest ecosystems have demonstrated positive correlations between tree species diversity and productivity, with a notable relationship between biodiversity and tree growth performance [2,4,7–10]. The positive impact of biodiversity on tree growth is considered a significant characteristic of the biodiversity effect, widely observed not only in large-scale studies [2] but also at finer scales, such as at the neighborhood level where individual trees interact with their diversely constituted neighboring trees [7–10]. These findings across scales underscore the robustness of the biodiversity–productivity relationship, highlighting the critical role of species diversity in fostering ecosystem productivity not only at the community level but also in influencing individual organism interactions within their immediate environments at a fine spatial scale [7–10].

Although the relationship between species diversity and ecosystem productivity is widely supported, it exhibits considerable variability across different spatial scales. At larger scales, such as landscapes or biomes, the BPR tends to show more consistent patterns, with higher biodiversity generally associated with increased productivity. At these broad scales, the impact of species diversity on productivity is influenced by factors such as species turnover (beta diversity) and the distribution of functional traits across the landscape, which promote more efficient resource use and enhance ecosystem stability [11–13]. However, the variability introduced by local density-dependent interactions at finer scales complicates the generalizability of the observed positive correlations at broader scales. Studies suggest that the strength and even direction of this relationship can vary significantly with spatial scale, with finer scales showing more variability [1,13]. This scale-dependent variability is attributed to localized resource competition and the specific ecological niches occupied by different species, which may not be as apparent in large-scale studies [7–9,14]. Given these insights, it is imperative to further investigate the mechanisms through which local intra- and interspecific competition affect the biodiversity–productivity relationship, particularly at the neighborhood scale [9,10,15–17]. Therefore, developing multiscale models that accurately capture the complex interplay of diversity's effects on ecosystem functioning across various scales is not only essential for advancing ecological theory and modeling but also crucial for understanding the more intricate neighborhood effects on the BPR [10,18–22].

At smaller scales, the variability in the species diversity–productivity relationship is significantly linked to neighbor effects, where both conspecific and heterospecific density dependence and resource competition may play a substantial role in modulating this relationship [7–9,15–17]. In forest communities, resources such as light and water are often limited, necessitating inevitable competition among individual trees with their neighboring conspecifics or heterospecifics as they grow [9–11]. The intensity of this competition is dependent on the density and size of adjacent trees [10]. On one hand, at the local scale, particularly at the fine scale of tree-to-tree interactions, neighboring tree individuals frequently experience intense intraspecific and interspecific competition [9–11,15–17], which is known as conspecific density dependence and has been widely documented in both tropical and subtropical forest communities [15–17]. On the other hand, the ability of plants of different sizes to acquire resources varies significantly. Larger trees are often more capable of capturing sunlight and accessing soil nutrients, thereby gaining a competitive advantage in resource acquisition [7–11]. Consequently, tree density and size markedly affect individual tree growth performance. However, the regulatory mechanisms and the relative importance of these effects on the species diversity–productivity relationship remain understudied. Therefore, incorporating the competitive effects of neighboring individuals into studies of the biodiversity–growth relationship is crucial, particularly at the neighborhood scale where these interactions are most pronounced.

In this context, our study quantified neighborhood species richness (NSR), conspecific and heterospecific neighborhood density (CND and HND), and conspecific and heterospecific resource competition indices (CNCI and HNCI) at the neighborhood scale (across five scale gradients from 2.5 m to 20 m). We linked these influencing factors with the individual trees' relative growth rate (RGR) to estimate the effect sizes of different factors on tree growth at specific scales. This approach enabled us to quantitatively analyze how species diversity and intra- and interspecific competition collectively regulate individual tree growth. To validate our hypotheses, we established a 9.6-hectare dynamic monitoring plot in a secondary forest that had undergone severe anthropogenic disturbance 20 years prior. We conducted a comprehensive survey and repeated measurements every five years. Our hypotheses are as follows (Figure 1): Hypothesis 1 asserts that there is a significant positive correlation between neighborhood scale biodiversity and relative growth rate (RGR) during the early stages of secondary succession, demonstrating a beneficial biodiversity–production relationship. Hypothesis 2 proposes that in forests undergoing early secondary recovery following disturbance, conspecific negative density dependence (CNDD) is prevalent, likely exerting a negative impact on individual RGR. Furthermore, the influence of heterospecific negative density dependence is proposed to be less pronounced than that of intraspecific competition. Hypothesis 3 suggests that neighborhood diversity, density, and resource competition effects collectively regulate the relationship between species diversity and tree growth, with the relative importance of various neighborhood effects varying across different testing scales. At the tree-to-tree neighborhood scale, conspecific negative density dependence may dominate, while the diversity effect tends to become relatively more important and exhibits a more pronounced positive effect on individual tree growth as spatial scale increases. Exploring the local density-dependent interactions at the neighborhood scale and their impact on species diversity and ecosystem productivity is crucial for ecological theory and ecosystem management practices.

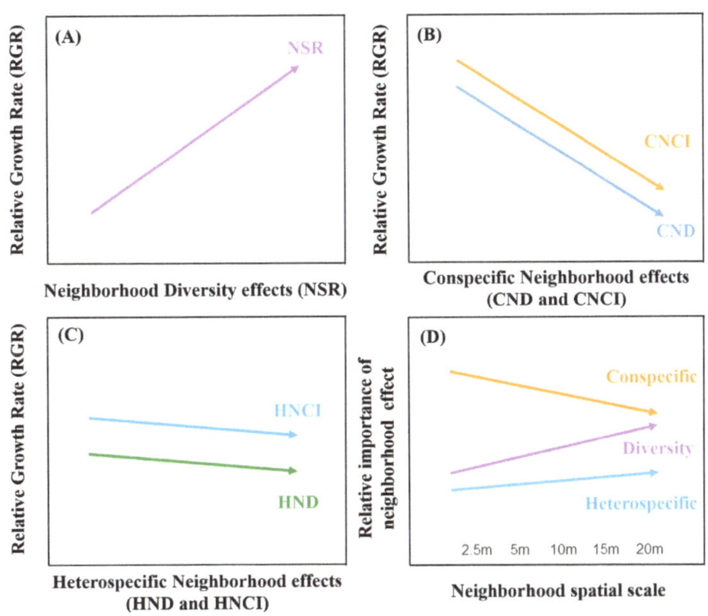

Figure 1. The research framework and scientific hypotheses. (i) In panel (**A**), Hypothesis 1 (H1) addresses the critical relationship between biodiversity effect and productivity. Specifically, we hypothesize a significant positive correlation between neighborhood species richness (NSR) and relative growth rate (RGR). (ii) In panels (**B**,**C**), Hypothesis 2 (H2) focuses on the roles of intraspecific

and interspecific competition in shaping tree growth patterns in secondary forests. This hypothesis quantifies conspecific and heterospecific neighborhood density (CND and HND), as well as conspecific and heterospecific resource competition indices (CNCI and HNCI) at the neighborhood scale. We hypothesize that conspecific neighborhood effects (CND and CNCI) exert a pronounced negative impact on RGR at finer scales, whereas heterospecific neighborhood effects (HND and HNCI) are generally insignificant. (iii) In panel (**D**), Hypothesis 3 (H3) examines the relative importance of neighborhood diversity effects, conspecific, and heterospecific neighborhood effects. Specifically, we hypothesize that neighborhood diversity, density, and resource competition collectively regulate the relationship between species diversity and tree growth. The relative importance of these neighborhood effects is expected to vary across different spatial scales. At the tree-to-tree neighborhood scale, conspecific negative density dependence may dominate, while the biodiversity effect is anticipated to become increasingly important and exhibit a more pronounced positive effect on individual tree growth as the spatial scale increases.

2. Results

2.1. Neighborhood Diversity Effects

Consistent with our initial hypothesis (Figure 1, H1), parameter estimates from linear mixed-effects models (LMMs) demonstrated a significant positive correlation between individual tree neighborhood diversity and their relative growth rate (Figure 2). Specifically, an increase in neighborhood diversity correlates with improved growth performance of individual trees, illustrating a positive biodiversity–productivity relationship at the scale of 5–20 m (positive significant relationship at the scale of 10–20 m). We observed a scale-dependent effect across different species (Figure 3), where the biodiversity–productivity relationship exhibited greater variability at smaller scales but gradually stabilized into a consistent positive correlation as spatial scale increased.

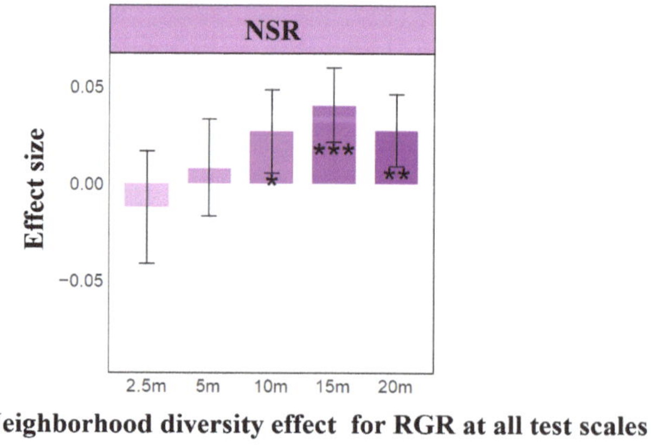

Figure 2. Parameter estimates of species diversity effects on relative growth rate (RGR) at neighborhood scales. The purple bar graphs depict the parameter estimation of neighborhood diversity richness (NSR) on the RGR of individual focal trees across different spatial scales. Positive values denote positive effects, while negative values signify negative effects. Significance levels are denoted by an asterisk (* $p < 0.05$; ** $p < 0.01$; *** $p < 0.001$).

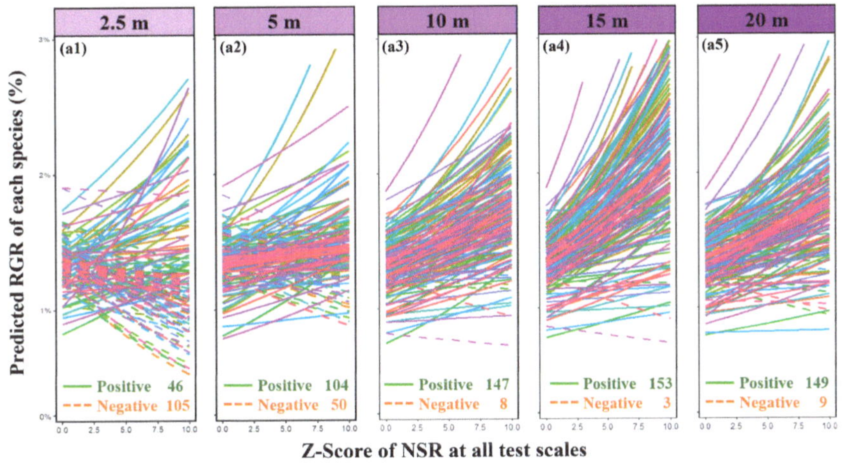

Figure 3. Multiscale relationship between species richness and relative growth rate (RGR) among all species. It displays annual diversity–RGR relationships for each of the 158 observed species at various spatial scales: 2.5 m (**a1**), 5 m (**a2**), 10 m (**a3**), 15 m (**a4**), 20 m (**a5**). Different species are represented by different colors in the lines (see Figure S1 in the supporting information), the solid line represents a positive correlation, and the dotted line represents a negative correlation. Predicted RGRs are back-transformed from the linear mixed model as described in the text, and all biodiversity effects were Z-score transformed at quantification. To enhance comparability and uniformity of the presentation results, we converted the Z-score values to positive in the figures; the untransformed original values can be found in Supplementary Figure S2.

2.2. Intraspecific and Interspecific Competition

Results from linear mixed-effects models (LMMs) reveal a significant and scale-varying quantitative relationship between intraspecific neighborhood effects and RGR. Specifically, at a small scale (2.5 m), both conspecific neighbor density (CND) and conspecific neighbor competition index (CNCI) consistently exhibited a significant negative impact on RGR (Figure 4a,b). Across scales from 5 m to 20 m, CND showed a positive but not significant correlation with RGR, with substantial variability among different species (Figure 5(a1–a5)). CNCI maintained a significant negative correlation with RGR across all scales (Figure 5(b1–b5)), displaying a uniform trend among 158 species within a 20 × 20 m scale. However, the detection of heterospecific neighborhood effects (HND and HNCI) was largely insignificant. Despite observing a significant negative correlation between HNCI and RGR at scales of 15 m and 20 m (Figure 6b), many species demonstrated specificity in their responses (Figure 7(b1–b5)). The relationship between HND and RGR was not significant across all scales (Figure 6a), with no discernible trend and considerable variability among different species (Figure 7(a1–a5)). As we mentioned in Hypothesis 2 (H2), it can be inferred that the intensity of interspecific competition in this study is significantly lower than that of intraspecific competition, and it may represent one of the mechanisms by which diversity in secondary forests rapidly recovers during early successional stages.

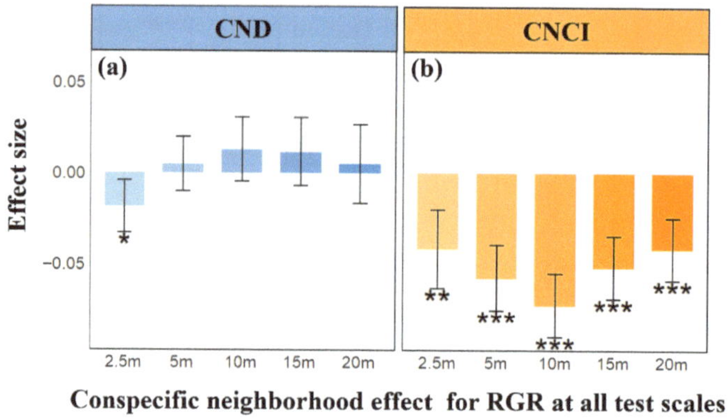

Figure 4. Parameter estimates for the effects of conspecific density and resource competition on relative growth rate (RGR) at neighborhood scales. The dark blue and orange bar graphs depict the parameter estimation of conspecific neighborhood density (CND) (**a**) and conspecific neighbor competition Index (CNCI) (**b**) on the RGR of individual focal trees across different spatial scales, respectively. Positive values denote positive effects, while negative values signify negative effects. Significance levels are denoted by an asterisk (* $p < 0.05$; ** $p < 0.01$; *** $p < 0.001$).

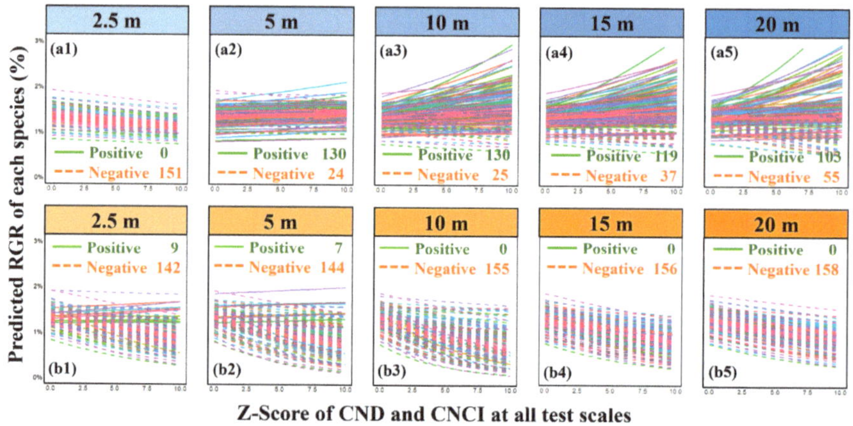

Figure 5. Multiscale relationship between conspecific density, resource competition, and relative growth rate (RGR) among all species. It displays annual conspecific neighborhood effect–RGR relationships for each of the 158 observed species at various spatial scales: 2.5 m (**a1**,**b1**), 5 m (**a2**,**b2**), 10 m (**a3**,**b3**), 15 m (**a4**,**b4**), 20 m (**a5**,**b5**). Different species are represented by different colors in the lines, the solid line represents a positive correlation, and the dotted line represents a negative correlation. Predicted RGR are back-transformed from the linear mixed model as described in the text, and all conspecific neighborhood effects were Z-score transformed at quantification. To enhance comparability and uniformity of the presentation results, we converted the Z-score values to positive in the figures; the untransformed original values can be found in Supplementary Figure S2.

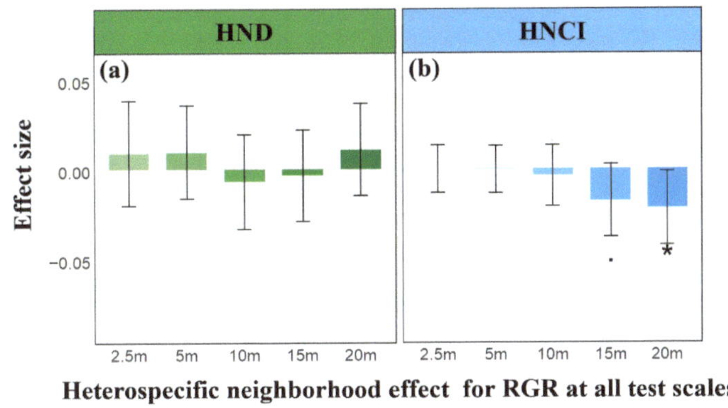

Figure 6. Parameter estimates for the effects of heterospecific density and resource competition on RGR at neighborhood scales. The green and light blue bar graphs depict the parameter estimation of heterospecific neighborhood density (HND) (**a**) and heterospecific neighborhood competition index (HNCI) (**b**) on the RGR of individual focal trees across different spatial scales, respectively. Positive values denote positive effects, while negative values signify negative effects. Significance levels are denoted by an asterisk (*) or a dot (·) (· $p < 0.05$; * $p < 0.1$).

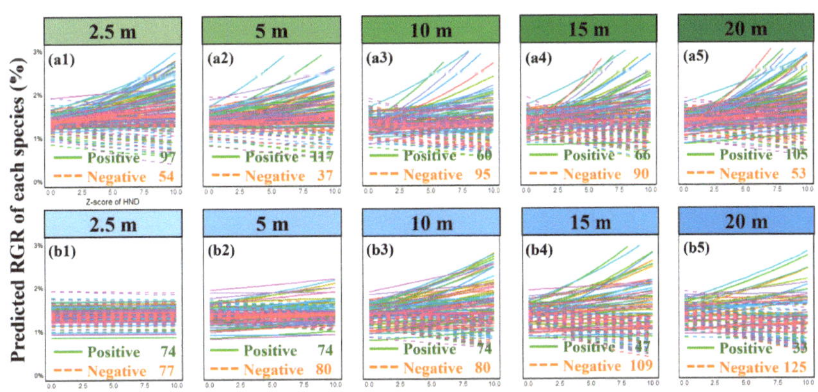

Figure 7. Multiscale relationship between heterospecific density, resource competition, and relative growth rate (RGR) among all species. It displays annual heterospecific neighborhood effect–RGR relationships for each of the 158 observed species at various spatial scales: 2.5 m (**a1**,**b1**), 5 m (**a2**,**b2**), 10 m (**a3**,**b3**), 15 m (**a4**,**b4**), 20 m (**a5**,**b5**). Different species are represented by different colors in the lines, the solid line represents a positive correlation, and the dotted line represents a negative correlation. Predicted RGR values are back-transformed from the linear mixed model as described in the text, and all heterospecific neighborhood effects were Z-score transformed at quantification. To enhance the comparability and uniformity of the presentation results, we converted the Z-score values to positive in the figures; the untransformed original values can be found in Supplementary Figure S2.

2.3. Relative Importance of Diversity and Density Effect across Fine Spatial Scales

As we expected in Hypothesis 3 (H3), neighborhood effects such as neighborhood species richness (NSR), conspecific negative density dependence (CNDD), neighborhood competition index (NCI), and heterospecific neighborhood competition index (HNCI) collectively regulate the relationship between species diversity and productivity. Among these, only NSR exhibits a positive effect on the growth of focal tree species (Figure 8).

However, the relative importance of various neighborhood effects varies with different testing scales. At a neighborhood scale of 5 m, we observed that neighborhood effects were overwhelmingly dominated by negative effects due to conspecific density or resource competition, making it difficult to discern a significant positive diversity–productivity relationship. However, as spatial scale increased, the relative importance of neighborhood species richness (NSR) became more pronounced at scales of 10 to 20 m, accounting for approximately 20% to 40% of the total effect and contributing to a positive diversity–productivity relationship. Additionally, at scales exceeding 15 m, interspecific resource competition led to a reduction in growth, representing 15% to 20% of the total effect, though conspecific negative density dependence remained predominant. In summary, the negative impact of conspecific density and resource competition on individual tree growth was predominant. However, as the test scale increased, the detectability of interspecific neighborhood effects became more significant, particularly the NSR effect at scales of 10 to 20 m, which facilitated a positive diversity–productivity relationship.

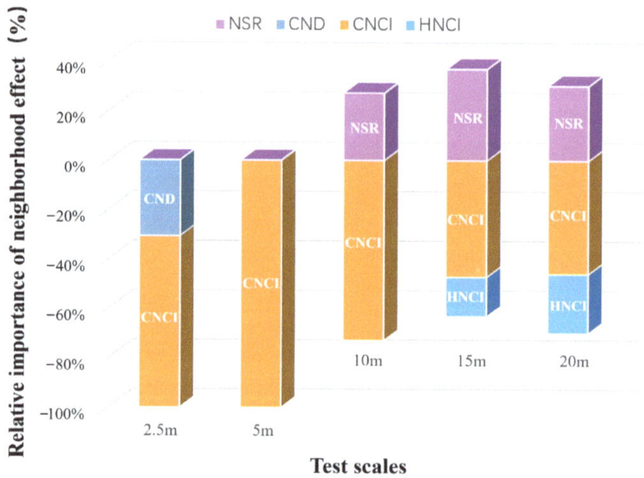

Figure 8. Relative importance of neighborhood effects at all test scales. This figure displays the relative effect sizes of neighborhood species richness (NSR), conspecific negative density (CND), conspecific neighborhood competition index (CNCI), and heterospecific neighborhood competition index (HNCI) on the relative growth rate. The graph only includes significant effects; since heterospecific negative density (HND) was not significant at any test scale in this study, it is not included in the calculations of relative importance. The relative effect of each predictor (neighborhood effect) and their interactions is calculated as the ratio of its parameter estimate to the sum of all parameter estimates.

3. Discussion

The intricate relationship between biodiversity and ecosystem productivity, as highlighted by our research within the context of secondary forests, underscores a pivotal consensus: species diversity enhances ecosystem productivity. This understanding, deeply rooted in the foundational studies by Tilman et al. (2014) and Cardinale et al. (2012), is extended by our findings, which emphasize the crucial role of neighborhood-scale diversity [1,5]. Similar to earlier research at broader scales, such as landscapes or biomes, the biodiversity–productivity relationship (BPR) tends to display more consistent patterns, with higher biodiversity generally linked to increased productivity [11,12,23,24]. Our study also observed a similar phenomenon, namely an increase in tree growth rates with higher neighborhood diversity, within a neighborhood scale of 10–20 m (Figure 2), thus affirming a positive correlation between biodiversity and productivity. This observation provides tangible evidence of biodiversity in boosting ecosystem productivity at the neighborhood scale.

However, at smaller neighborhood scales (below 2.5 m), our research observes that the consistently positive diversity–productivity relationship noted at 10–20 m scales becomes more complex (Figure 3) and even exhibits a negative diversity–growth relationship at a scale of 2.5 m (Figure 2), although it is not statistically significant. We attribute this high variability and the non-significant negative relationship between plant diversity and growth at these smaller scales to the intricate interactions between the focal tree species and their neighboring individuals, particularly concerning conspecific density dependence and resource competition [7–10,15–17]. Additionally, the inclusion of random slopes for species enabled us to capture species-specific response variability, which is crucial for understanding the nuanced interactions at smaller scales. Our analysis of conspecific neighborhood effects (CND and CNCI) supports this view, particularly at very small scales below 2.5 m, where we found that neighborhood effects were overwhelmingly dominated by negative effects due to intraspecific competition (Figure 8), as indicated by the conspecific neighbor density (CND) and conspecific neighbor competition index (CNCI), which had a significant negative impact on RGR across various scales (Figure 4). This suggests that individuals of the same species exert a stronger competitive pressure on each other, likely due to direct competition for identical resources (light, water, and nutrients) and space [7,8,10,15–17,25,26]. Especially at very small scales, the limited space leads to the dominance of negative density effects among neighboring tree individuals (Figure 8), making it challenging to detect a positive diversity–growth relationship. On the other hand, interspecific competition, as measured by the heterospecific neighbor density (HND) and heterospecific neighbor competition index (HNCI), showed a less significant impact on RGR (Figure 6). This finding highlights the importance of niche differentiation in mediating competition among coexisting species [7,9,10,17]. According to niche theory, species coexistence is facilitated by differences in resource use and habitat preferences, which reduce direct competition and allow for a more equitable distribution of resources among species [27–30].

Interestingly, as spatial scale increases, the relative importance of neighborhood species richness (NSR) at scales of 10 m to 20 m increases, accounting for up to approximately 40% of the total neighborhood effect, thereby contributing to a positive biodiversity–productivity relationship. These results underscore the complexity of ecological interactions and the role of spatial context in mediating these interactions, indicating that at smaller scales, intense competition for resources may overshadow the positive effects of species diversity on productivity [7–10,17,31]. In contrast, at broader scales, the benefits of species diversity, possibly through mechanisms such as niche complementarity and reduced competition, become more apparent. Our research further elucidates the scale-dependent dynamics of the biodiversity–productivity relationship, a topic increasingly emphasized in recent ecological studies. Huang et al. (2018) and Liang et al. (2016) demonstrated that the impacts of biodiversity on ecosystem functioning can vary significantly across large spatial scales [4,32]. This variability implies that the mechanisms through which biodiversity influences ecosystem productivity—such as niche differentiation and resource partitioning—may manifest differently depending on the spatial scale under investigation [7,9,10,33]. Our findings unequivocally support this perspective, demonstrating that the positive effects of biodiversity are not consistent across different tree species but rather are modulated by the spatial scale at which they are examined.

Additionally, we observed a scale-dependent effect among different species, whereby at smaller scales, the biodiversity–productivity relationship exhibits greater interspecific variation (Figure 3a1), but as spatial scale increases, this relationship gradually stabilizes into a consistent positive correlation (Figure 3a5). This suggests that different species may respond differently to neighborhood effects—although neighborhood tree species richness generally promotes individual tree productivity, species with different resource utilization and competition strategies may exhibit varying responses [7,8]. Specifically, resource-acquisitive species are often more susceptible to reductions in individual growth due to neighborhood competition at smaller spatial scales, while species with more conservative resource utilization strategies may benefit more from diversity effects in individual growth [7,9,10].

In summary, our research extends the scope of previous studies on the plant species diversity–productivity relationship by demonstrating that at very small scales, the combined effects of conspecific and heterospecific density dependence and resource competition can alter the previously stable positive biodiversity–productivity relationship observed at larger scales. Specifically, at tree-to-tree neighborhood distances of less than 2.5 m, tree growth is predominantly influenced by conspecific neighborhood effects, which adversely affect growth. Traditional views have emphasized the role of interspecific competition in driving community assembly and species distribution patterns [28]. However, our findings, along with recent studies [29,30], suggest that intraspecific competition may play an equally, if not more, significant role in influencing plant community dynamics and ecosystem functioning. Understanding the differential impacts of intraspecific and interspecific competition on tree growth and the biodiversity–productivity relationship is crucial for gaining insights into the dynamics of secondary forest ecosystems. Moreover, the dominance of intraspecific competition, particularly in the early stages of forest succession, may significantly influence patterns of species recruitment, growth, and mortality. This, in turn, shapes the trajectory of forest development and recovery [9–11,31,32]. Therefore, recognizing the importance of intraspecific competition provides a more comprehensive understanding of forest ecology and the factors driving ecosystem resilience and productivity. Our research also confirms that the positive effects of biodiversity are modulated by spatial scale and interspecific variability, emphasizing the importance of scale in understanding ecological phenomena and suggesting that processes observed at one scale may not be directly extrapolated to another [14,34]. This is critical for understanding biodiversity's effects and underscores the importance of considering spatial scale in ecological research and ecosystem management practices. Recognizing the importance of intraspecific competition and the scale-dependent nature of biodiversity effects provides a more comprehensive understanding of forest ecology and the factors driving ecosystem resilience and productivity.

4. Materials and Methods

4.1. Data Collection

The research area is situated in the Wuyishan National Park, located in the northwestern part of Fujian Province, China. This region experiences an average annual temperature of 19.2 °C and receives about 1600 mm of rainfall yearly (Figure S3 in the Supplementary Files). It enjoys an average annual sunshine of 1910.2 h, with a frost-free season lasting between 227 and 246 days. The dominant natural vegetation in this locale is the subtropical evergreen broad-leaved forest [35], although extensive commercial logging has historically transformed many primary forests into secondary forests [9,36].

For our research, we established a 9.6-hectare (400 m × 240 m) dynamic observation plot (27°35′24.23″ N, 117°45′55.43″ E) within the subtropical evergreen broad-leaved secondary forest, covering dimensions of 400 m by 240 m (Figure S4 in the Supplementary Material). This plot lies at an altitude that varies from 450 to 580 m, exhibiting minimal topographical variation. The long axis of the plot runs parallel to the main ridge in a northeast–southwest orientation. Approximately two-thirds of the plot area is on the southeast slope, with the remainder on the northwest slope. Predominant tree species within the plot include evergreen broad-leaved species and subspecies like *Castanopsis carlesii*, *Castanopsis eyrei*, and *Schima superba* [9,35,36].

In accordance with the CTFS (Center for Tropical Forest Science) survey protocols, the entire plot was divided into 240 large quadrats (20 m × 20 m), and each large quadrat was further subdivided into 16 smaller plots (5 m × 5 m), totaling 3840 small plots. These smaller quadrats were used as work units to measure the relative position, DBH (diameter at breast height), and other individual attributes of all trees. From October to December 2013, during the first survey, we recorded species, relative position, DBH, height, and crown base height for all tree individuals with DBH $\geq$ 1 cm. Among these, one 5 m $\times$ 5 m or 1 m $\times$ 1 m subplot was selected in each large quadrat to survey shrubs, herbaceous plants, and lianas, recording their species, abundance, average height and cover (for shrubs and herbaceous plants), as well as basal diameter and length (for lianas). Specifically, for shrubs less than 1.3 m in height (the height at which DBH is measured) or with DBH < 1 cm, we only measured their average height and cover. The species listed in our study are exclusively woody plants. While we have also collected data on shrubs and herbaceous plants, these data were not included in this study. Our analysis only considers woody plants with DBH $\geq$ 1 cm.

The first census showed a total of 68,336 tree individuals (including branches and sprouts) with DBH $\geq$ 1 cm, belonging to 173 species, 88 genera, and 48 families. The co-dominant families included Fagaceae, Ericaceae, and Elaeocarpaceae, with co-dominant species including *Castanopsis carlesii, Castanopsis fordii, Castanopsis eyrei, Engelhardia roxburghiana, Syzygium buxifolium*, and *Schima superba*. No single species was overwhelmingly dominant (Table S1 in the Supplementary Files), and the stand structure indicated that the forest community in our study was still in the early stage of secondary succession because most tree individuals were saplings [9,36]. The second survey was conducted from September to December 2018. A total of 63,897 live trees (Table S2 in the Supplementary Files) were surveyed, including newly recruited individuals. Additionally, we noted that a total of 148 tree species (10.87% of the total number of tree) had died between 2013 and 2018 [9,36].

4.2. Relative Growth Rate

To evaluate tree productivity, we utilized the relative growth rate (RGR) of the tree's wood volume. For each target tree, we calculated the wood volume (V) by employing a form factor of 0.5, which represents an average for young subtropical trees, where $V = (\pi \cdot d^2/4)hf$, d being the diameter at breast height (DBH), h the height of the tree, and f the form factor representing a cylinder [7,8]. The RGR of wood volume was determined using the following formula:

$$\text{RGR} = \frac{log(V_2/V_1)}{(t_2 - t_1)} \quad (1)$$

where V_1 and V_2 represent the volumes of tree wood at the start t_2 and end t_1 of the study period from 2013 to 2018 (Figure S5 in the supporting information). We opted for RGR over the absolute growth rate due to the significant variation in the initial sizes of the trees under observation. RGR is a more reliable measure that is less influenced by the initial size differences among trees [7,37].

4.3. Neighborhood Diversity, Intraspecific/Interspecific Competition, and Test Scale

In our analysis, we developed a framework to elucidate the relationships between neighborhood diversity (NSR) and intraspecific (both conspecific neighborhood density, CND, and conspecific neighborhood competition index, CNCI) and interspecific competition represented by heterospecific neighborhood density, HND, and heterospecific neighborhood competition indices, HNCI, respectively, with the relative growth rate (RGR) of trees [7,38]. This methodology aimed to explore how these factors collectively influence tree growth [7–10,15,16].

Neighborhood species richness (NSR) was determined by counting the number of distinct tree species within a defined vicinity of each focal tree (Figure S6 in the Supplementary Files). Conspecific negative dependence (CND) and heterospecific negative dependence (HND) were assessed by examining the density of same-species and different-species trees surrounding a focal tree (Figure S7 in the Supplementary Files), respectively. The conspecific and heterospecific neighborhood competition indices (CNCI and HNCI) quantified the extent of resource competition, calculated by evaluating the DBH (diameter at breast height) area of neighboring trees of the same and different species (Figure S8 in the Supplementary Files), respectively. These indices, serving as a gauge for the abundance of competitors, were formulated as $NCI = \sum_{j \neq i} \frac{\pi D_j^2}{4}$, where D_j represents the diameter at breast height (DBH) of neighboring trees [7]. CNCI includes only conspecific trees. For a given focal tree, CNCI is calculated as the sum of the DBH of all conspecific neighboring trees within a specified radius (from 5 m to 20 m). HNCI includes only heterospecific trees. Similarly, HNCI is calculated as the sum of the DBH areas of all heterospecific neighboring trees within a specified radius.

Neighborhood diversity and intra-/interspecific competition are highly spatial scale-dependent and closely related to the scale of the sampling radius [7,9,39]. In this study, the "neighborhood scale or local scale" (i.e., the test scale) was defined as the range with the focal tree species as the center and a radius less than 20 m to comprehensively evaluate the strength of the NSR effect at different spatial scales. In addition, at this neighborhood scale, our results could fully reflect the biological interaction relationship between species and avoid being confounded by the influence of habitat heterogeneity factors in the plot [17,33,40–42]. We calculated the NSR/CND/HND/CNCI/HNCI of focal tree species at different neighborhood scales in R-Studio (R 4.05, Boston, MA, USA)

4.4. Multiscale Neighborhood Effect Models for Tree Relative Growth Rate

To establish the relationship between neighborhood effects (NSR, CND, CNCI, HND, and HNCI) and relative growth rate (RGR), we employed linear mixed-effects models (LMMs). These models are tailored to elucidate the complex interactions influencing the annual growth rate of wood volume within a tree's neighborhood. This approach offers a detailed understanding of how diversity, density dependence, and resource competition impact growth, enhancing our ability to predict growth dynamics across various species and environmental conditions.

In the linear mixed-effects model (LMM), α represents the intercept, indicating the baseline relative growth rate (RGR), and the β coefficients represent the fixed effects of the predictors (NSR, CND, CNCI, HND, and HNCI) on RGR. The random effects include species identity (random intercepts and slopes) and plot identity (random intercepts). Specifically, c_s represents the random intercept for species s, accounting for the variability in the average growth rate across different species. $u_{1s}, u_{2s}, \ldots, u_{5s}$ are the species-specific random slopes for the predictors NSR, CND, CNCI, HND, and HNCI, respectively. These slopes capture the species-specific responses to the predictors, such as initial tree height, neighborhood competition, and neighborhood species richness, thereby accounting for inherent variability among species. t_p represents the random effect of plot identity p, accounting for environmental heterogeneity among plots. These random effects account for intrinsic variability, thereby enhancing the robustness of the analysis. The error term (ε) is assumed to be normally distributed. The specific model structure is as follows:

$$
\begin{aligned}
RGR_{i,s,p} = \alpha + (\beta_1 + u_{1s})NSR_{i,s,p} + (\beta_1 + u_{2s})CND_{i,s,p} + (\beta_1 + u_{3s})CNCI_{i,s,p} + (\beta_1 + u_{4s})HND_{i,s,p} \\
+(\beta_1 + u_{5s})HNCI_{i,s,p} + DBH_{i,s,p} + c_s + t_p + \epsilon_{i,s,p}
\end{aligned} \tag{2}
$$

We accounted for variation in abiotic growing conditions and species-specific effects by incorporating plot (quadrats), species identity, and neighborhood species effects (NSR, CND, CNCI, HND, and HNCI) into the random structure of our analysis. Specifically, a linear mixed-effects model (LMM) was employed, featuring both random intercepts and random slopes to account for variability among species and random intercepts alone to account for variability among plots [7,8,15,16]. The smallest plot scale of 5 × 5 m was deliberately chosen to effectively capture and control for tree dependencies. By including random intercepts for plots, we managed spatial dependencies and environmental heterogeneity within plots. Furthermore, by incorporating random slopes for species, we captured species-specific response variability, which further mitigated the issue of tree independence.

We utilized the "lme4 1.1-31" package for fitting LMMs [7,15,16]. According to the definitions of test scales in Section 2.3 of our study, we examined the effects of NSR, CND, CNCI, HND, and HNCI at distances of 2.5 m, 5 m, 10 m, 15 m, and 20 m from the focal tree species on RGR, thus establishing a multi-scale neighborhood effects model. The relative effect of each predictor (neighborhood effect) and their interactions is calculated as the ratio of its parameter estimate to the sum of all parameter estimates, expressed as a percentage [43]. Graphical and stand structural analyses were conducted using Excel and R-studio, utilizing R version 4.05 with packages, 'vegan' 2.5-7, and 'ads' 1.5-5.

5. Conclusions

The elucidation of the relationship between species diversity and ecosystem productivity has remained a cornerstone in ecological research, with significant implications for biodiversity conservation and ecosystem management strategies [1,44,45]. Our study observed a significant positive correlation between neighborhood-scale species diversity and the relative growth rate (RGR) of individual trees within a secondary forest ecosystem. Confirming our initial hypothesis (H1), our findings underscore a significant positive correlation between neighborhood diversity and individual tree growth, reaffirming the importance of biodiversity in fostering ecosystem productivity. Notably, we observed a scale-dependent effect, wherein the biodiversity–productivity relationship exhibited greater variability at smaller scales but stabilized into a consistent positive correlation as spatial scale increased. Our investigation into intraspecific and interspecific competition further elucidates the mechanisms driving the observed patterns. At smaller scales, intraspecific competition, as indicated by the conspecific neighbor density (CND) and conspecific neighbor competition index (CNCI), exerted a notable negative impact on relative growth rate (RGR), reflecting the dominance of conspecific density-dependent effects. In contrast, interspecific competition, represented by the heterospecific neighbor density (HND) and heterospecific neighbor competition index (HNCI), exhibited insignificance or a lesser influence on RGR, suggesting a lower intensity compared to intraspecific competition. Combining these insights, it becomes evident that intraspecific competition plays a pivotal role in shaping individual tree growth, especially at smaller scales, while interspecific competition contributes less significantly. However, as spatial scale increases, the relative importance of neighborhood species richness (NSR) becomes more pronounced, contributing to a positive diversity–productivity relationship. This shift highlights the complex interplay of local density-dependent interactions and spatial scale in modulating ecosystem functioning.

It is important to note that diameter at breast height (DBH) is a widely used and accepted method for quantifying competition, especially due to its robust and reliable nature in large-scale surveys [7–10,15–17]. Moreover, DBH is one of the most precise measurable variables obtainable through traditional, non-automated field surveys. Other variables, like tree height and crown width, often encounter significant error due to factors like stand density and individual tree occlusion, making them less reliable, especially in expansive surveys. However, we acknowledge the potential limitations of using only DBH to evaluate resource competition in diverse ecosystems, particularly in multispecies evergreen forest communities. Future forest surveys should consider employing more precise instruments, such as LiDAR (Light Detection and Ranging), to obtain detailed forest structure data. Such advanced technologies can offer more precise measurements of tree height, canopy volume, and spatial distribution, thereby improving our ability to quantify competition and enhancing our understanding of biodiversity and ecosystem productivity relationships.

Overall, our study contributes to a deeper understanding of how biodiversity influences ecosystem productivity across different scales, emphasizing the need for a nuanced approach in ecosystem management and conservation efforts. By unraveling the intricate mechanisms underlying the biodiversity–productivity relationship, our findings offer insights for fostering sustainable stewardship of natural resources and guiding forest restoration initiatives in degraded ecosystems.

Supplementary Materials: The following supporting information can be downloaded at https://www.mdpi.com/article/10.3390/plants13141994/s1, Figure S1. Relationship between species richness and relative growth rate (RGR) among all species. It displays annual diversity–RGR relationships for each of the 158 observed species at 20 m. Different species are represented by different colors in the lines. Predicted RGR are back-transformed from the linear mixed model as described in the text, and all diversity effects were Z-score transformed at quantification. Figure S2. Multiscale relationship between neighborhood effect (NSR, CND, CNCI, HND, and HNCI) and relative growth rate (RGR) among all species. The annual neighborhood effect–RGR relationship across the 158 observed species is represented by lines of different colors (refer to Figure S1), with solid lines indicating a positive relationship and dashed lines indicating a negative one. The predicted values of RGR are obtained by back-transforming from the linear mixed models, with all diversity effects quantified by Z-score transformation. In this Figure, it is noted that Figures 3, 5 and 7 in the main text only present the prediction results for positive Z-scores, whereas the original results are displayed here. Figure S3. Schematic representation of the location of the study area. The red dots in the figure represent the location of the subtropical evergreen broad-leaved forest plot in the Wuyi Mountains, China. Figure S4. The spatial pattern of all tree individuals in the evergreen broad-leaved secondary forest dynamic observation site. Circles represent the locations of individual trees, whose sizes are proportional to the tree DBH (diameter at breast height); dark green circles represent tree individuals in first census in 2013, and steel blue circles represent tree individuals in 2018. Figure S5. The spatial pattern of relative growth rate (RGR) in the evergreen broad-leaved secondary forest dynamic observation site. This figure represents the annual relative growth rate (RGR) of tree individuals during 2013–2018. RGR decreases gradually as the color transitions from red to blue. The spatial intensity was estimated using an Epanechnikov kernel with a bandwidth of 10 m. Figure S6. The spatial pattern of neighborhood diversity (NSR). This figure represents the neighborhood species richness (NSR) of tree individuals in 2018. NSR decreases gradually as the color transitions from red to blue. The spatial intensity was estimated using an Epanechnikov kernel with a bandwidth of 10 m. Figure S7. The spatial intensities of neighborhood density. This figure represents the intensity of conspecific and heterospecific neighborhood densities of tree individuals in 2018. The intensity decreases gradually as the color transitions from red to blue. The spatial intensity was estimated using an Epanechnikov kernel with a bandwidth of 10 m. Figure S8. The spatial pattern of neighborhood size (DBH). This figure represents the intensity of neighborhood size (DBH) of tree individuals in 2018. DBH decreases gradually as the color transitions from red to blue. The spatial intensity was estimated using an Epanechnikov kernel with a bandwidth of 10 m. Table S1. Basic characteristics of the 28 co-dominant tree species in the subtropical evergreen broad-leaved forest plot in the Wuyi Mountains, China. Table S2. Forest dynamic in the subtropical evergreen broad-leaved forest plot in the Wuyi Mountains, China for the years 2013 and 2018.

Author Contributions: Conceptualization, H.Z.; funding acquisition, H.D. and H.Z.; investigation, H.Z., X.Z. (Xiao Zheng), Y.H., X.Z. (Xu Zhou), Y.F., L.X. and Y.L.; writing—original draft, H.Z.; writing—review and editing, H.Z. and Y.G. Project administration, H.D. and Y.F. All authors have read and agreed to the published version of the manuscript.

Funding: This work was funded by the Innovative Team Project of Nanjing Institute of Environmental Sciences MEE (ZX2023QT022); The Central Public Welfare Scientific Institution Basal Research Fund, MEE (GYZX240104, GYZX210302); Sino-German Environmental Partnership Phase III: Strengthening Chinese environmental and nature conservation policy through dialogue (SGEP2022-2027); Biodiversity Survey, Observation and Assessment Program of Ministry of Ecology and Environment of China (BSOA2019-2024); The National Environmental Protection Public Welfare Industry Targeted Research Fund: Research and demonstration of key technologies for dynamic supervision of nature reserves (201509042); The Special Foundation for National Science and Technology Basic Resources Investigation of China (2019FY202300); and The Jiangsu Postdoctoral Research Funding Program (2021K038A).

Data Availability Statement: The forest census data are available from the corresponding author on reasonable request.

Acknowledgments: We are grateful to the people who helped to collect field survey data.

Conflicts of Interest: The authors declare no conflicts of interest.

References

1. Tilman, D.; Reich, P.B.; Isbell, F. Biodiversity impacts ecosystem productivity as much as resources, disturbance, or herbivory. *Proc. Natl. Acad. Sci. USA* **2014**, *109*, 10394–10397. [CrossRef] [PubMed]
2. LaManna, J.A.; Halpern, C.B.; Tappeiner, J.C. Plant diversity increases with the strength of negative density dependence at the global scale. *Science* **2017**, *356*, 1389–1392. [CrossRef] [PubMed]
3. Schnabel, F.; Schwarz, J.A.; Dănescu, A.; Fichtner, A.; Nock, C.A.; Bauhus, J.; Potvin, C. Drivers of productivity and its temporal stability in a tropical tree diversity experiment. *Glob. Chang. Biol.* **2019**, *25*, 4257–4272. [CrossRef] [PubMed]
4. Liang, J.; Crowther, T.W.; Picard, N.; Wiser, S.; Zhou, M.; Alberti, G.; Schulze, E.D.; McGuire, A.D.; Bozzato, F.; Pretzsch, H.; et al. Positive biodiversity-productivity relationship predominant in global forests. *Science* **2016**, *354*, aaf8957. [CrossRef] [PubMed]
5. Cardinale, B.J.; Duffy, J.E.; Gonzalez, A.; Hooper, D.U.; Perrings, C.; Venail, P.; Narwani, A.; Mace, G.M.; Tilman, D.; Wardle, D.A.; et al. Biodiversity loss and its impact on humanity. *Nature* **2012**, *486*, 59–67. [CrossRef] [PubMed]
6. Tilman, D.; Reich, P.B.; Knops, J.; Wedin, D.; Mielke, T.; Lehman, C. Diversity and productivity in a long-term grassland experiment. *Science* **2001**, *294*, 843–845. [CrossRef] [PubMed]
7. Fichtner, A.; Härdtle, W.; Li, Y.; Bruelheide, H.; Kunz, M.; von Oheimb, G. From competition to facilitation: How tree species respond to neighbourhood diversity. *Ecol. Lett.* **2017**, *20*, 892–900. [CrossRef] [PubMed]
8. Fichtner, A.; Härdtle, W.; Li, Y.; Bruelheide, H.; Kunz, M.; von Oheimb, G. Neighbourhood diversity mitigates drought impacts on tree growth. *J. Ecol.* **2020**, *108*, 865–875. [CrossRef]
9. Zhang, H.N.; Chen, S.; Zheng, X.; Ge, X.; Li, Y.; Fang, Y.; Cui, P. Neighborhood diversity structure and neighborhood species richness effects differ across life stages in a subtropical natural secondary forest. *For. Ecosyst.* **2022**, *9*, 100075. [CrossRef]
10. Zhang, H.N.; Xu, J.; Meng, W.; Qi, C.; Xie, J.; Xie, Y.; Zhang, J.; Zhu, J.; Deng, Y.; Zhu, B. Positive interactions in shaping neighborhood diversity during secondary forests recovery: Revisiting the classical paradigm. *For. Ecol. Manag.* **2024**, *552*, 121586. [CrossRef]
11. Loreau, M.; Hector, A. Partitioning selection and complementarity in biodiversity experiments. *Nature* **2001**, *412*, 72–76. [CrossRef] [PubMed]
12. Isbell, F.; Craven, D.; Connolly, J.; Loreau, M.; Schmid, B.; Beierkuhnlein, C.; Bezemer, T.M.; Bonin, C.; Bruelheide, H.; de Luca, E.; et al. Biodiversity increases the resistance of ecosystem productivity to climate extremes. *Nature* **2015**, *526*, 574–577. [CrossRef]
13. Gonzalez, A.; Germain, R.M.; Srivastava, D.S.; Filotas, E.; Dee, L.E.; Gravel, D.; Thompson, P.L.; Isbell, F.; Wang, S.; Kéfi, S.; et al. Scaling-up biodiversity-ecosystem functioning research. *Ecol. Lett.* **2020**, *23*, 757–776. [CrossRef] [PubMed]
14. Levin, S.A. The problem of pattern and scale in ecology: The Robert H. MacArthur Award Lecture. *Ecology* **1992**, *73*, 1943–1967. [CrossRef]
15. Zhu, Y.; Comita, L.S.; Hubbell, S.P.; Ma, K. Conspecific and phylogenetic density-dependent survival differs across life stages in a tropical forest. *J. Ecol.* **2015**, *103*, 957–966. [CrossRef]
16. Zhu, Y.; Queenborough, S.A.; Condit, R.; Hubbell, S.P.; Ma, K.P.; Comita, L.S. Density-dependent survival varies with species life-history strategy in a tropical forest. *Ecol. Lett.* **2018**, *21*, 506–515. [CrossRef] [PubMed]
17. Zhang, H.N.; Chen, S.; Xia, X.; Ge, X.; Zhou, D.; Wang, Z. The competitive mechanism between post-abandonment Chinese fir plantations and rehabilitated natural secondary forest species under an in situ conservation policy. *For. Ecol. Manag.* **2021**, *502*, 119725. [CrossRef]
18. Bardgett, R.D.; Wardle, D.A. *Aboveground-Belowground Linkages: Biotic Interactions, Ecosystem Processes, and Global Change*; Oxford University Press: Oxford, UK, 2010.
19. Luo, W.; Liang, J.; Gatti, R.C.; Zhao, X.; Zhang, C. Parameterization of biodiversity–productivity relationship and its scale dependency using georeferenced tree-level data. *J. Ecol.* **2019**, *107*, 238–248. [CrossRef]
20. Hooper, D.U.; Adair, E.C.; Cardinale, B.J.; Byrnes, J.E.K.; Hungate, B.A.; Matulich, K.L.; Gonzalez, A.; Duffy, J.E.; Gamfeldt, L.; O'Connor, M.I. A global synthesis reveals biodiversity loss as a major driver of ecosystem change. *Nature* **2012**, *486*, 105–108. [CrossRef]
21. Srivastava, D.S.; Cadotte, M.W.; MacDonald, A.A.M.; Marushia, R.G.; Mirotchnick, N. Phylogenetic diversity and the functioning of ecosystems. *Ecol. Lett.* **2012**, *15*, 637–648. [CrossRef]
22. Thompson, P.L.; Benjamin, R.; McKenzie, J.S.; Woodhouse, I.H.; Wernberg, T.; Van Oppen, M.J.H. Loss of biodiversity and ecosystem functioning in Indo-Pacific coral reefs. *Sci. Adv.* **2018**, *4*, eaap9650. [CrossRef]
23. Wright, A.J.; Ebeling, A.; de Kroon, H.; Roscher, C.; Weigelt, A.; Buchmann, N.; Buchmann, T.; Fischer, C.; Hacker, N.; Hildebrandt, A.; et al. Flooding disturbances increase resource availability and productivity but reduce stability in diverse plant communities. *Nat. Commun.* **2015**, *6*, 6092. [CrossRef]
24. Mori, A.S.; Lertzman, K.P.; Gustafsson, L. Biodiversity and ecosystem services in forest ecosystems: A research agenda for applied forest ecology. *J. Appl. Ecol.* **2017**, *54*, 12–27. [CrossRef]
25. Tilman, D. *Resource Competition and Community Structure*; Princeton University Press: Princeton, NJ, USA, 1982.
26. Connell, J.H. On the prevalence and relative importance of interspecific competition: Evidence from field experiments. *Am. Nat.* **1983**, *122*, 661–696. [CrossRef]
27. Chesson, P. Mechanisms of maintenance of species diversity. *Annu. Rev. Ecol. Syst.* **2000**, *31*, 343–366. [CrossRef]
28. MacArthur, R.; Levins, R. The limiting similarity, convergence, and divergence of coexisting species. *Am. Nat.* **1967**, *101*, 377–385. [CrossRef]

29. Levine, J.M.; HilleRisLambers, J. The importance of niches for the maintenance of species diversity. *Nature* **2009**, *461*, 254–257. [CrossRef] [PubMed]
30. Kunstler, G.; Falster, D.; Coomes, D.A.; Hui, F.; Kooyman, R.M.; Laughlin, D.C.; Poorter, L.; Vanderwel, M.; Vieilledent, G.; Wright, S.J.; et al. Plant functional traits have globally consistent effects on competition. *Nature* **2016**, *529*, 204–207. [CrossRef] [PubMed]
31. Zhang, H.N.; Yang, Q.; Zhou, D.; Xu, W.; Gao, J.; Wang, Z. How evergreen and deciduous trees coexist during secondary forest succession: Insights into forest restoration mechanisms in Chinese subtropical forest. *Glob. Ecol. Conserv.* **2021**, *25*, e01418. [CrossRef]
32. Huang, Y.; Chen, Y.; Castro-Izaguirre, N.; Baruffol, M.; Brezzi, M.; Lang, A.; Li, Y.; Härdtle, W.; von Oheimb, G.; Yang, X.; et al. Impacts of species richness on productivity in a large-scale subtropical forest experiment. *Science* **2018**, *362*, 80–83. [CrossRef]
33. Zhang, H.N.; Xue, J.H. Spatial pattern and competitive relationships of moso bamboo in a native subtropical rainforest community. *Forests* **2018**, *9*, 774. [CrossRef]
34. Chase, J.M.; Leibold, M.A. Spatial scale dictates the productivity-biodiversity relationship. *Nature* **2002**, *416*, 427–430. [CrossRef]
35. Wu, Z.Y. *Vegetation of China*; Science Press: Beijing, China, 1980; pp. 823–888. (In Chinese with English abstract)
36. Ding, H.; Chen, S.F.; Xu, H.; Luo, X.Q.; Li, Y.; Xu, K.; Wu, Y.; Wu, Y.Q.; Yong, F.; Zhou, W.L.; et al. Dynamics of arbor layer in the Subtropical Evergreen Broad-leaved Forest in the Wuyi Mountains, Fujian Province, southeastern China in 2013–2018. *Acta Ecol. Sin.* **2022**, *42*, 3458–3469. (In Chinese with English abstract) [CrossRef]
37. Mencuccini, M.; Martínez-Vilalta, J.; Vanderklein, D.; Hamid, H.A.; Korakaki, E.; Lee, S.; Michiels, B. Size-mediated ageing reduces vigour in trees. *Ecol. Lett.* **2005**, *8*, 1183–1190. [CrossRef]
38. Zhang, H.N.; Zhang, X.S.; Lv, Y.Y.; Zeng, Y.; Qi, C.; Xie, J.; Xie, Y.; Zhang, J.; Zhu, J.; Deng, Y.; et al. How topography and neighbor shape the fate of trees in subtropical forest restoration: Environmental filtering and resource competition drive natural regeneration. *For. Ecosyst.* **2024**, *11*, 100169. [CrossRef]
39. Rayburn, A.; Wiegand, T. Individual species–area relationships and spatial patterns of species diversity in a Great Basin, semi-arid shrubland. *Ecography* **2012**, *35*, 341–347. [CrossRef]
40. Baddeley, A.; Rubak, E.; Turner, R. *Spatial Point Patterns: Methodology and Applications with R*, 1st ed.; Chapman and Hall/CRC: New York, NY, USA, 2015. [CrossRef]
41. Getzin, S.; Dean, C.; He, F.; Trofymow, J.; Wiegand, K.; Wiegand, T. Spatial patterns and competition of tree species in a Douglas-fir chronosequence on Vancouver Island. *Ecography* **2006**, *29*, 671–682. [CrossRef]
42. Hao, Z.; Zhang, J.; Song, B.; Ye, J.; Li, B. Vertical structure and spatial associations of dominant tree species in an old-growth temperate forest. *For. Ecol. Manag.* **2007**, *252*, 1–11. [CrossRef]
43. García-Palacios, P.; Gross, N.; Gaitán, J.; Maestre, F.T. Climate mediates the biodiversity–ecosystem stability relationship globally. *Proc. Natl. Acad. Sci. USA* **2018**, *115*, 8400–8405. [CrossRef]
44. Liang, M.; Crowther, T.W.; Picard, N.; Wiser, S.; Zhou, M.; Alberti, G.; Reich, P.B. Consistent stabilizing effects of plant diversity across spatial scales and climatic gradients. *Nat. Ecol. Evol.* **2022**, *6*, 1669–1675. [CrossRef]
45. Schnabe, F.; Gessler, A.; He, J.S.; Keitel, C.; Kreft, H.; Nadrowski, K.; Scherer-Lorenzen, M. Species richness stabilizes productivity via asynchrony and drought-tolerance diversity in a large-scale tree biodiversity experiment. *Sci. Adv.* **2021**, *7*, eabk1643. [CrossRef] [PubMed]

plants

MDPI

Article

Genetic Diversity and Structure of *Terminalia bellerica* (Gaertn. Roxb.) Population in India as Revealed by Genetic Analysis

Subramani Umesh Kanna [1], Kalappan Thangamuthu Parthiban [2], Kandasamy Senthilraja [3], Subramanian Venkatesan [3], Dhandayuthapani Udhaya Nandhini [4], Shanmugam Mohan Kumar [5], Manickam Dhasarathan [5], Palaniyappan Kumaresan [6], Makkena Jaswanth Sai [2], Muthurajan Raveendran [3] and Vellingiri Geethalakshmi [5,*]

1. Directorate of Planning and Monitoring, Tamil Nadu Agricultural University, Coimbatore 641 003, Tamil Nadu, India; umeshkanna.s@tnau.ac.in
2. Forest College and Research Institute, Tamil Nadu Agricultural University, Mettupalayam 641 301, Tamil Nadu, India; ktparthi2001@gmail.com (K.T.P.); jaswanth.fcri@gmail.com (M.J.S.)
3. Directorate of Research, Tamil Nadu Agricultural University, Coimbatore 641 003, Tamil Nadu, India; senthilrajaens@gmail.com (K.S.); svengat95@gmail.com (S.V.); raveendrantnau@gmail.com (M.R.)
4. Centre of Excellence in Sustaining Soil Health, Anbil Dharmalingam Agricultural College and Research Institute, Trichy 620 027, Tamil Nadu, India
5. Agro-Climatic Research Centre, Tamil Nadu Agricultural University, Coimbatore 641 003, Tamil Nadu, India; psmohan13@gmail.com (S.M.K.); plantdr.dhasarathan@gmail.com (M.D.)
6. Centre for Water and Geospatial Studies, Tamil Nadu Agricultural University, Coimbatore 641 003, Tamil Nadu, India; acfkumagri@gmail.com

* Correspondence: geetha@tnau.ac.in

Abstract: In this study, an extensive exploration survey of wild progeny was conducted which yielded 18 candidate plus trees (CPTs) of *Terminalia bellerica*. Seeds of these CPTs were collected from diverse locations between 10°54′ and 28°07′ E longitude, and 76°27′ and 95°32′ N latitude, covering 18 different locations across 5 states of the Indian subcontinent. The objective of the progeny trial was to assess genetic associations and variability in growth and physio-chemical characteristics. Significant variations ($p < 0.05$) were observed among the growth traits, encompassing plant height, basal diameter, girth at breast height and volume, as well as physio-chemical characteristics such as leaf length, width, area and chlorophyll content, carotenoids, and protein in the progeny trial. Broad-sense heritability (h^2_b) estimates were consistently high, exceeding 80% for all growth and physiological related traits under investigation except for plant height, leaf length, and girth at breast height. A correlation study revealed that selecting based on plant height, leaf area, and girth at breast height effectively enhances *T. bellerica* volume. A moderate genetic advance in percent of the mean (GAM) was observed for most traits, except leaf length, leaf width, girth at breast height, and plant height. Across all 13 traits, phenotypic coefficient of variation (PCV) surpassed genotypic coefficient of variation (GCV). Utilizing principal component analysis (PCA) and dendrogram construction categorized the genotypes into seven distinct groups. In conclusion, the study has demonstrated that targeting girth at breast height and plant height would be a highly effective strategy for the establishment of elite seedling nurseries and clonal seed nurseries for varietal and hybridization programs in the future.

Keywords: *Terminalia bellerica*; genetic diversity; heritability; clustering; progeny trial

Citation: Umesh Kanna, S.; Parthiban, K.T.; Senthilraja, K.; Venkatesan, S.; Udhaya Nandhini, D.; Mohan Kumar, S.; Dhasarathan, M.; Kumaresan, P.; Sai, M.J.; Raveendran, M.; et al. Genetic Diversity and Structure of *Terminalia bellerica* (Gaertn. Roxb.) Population in India as Revealed by Genetic Analysis. *Plants* **2024**, *13*, 470. https://doi.org/10.3390/plants13040470

Academic Editors: Xugao Wang, Zuoqiang Yuan and Chengjin Chu

Received: 27 November 2023
Revised: 20 December 2023
Accepted: 29 December 2023
Published: 6 February 2024

1. Introduction

T. bellerica (Gaertn.) Roxb. is an impressive deciduous tree known for its rapid growth and substantial size. With its expansive and spherical crown, it can reach remarkable heights of up to 50 m in its native environment, although it generally grows smaller when nurtured. This tree often showcases prominent buttresses and keeps its branches absent for the initial 20 m of its trunk. The bark of *Terminalia* exhibits a distinctive ashy grey color, highlighted by delicate longitudinal cracks. Additionally, the inner bark showcases a subtle yellowish tint. *Terminalia* can be commonly found in native woodlands across several regions of the Indian subcontinent, including West Bengal, Madhya Pradesh, Uttar Pradesh, Maharashtra, Assam, Tamil Nadu, Rajasthan, Karnataka, Kerala, and Punjab [1].

The wood derived from *T. bellerica* is known for its exceptional hardness and can be utilized in various applications. These include minor construction, the creation of grain measurement tools, boat side planks, fodder production, providing food for Tasar silkworms, soap manufacturing, as well as the extraction of gum with demulcent and purgative properties. Additionally, this wood is also used in the production of ayurvedic medicines such as Triphala. Hence, *T. bellerica* has been selected for the ongoing study. Candidate plus trees (CPTs) have been carefully chosen based on their exceptional morphometric characteristics from diverse geographical regions. The objective is to identify offspring with enhanced productivity through systematic tree improvement initiatives.

The establishment and productivity of forest tree plantations heavily depend on the selection of species and seed sources within those species [2]. Thus, it is crucial to grasp the variation within these seed sources to ensure effective tree improvement programs [3]. In order to obtain superior genetic material, tree breeders must evaluate the traits that require enhancement while taking into account their variability in terms of both morphological and biochemical characteristics [4]. Conducting variability studies is an essential requirement for any tree improvement program [5], however, in the case of *T. bellerica*, such studies are still in the early stages of development.

Besides variability, the role of heritability in estimating potential gains from selection programs is of the utmost importance [6]. Understanding the heritability of selected traits is essential [7], hence it is valuable to assess the genetic analysis to determine the heritable components. At present, there is a lack of information about *T. bellerica* in this field.

Evaluating the extent and type of variation in the initial population is essential for enhancing both qualitative and quantitative progress. Historically, assessing genetic diversity in trees involved conducting provenance/progeny tests and utilizing the Mahalanobis D2 statistic [8]. By clustering genotypes, distantly related clusters can be identified for hybridization, leading to improved segregation to facilitate selection of superior groups or individuals. The individuals or groups that demonstrate enhanced energy and enthusiasm can be effectively utilized in planting programs to enhance productivity [9–13].

Previously, the evaluation of genetic diversity relied on investigations that focused on comparative anatomy, morphology, physiology, and biochemistry [3]. However, the introduction of molecular marker techniques has revolutionized our comprehension of tropical tree population genetics. These techniques enable the analysis of protein or DNA polymorphism and have been instrumental in advancing our understanding in this field [4,14–16]. The utilization of DNA marker studies in tropical trees has proven to be effective in various applications. These include understanding variations in origin, determining genotypic identity, characterizing germplasm at a molecular level [17], identifying quantitative trait loci [18,19], studying molecular systematics [20,21], and evaluating genetic diversity [12,13,22]. However, there is a noticeable lack of research on these aspects specifically related to *T. bellerica*. In light of the aforementioned information, this document presents a research study that seeks to explore the potential of *T. bellerica* as a feasible alternative to secondary timber genetic resources. The main aim is to tackle the growing demand for raw wood materials within industries that heavily depend on forest resources.

2. Results

2.1. Growth Traits

The analysis of variances indicated significant variation among all the 18 accessions studied in the measured growth traits at a significance level of $p < 0.05$ (Figure S1). Significant variation in plant height was observed during the growth of progeny from 18 accessions in the *T. bellerica* (Table 1). Compared to other accessions, FCRITB17 (7.19 m) and FCRITB16 (7.17 m) exhibited significantly higher values for plant height. Conversely, a group of seven accessions exhibited lower heights, measuring less than 6.50 m which included FCRITB18 (6.41 m), FCRITB06 (6.39 m), FCRITB04 (6.29 m), and FCRITB112 (6.23 m).

Table 1. Mean performance of selected genotypes for growth and physiological traits in *T. bellerica*.

Accession Name	Plant Height (m)	Basal Diameter (cm)	Girth at Breast Height (m)	Volume (m^3)	Leaf Length (cm)	Leaf Width (cm)	Leaf Area (cm^2)
FCRITB01	6.50 ± 0.14 ^{efgh}	44.2 ± 0.63 ^{d}	0.09 ± 0.001 ^{b}	0.692 ± 0.00 ^{ef}	26.5 ± 0.10 ^{defg}	11.6 ± 0.01 ^{de}	208 ± 1.88 ^{ef}
FCRITB02	6.67 ± 0.11 ^{cdef}	43.6 ± 0.00 ^{de}	0.09 ± 0.000 ^{b}	0.690 ± 0.00 ^{de}	25.9 ± 0.54 ^{efg}	11.0 ± 0.17 ^{gh}	194 ± 0.58 ^{jk}
FCRITB03	6.73 ± 0.16 ^{cde}	50.9 ± 0.81 ^{a}	0.09 ± 0.000 ^{b}	0.631 ± 0.00 ^{hij}	26.6 ± 0.26 ^{cdef}	11.2 ± 0.21 ^{fgh}	201 ± 0.35 ^{fgh}
FCRITB04	6.29 ± 0.09 ^{gh}	41.6 ± 0.40 ^{gh}	0.09 ± 0.001 ^{b}	0.625 ± 0.01 ^{ij}	27.5 ± 0.30 ^{ab}	11.9 ± 0.09 ^{abcd}	222 ± 1.51 ^{bc}
FCRITB05	6.53 ± 0.03 ^{defg}	43.4 ± 0.73 ^{def}	0.09 ± 0.001 ^{b}	0.627 ± 0.00h ^{ij}	25.9 ± 0.38 ^{fg}	11.1 ± 0.00 ^{fgh}	195 ± 2.13 ^{ij}
FCRITB06	6.39 ± 0.09 ^{fgh}	41.9 ± 0.00 ^{fg}	0.09 ± 0.001 ^{b}	0.594 ± 0.00 ^{k}	25.6 ± 0.01 ^{g}	10.7 ± 0.01 ^{i}	186 ± 0.52 ^{k}
FCRITB07	6.65 ± 0.07 ^{cdef}	44.4 ± 0.85 ^{d}	0.10 ± 0.000 ^{a}	0.775 ± 0.00 ^{b}	27.1 ± 0.39 ^{bcd}	11.8 ± 0.03 ^{bcd}	218 ± 2.10 ^{cd}
FCRITB08	6.53 ± 0.09 ^{defg}	44.4 ± 0.06 ^{d}	0.09 ± 0.002 ^{b}	0.720 ± 0.00 ^{c}	26.7 ± 0.01 ^{bcde}	11.3 ± 0.10 ^{efg}	205 ± 3.99 ^{fg}
FCRITB09	6.80 ± 0.09 ^{bcd}	47.6 ± 0.18 ^{bc}	0.09 ± 0.001 ^{b}	0.711 ± 0.02 ^{cd}	26.8 ± 0.11 ^{cdef}	11.4 ± 0.12 ^{ef}	207 ± 3.73 ^{ef}
FCRITB10	7.04 ± 0.14 ^{ab}	49.5 ± 0.15 ^{b}	0.09 ± 0.000 ^{b}	0.663 ± 0.01 ^{fg}	26.1 ± 0.10 ^{fg}	11.3 ± 0.05 ^{fgh}	199 ± 2.18 ^{ghi}
FCRITB11	6.63 ± 0.15 ^{def}	46.7 ± 0.19 ^{c}	0.09 ± 0.002 ^{b}	0.648 ± 0.01 ^{gh}	25.8 ± 0.25 ^{fg}	10.9 ± 0.03 ^{hi}	192 ± 0.70 ^{jk}
FCRITB12	6.23 ± 0.11 ^{h}	40.3 ± 0.51 ^{h}	0.09 ± 0.001 ^{b}	0.613 ± 0.01 ^{jk}	26.4 ± 0.11 ^{cdefg}	11.2 ± 0.22 ^{fgh}	201 ± 0.60 ^{ghi}
FCRITB13	6.55 ± 0.06 ^{defg}	41.7 ± 0.52 ^{fg}	0.09 ± 0.000 ^{b}	0.691 ± 0.01 ^{e}	27.1 ± 0.14 ^{bc}	11.6 ± 0.22 ^{cde}	213 ± 4.08 ^{de}
FCRITB14	6.94 ± 0.03 ^{abc}	39.2 ± 0.25 ^{i}	0.09 ± 0.001 ^{b}	0.616 ± 0.01 ^{ijk}	26.0 ± 0.61 ^{efg}	11.1 ± 0.02 ^{fgh}	196 ± 0.06 ^{hij}
FCRITB15	6.73 ± 0.13 ^{cde}	42.1 ± 0.06 ^{fg}	0.09 ± 0.001 ^{b}	0.639 ± 0.00 ^{hi}	28.3 ± 0.34 ^{a}	12.3 ± 0.04 ^{a}	237 ± 0.75 ^{a}
FCRITB16	7.17 ± 0.07 ^{a}	42.5 ± 0.45 ^{efg}	0.10 ± 0.002 ^{a}	0.819 ± 0.00 ^{a}	27.7 ± 0.11 ^{b}	11.9 ± 0.00 ^{abc}	220 ± 4.81 ^{bcd}
FCRITB17	7.19 ± 0.01 ^{a}	41.9 ± 0.26 ^{fg}	0.09 ± 0.000 ^{b}	0.708 ± 0.00 ^{cde}	28.3 ± 0.20 ^{a}	12.2 ± 0.23 ^{ab}	234 ± 2.13 ^{a}
FCRITB18	6.41 ± 0.02 ^{fgh}	43.7 ± 0.48 ^{de}	0.10 ± 0.001 ^{a}	0.731 ± 0.02 ^{c}	27.3 ± 0.14 ^{b}	12.2 ± 0.04 ^{ab}	225 ± 0.21 ^{b}

Note: Data are the mean values of three replicates with ± standard error. Means followed by the same letter within each column are not significantly different at the 0.05 level.

Significant variation was noted among the accessions in terms of the basal diameter, with an observed mean of 43.9 cm occurring. The accession with the highest basal diameter (50.9 cm) was FCRITB03, followed by FCRITB06 (49.5 cm). The FCRITB14 with a basal diameter of 39.2 cm was apparently distinct, and its measurement was significantly lower than all other accessions.

The range of volumes observed in this study varied from 0.1653 m^3 to 0.2251 m^3. Among the nine progenies analyzed, namely FCRITB07 (0.2088 m^3), FCITB 10 (0.1791 m^3), FCRITB14 (0.1765 m^3), FCRITB16 (0.2251 m^3), FCRITB17 (0.1829 m^3), and FCRITB18 (0.2013 m^3) exhibited higher volumes compared to the average value obtained from the overall sample. It is worth mentioning that among the identified progeny, FCRITB16 exhibited the highest recorded volume of 0.2251 m^3, whereas progeny FCRITB12 achieved the lowest volume of 0.1585 m^3. The length of the leaves was observed to range from 25.83 cm to 28.30 cm, with an overall average of 26.75 cm. Among the 18 progenies, progeny FCRITB15 and FCRITB17 recorded the maximum leaf length at 28.30 cm, while progeny FCRITB11 exhibited the minimum leaf length at 25.83 cm. The leaf area value exhibited variation ranging from 236.63 cm^2 to 185.99 cm^2, with an overall mean value of 208.46 cm^2. Out of the 18 progenies of *T. bellerica*, 7 progenies—FCRITB04, FCRITB07, FCRITB13, FCRITB15, FCRI TB 16, FCRITB17, and FCRITB18—displayed higher leaf areas compared to the general mean. The progeny FCRITB15 recorded the maximum leaf area of (236.63 cm^2), while the progeny FCRITB06 exhibited the smallest leaf area of (185.99 cm^2) (Table 1).

2.2. Biochemical Traits

The ANOVA revealed prominent variation ($p < 0.05$) among the 18 studied accessions across all the biochemical traits (Figure S2). Among the 18 progeny, FCRITB05 exhibited higher chlorophyll 'a' content with a measurement of 1.127 mg/g, while progeny FCRITB18 had the lowest chlorophyll 'a' content at 0.489 mg/g. The overall average for chlorophyll 'a' content was recorded as 0.743 mg/g. In terms of chlorophyll 'b' content, the general mean was determined to be 0.471 mg/g, ranging from 0.213 mg/g to 0.979 mg/g. Seven specific progenies—FCRITB05 (0.879 mg/g), FCRITB06 (0.613 mg/g), FCRITB07 (0.701 mg/g), FCRITB09 (0.594 mg/g), FCRITB10 (0.534 mg/g), FCR IT B11 (0.497 mg/g), and FCRITB13 (0.293 mg/g)—exhibited higher levels of chlorophyll 'b' content compared to the overall average value. The total chlorophyll content varied between 0.501 mg/g and 1.943 mg/g. Among the seven progenies—FCRITB05 (1.943 mg/g), FCRITB06 (0.891 mg/g), FCRITB07 (1.012 mg/g), FCRITB09 (0.904 mg/g), FCRITB10 (0.897 mg/g), FCRITB11 (0.919 mg/g), and FCRITB13 (1.009 mg/g)—the maximum total chlorophyll content was observed when compared to the overall average value. Carotenoid content varied from 0.886 mg/g to 0.372 mg/g, with an average value of 0.581 mg/g. Specifically, the progenies FCRITB02 (0.645 mg/g), FCRITB06 (0.663 mg/g), FCRITB08 (0.604 mg/g), FCRITB10 (0.831 mg/g), FCRITB14 (0.618 mg/g), FCRITB15 (0.818 mg/g), and FCRITB17 (0.886 mg/g) exhibited higher carotenoid content compared to the overall average value. Out of the 18 progenies of *T. bellerica*, 10 specific progenies, including FCRITB03, FCRITB04, FCRITB06, FCRITB10, FCRITB11, FCRITB12, FCRITB13, FCRITB14, and FCRITB17, have demonstrated a higher crude protein content in comparison to the overall average (Table 2).

Table 2. Mean performance of selected genotypes for biochemical traits in *T. bellerica*.

Accession Name	Chlorophyll a (mg/g)	Chlorophyll b (mg/g)	Chlorophyll a and b	Total Chlorophyll (mg/g)	Carotenoid mg/g	Crude Protein%
FCRITB01	0.628 ±0.01 ^{g}	0.435 ±0.00 ^{i}	1.063 ±0.00 ^{i}	0.754 ±0.00 ^{e}	0.561 ±0.00 ^{f}	11.59 ±0.19 ^{f}
FCRITB02	0.552 ±0.01 ^{h}	0.213 ±0.00 ^{n}	0.765 ±0.01 ^{l}	0.690 ±0.00 ^{f}	0.645 ±0.00 ^{d}	21.11 ±0.31 ^{d}
FCRITB03	0.515 ±0.00 ^{i}	0.278 ±0.00 ^{l}	0.793 ±0.00 ^{k}	0.654 ±0.01 ^{g}	0.574 ±0.01 ^{f}	27.78 ±0.08 ^{f}
FCRITB04	0.771 ±0.02 ^{f}	0.398 ±0.00 ^{j}	1.169 ±0.01 ^{h}	0.798 ±0.01 ^{d}	0.513 ±0.00 ^{g}	26.26 ±0.00 ^{g}
FCRITB05	1.127 ±0.02 ^{a}	0.979 ±0.01 ^{a}	2.106 ±0.00 ^{a}	1.943 ±0.04 ^{a}	0.518 ±0.00 ^{g}	39.88 ±0.07 ^{g}
FCRITB06	0.826 ±0.00 ^{e}	0.613 ±0.01 ^{d}	1.439 ±0.02 ^{e}	0.891 ±0.00 ^{c}	0.663 ±0.00 ^{d}	29.18 ±0.10 ^{d}
FCRITB07	0.926 ±0.01 ^{c}	0.701 ±0.01 ^{c}	1.627 ±0.03 ^{c}	1.012 ±0.00 ^{b}	0.513 ±0.01 ^{g}	12.15 ±0.14 ^{g}
FCRITB08	0.519 ±0.01 ^{i}	0.298 ±0.00 ^{k}	0.817 ±0.02 ^{k}	0.649 ±0.00 ^{g}	0.604 ±0.01 ^{e}	13.88 ±0.09 ^{e}
FCRITB09	0.892 ±0.01 ^{d}	0.594 ±0.01 ^{e}	1.486 ±0.00 ^{d}	0.904 ±0.01 ^{c}	0.416 ±0.00 ^{h}	11.73 ±0.02 ^{h}
FCRITB10	0.836 ±0.00 ^{e}	0.534 ±0.01 ^{f}	1.370 ±0.00 ^{f}	0.897 ±0.01 ^{c}	0.831 ±0.01 ^{b}	39.39 ±0.13 ^{b}
FCRITB11	0.823 ±0.02 ^{e}	0.497 ±0.00 ^{g}	1.320 ±0.01 ^{g}	0.919 ±0.01 ^{c}	0.498 ±0.00 ^{g}	31.40 ±0.05 ^{g}
FCRITB12	0.562 ±0.00 ^{h}	0.301 ±0.00 ^{k}	0.863 ±0.00 ^{j}	0.619 ±0.00 ^{h}	0.364 ±0.00 ^{i}	22.16 ±0.12 ^{i}
FCRITB13	0.997 ±0.01 ^{b}	0.923 ±0.00 ^{b}	1.920 ±0.01 ^{b}	1.009 ±0.00 ^{b}	0.574 ±0.00 ^{f}	39.88 ±0.14 ^{f}
FCRITB14	0.615 ±0.01 ^{g}	0.464 ±0.00 ^{h}	1.079 ±0.01 ^{i}	0.710 ±0.00 ^{f}	0.618 ±0.01 ^{e}	28.45 ±0.30 ^{e}
FCRITB15	0.436 ±0.00 ^{j}	0.254 ±0.00 ^{m}	0.690 ±0.00 ^{m}	0.593 ±0.00 ^{h}	0.818 ±0.00 ^{c}	13.88 ±0.23 ^{c}
FCRITB16	0.513 ±0.00 ^{i}	0.294 ±0.00 ^{kl}	0.807 ±0.00 ^{k}	0.604 ±0.00 ^{h}	0.498 ±0.01 ^{g}	13.64 ±0.02 ^{g}
FCRITB17	0.612 ±0.00 ^{g}	0.454 ±0.01 ^{hi}	1.066 ±0.01 ^{i}	0.784 ±0.01 ^{d}	0.886 ±0.01 ^{a}	29.09 ± 0.07 ^{a}
FCRITB18	0.489 ±0.00 ^{i}	0.254 ±0.00 ^{m}	0.743 ±0.02 ^{l}	0.501 ±0.01 ^{i}	0.372 ±0.00 ^{i}	21.31 ± 0.06 ^{i}

Traits marked with the same superscript letter are not statistically significantly different at a significance level of $p = 0.05$.

2.3. Heritability

In the conducted study, it was observed that all 13 traits demonstrated a significant level of heritability, ranging from 68.11% to 99.94%, as indicated in Table 3. Among the studied traits, crude protein exhibited the highest level of heritability at 99.94%. This was closely followed by chlorophyll a, b, ratio to chlorophyll a, b (99. 76%, 99. 74%, and 99.31% respectively), as well as carotenoid (99.36%).

Table 3. Genetic estimates of selected progeny traits in *T. bellerica*.

	Traits	Phenotypic Coefficient of Variation	Genotypic Coefficient of Variation	Heritability Broad Sense (%)	GA (%) of Mean
Growth traits	Plant height (m)	4.70	3.88	68.11	6.59
	Basal diameter (cm)	6.94	6.71	93.37	13.35
	Girt at breast height (m)	4.69	4.16	78.68	7.61
	Volume (m^3)	9.01	8.76	94.54	17.54
Physiological traits	Leaf length (cm)	3.47	2.92	71.00	5.07
	Leaf Width (cm)	4.21	3.80	81.65	7.08
	Leaf area (cm^2)	7.34	7.08	92.98	14.06
Biochemical traits	Chlorophyll a (mg/g)	28.55	28.45	99.31	58.41
	Chlorophyll b (mg/g)	47.80	47.74	99.76	98.24
	Chl a/Chl b	35.65	35.60	99.74	73.25
	Total Chlorophyll (mg/g)	37.95	37.88	99.63	77.89
	Carotenoid (mg/g)	25.36	25.28	99.36	51.91
	Crude Protein (%)	41.49	41.48	99.94	85.42

2.4. Genotypic and Phenotypic Variation

The presence of high variability in chlorophyll a, chlorophyll b, chl a/chl b, total chlorophyll, carotenoid, and crude protein is indicated by the highest values of GCV and PCV. The magnitude of phenotypic coefficient of variation (PCV) was greater than the respective genotypic coefficient of variation (GCV) for all the studied traits, albeit with only a slight difference.

In Table 3, the analysis of the genotypic coefficient of variance (GCV) and phenotypic coefficient of variance (PCV) for multiple traits is presented. The findings reveal that, in the majority of cases, the phenotypic coefficient variances (PCVs) are slightly higher than the genotypic coefficient variances (GCVs). This indicates that the traits being studied are relatively less affected by environmental factors. Notably, the highest values for both GCV and PCV were observed in chlorophyll b (47.80% and 47.74%), crude protein (41.49% and 41.48%), total chlorophyll (37.95% and 37.88%), Chl a/Chl b (35. 65% and 35.60%), chlorophyll a (28.55% and 28.45%), and carotenoid (25.36% and 25.28%). On the other hand, the traits of plant height, basal diameter, girth at breast height, volume, leaf length, leaf width, and leaf area showed limited variability as indicated by their low GCV and PCV values.

2.5. Genetic Advance

The findings of this investigation revealed all three kinds of genetic advances (low, moderate, and high). Some traits, including leaf length, leaf width, girth at breast height, and plant height displayed genetic advances of less than 10% in Table 3. In contrast, basal diameter, leaf area, and volume exhibited moderate genetic advances ranging from 13.35% to 17.54%. Particularly, biochemical traits demonstrated a high genetic advance, exceeding

20% with a maximum of 98.24% by chlorophyll b. This study uncovers biochemical traits with both high heritability and genetic advance as a percentage of the mean (>50).

2.6. Correlation among the Traits

Correlation analysis revealed several significant relationships between mean progenies traits (Table 4). Plant height of provinces was significantly correlated (r = 0.545 *) with carotenoid (Figure S2) and positively correlated with basal diameter (r = 0.194), girth at breast height (r = 0.158), leaf area (r = 0.248), volume (r = 0.413), leaf length (r = 0.205), leaf width (r = 0.194), and crude protein (r = 0.025). Progenies heights were negatively correlated with chlorophyll content.

2.7. Principle Component Analysis

Principle Component Analysis (PCA) was executed to facilitate the visualization of the entire dataset through a condensed dimension plot. The application of PCA was for determining genetic relationships among progenies and exploring correlations among growth, physiological, and biochemical traits. In this study, the performed PCA revealed that over 99% of the observed variances could be accounted for by the initial three principal components (Figure S3). Specifically, PC1, PC2, and PC3 contributed 73.1%, 24.1% and 2.6% to the total variability, respectively (Figure 1. PC1 predominantly represents leaf area and crude protein, PC2 explains the same, and PC3 primarily contributes to basal diameter.

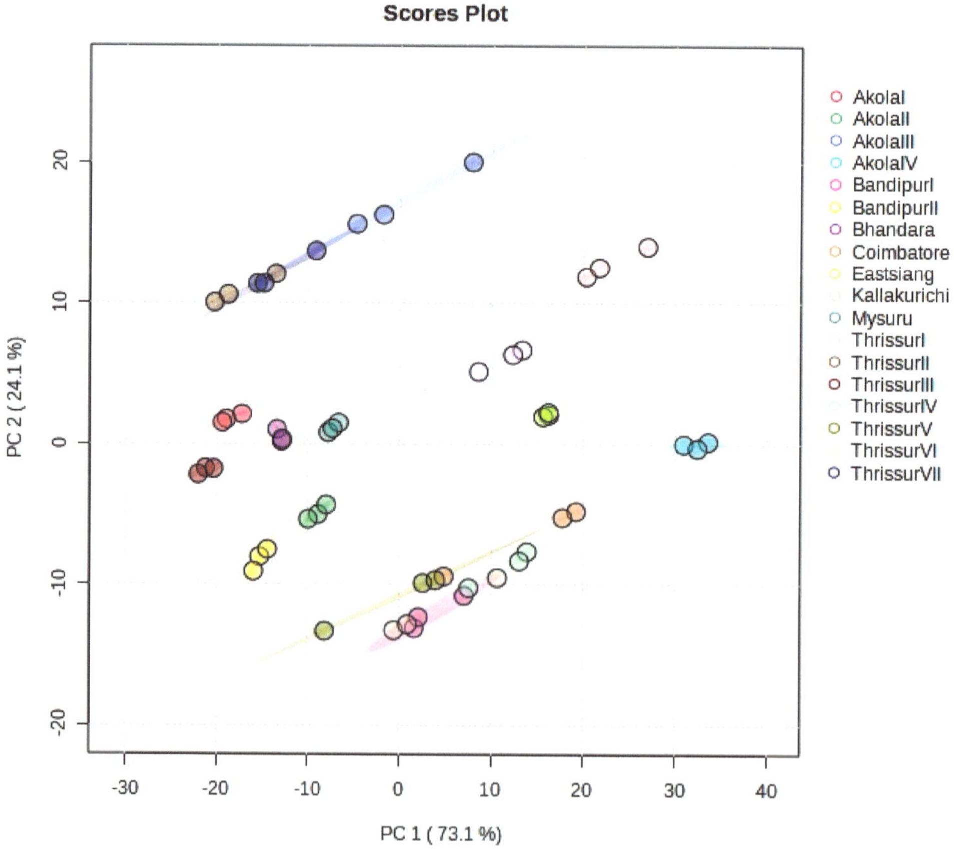

Figure 1. PCA showed the variation in the 18 provinces.

Table 4. Pearson Correlation Coefficient among different characters studied.

	PH		GBH	LA	Volume	Leaf Length	Leaf Width	Chl a	Chl b	Chl a/Chl b	Total Chlorophyll	Carotenoid	Crude Protein
PH	1	0.194	0.158	0.248	0.413	0.205	0.194	−0.136	−0.064	−0.096	−0.087	0.545 *	0.025
BD		1	−0.020	−0.188	0.119	−0.284	−0.231	0.088	−0.044	0.026	0.036	0.007	0.023
GBH			1	0.383	0.747 **	0.317	0.489 *	−0.096	−0.082	−0.088	−0.172	−0.312	−0.321
LA				1	0.416	0.962 **	0.973 **	−0.313	−0.203	−0.257	−0.304	0.220	−0.336
Volume					1	0.366	0.523 *	−0.105	−0.075	−0.086	−0.176	−0.150	−0.504 *
Leaf length						1	0.932 **	−0.386	−0.272	−0.331	−0.358	0.215	−0.302
Leaf Width							1	−0.323	−0.223	−0.273	−0.322	0.072	−0.410
Chl a								1	0.938 **	0.983 **	0.852 **	−0.122	0.514 *
Chl b									1	0.985 **	0.850 **	−0.034	0.529 *
Chl a/Chl b										1	0.866 **	−0.082	0.523 *
Total Chlorophyll											1	−0.042	0.513 *
Carotenoid												1	0.256
Crude Protein													1

**. Correlation is significant at the 0.01 level (two-tailed) *. Correlation is significant at the 0.05 level (two-tailed). Note: PH—Plant height; BD—Basal diameter; GBH—Girth at breast height; LA—Leaf area.

Component scores for the 18 studied progenies are shown in Table S1. Positive values for PC1 indicate progenies with plant height, girth at breast height, leaf area, volume, leaf length, leaf width, and carotenoids in general. FCRITB10, FCRITB16, and FCRITB17 belong to this group. The lowest values for PC1 indicate basal diameter such as FCRITB12 and FCRITB14. The highest values for PC2 indicate all the parameters except for basal diameter and volume. The scatter biplot in Figure S4 shows the relationship between studied genotypes and depicts a clear pattern of the grouping of provincesprovinces. All the provincesprovinces were scattered widely in different quarters.

2.8. Heatmap Clustering

Figure 2 illustrates K-Means hierarchical clustering for growth characteristics, physiological traits, and biochemical traits in *T. bellerica* accessions. A total of 18 *Terminalia* progenies were categorized into 7 clusters using K-Means clustering, with cluster V and VII holding the highest number of accessions (4) and others sharing 2 progenies under each.

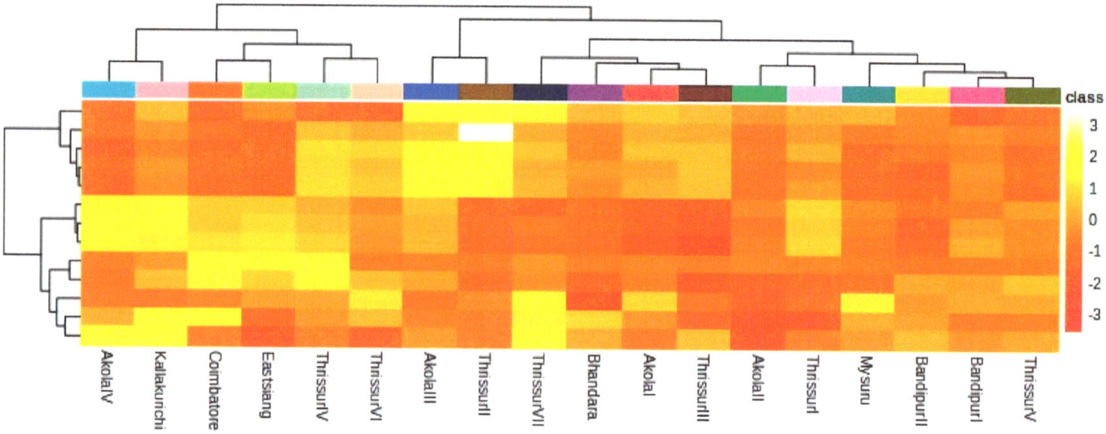

Figure 2. Heatmap dendrogram showed the variations among the 18 progenies.

Interestingly, cluster VII consisting of FCRITB01, FCRITB02, FCRITB03, and FCRITB08 exhibited similar mean values for the growth and biophysiological characters. FCRITB15nd FCRITB17 were placed under cluster I and showed similar values for the traits. Despite having the highest number of clustering, progenies did not exhibit high mean values for any of the measured traits amidst the different climatic provincesprovinces.

3. Discussion

After a thorough analysis of 18 progenies of *T. bellerica*, it was observed that progenies FCRITB02 and FCRITB13 showed initial superiority based on their biometric attributes. However, progenies FCRITB05, FCRITB07, and FCRITB09 demonstrated significant superiority in more than three biometric traits including plant height, basal diameter, leaf length, leaf width, and leaf area. This highlights the potential of these specific progenies for further study or utilization in future breeding programs.

The progenies, specifically FCRITB03, FCRITB10, FCRITB16, and FCRITB17 displayed significantly higher measurements in various biometric characteristics including height, basal diameter, leaf length, leaf width, and leaf area. The outcomes are consistent with *Neolamarckia cadamba*, *Casuarina* clones, *Ailanthus excelsa*, *Santalum album*, *Dalbergia sissoo*, *Pongamia pinnata*, *Acacia* species, Salix species, *Aquilaria malaccensis*, *Melia azedarach*, *Leucaena leucocephala*, and *Toona ciliata* [7,9,11,23–32]. Furthermore, the comparable height observed in FCRITB16 and FCRITB17 can be attributed to the similarity in weather parameters, with both provincesprovinces experiencing an average temperature of 17.3 °C and annual rainfall of 2036 mm.

Leaf area is the most essential characteristic when it comes to biomass production. The progenies displayed a significant amount of variation in terms of leaf traits, indicating that these traits can be effectively utilized for selection purposes. The conducted study revealed notable variations in the investigated leaf-related characteristics, including leaf length, leaf breadth, and leaf area. It was observed that among the 18 progenies examined, FCRITB17 demonstrated superiority in all of the analyzed leaf parameters. This could be attributed to its remarkable growth and volume, potentially explaining its exceptional performance. The current investigation is supported by prior evidence showing variation in leaf features and their correlation to production in various plant species such as *Toona ciliata* [33], *N. cadamba* [34,35], *Ficus carica* [36], *Acacia* species [7], *Pongamia pinnata* [37], *Aquilaria malaccensis* [29], Poplar [38], *Dalbergia sissoo* [27], and *Acacia catechu* [39].

Following the analysis of biochemical data from a study on 18 offspring of *T. bellerica*, it was determined that one particular offspring—FCRITB10—consistently displayed significantly elevated levels for all six studied biochemical parameters. Additionally, two other offspring, namely FCRITB12 and FCRITB13, demonstrated superiority in five parameters: chlorophyll 'a', chlorophyll 'b', chlorophyll a/b ratio, total chlorophyll, and carotenoid levels.

In the present study, the biochemical characteristics of 18 progenies of *T. bellerica* were boserved. Out of the six examined biochemical properties, it is noteworthy that only one progeny, namely FCRITB05, consistently demonstrated superior performance compared to the other progenies. This superiority was observed to be significantly remarkable. Three progenies—FCRITB07, FCRITB10, and FCRITB13—have exhibited their superiority in five biochemical characteristics. These characteristics include chlorophyll 'a', chlorophyll 'b', chlorophyll a/b ratio, total chlorophyll, and carotenoids. Previous studies have been conducted on various plant species such as *L. leucocephala*, *N. cadamba*, *Ailanthus excelsa*, *Albizia lebbeck*, *Acacia catechu*, *Bassia latifolia*, *Mangifera indica*, and *Ulmis pumila* [40–45]. These investigations have shown that these plants exhibit similar variations in terms of their biochemical attributes. Therefore, the findings from previous studies provide support for the conclusions made in the current investigation. Due to its superior performance in a wide range of biometric and biochemical characteristics, FCRITB05 outperformed other progenies of *T. bellerica*. As a result, it is currently being evaluated for prompt integration into future breeding programs.

Heritability serves as a reliable indicator of the transmission of traits from parents to their progeny, categorized as low (below 30%), medium (30–60%), and high (above 60%). The concept of heritability plays a pivotal role in the field of plant breeding, assisting breeders in the selection of genotypes from a wide range of genetic populations. High heritability values are particularly valuable as they enable the effective selection of specific traits. In the conducted study, it was observed that all 13 traits demonstrated a significant level of heritability, ranging from 68.11% to 99.94% as indicated in Table 5. Among the studied traits, crude protein exhibited the highest level of heritability at 99.94%.

The presence of high variability in chlorophyll a, chlorophyll b, chl a/chl b, total chlorophyll, carotenoid, and crude protein is indicated by the highest values of GCV and PCV. The magnitude of phenotypic coefficient of variation (PCV) was greater than the respective genotypic coefficient of variation (GCV) for all the studied traits, albeit with only a slight difference. Similar findings were reported by Rao et al. [46]. In the current study, there was minimal disparity between genotypic and phenotypic coefficients of variation for all the studied traits except plant height. This implies that these traits are less susceptible to environmental influences. Comparable results were reported in *Populus deltoids* [47] and in willow trees [48]. The marginal difference between PCV and GCV of almost all the characters studied in all the traits suggested that there was high heritability of variation among the characters.

Table 5. Provinces selected from the Indian subcontinent and the geographical location details.

S.No	Sources	District	State	Latitude	Longitude	Assigned Name
1.	Bandipur Tiger Reserve and National Park	Bandipur	Karnataka	11.664547	76.626421	FCRITB1
2.	Bandipur Tiger Reserve and National Park	Bandipur	Karnataka	11.664571	76.626418	FCRITB2
3.	Mysuru Zoo	Mysuru	Karnataka	12.30053	76.669647	FCRITB3
4.	Kerala Agricultural University	Thrissur	Kerala	10.3832	76.3296	FCRITB4
5.	Kerala Agricultural University	Thrissur	Kerala	10.3836	76.3299	FCRITB5
6.	Vellanikkara	Thrissur	Kerala	10.548235	76.278912	FCRITB6
7.	Vellanikkara	Thrissur	Kerala	10.548007	76.278745	FCRITB7
8.	Vellanikkara	Thrissur	Kerala	10.54824	76.278874	FCRITB8
9.	Bentham and Hooker Garden	Thrissur	Kerala	10.547665	76.278592	FCRITB9
10.	Bentham and Hooker Garden	Thrissur	Kerala	10.550524	76.280483	FCRITB10
11.	Akola	Akola	Maharashtra	20.703063	77.069286	FCRITB11
12.	Akola	Akola	Maharashtra	20.703088	77.069316	FCRITB12
13.	Akola	Akola	Maharashtra	20.703003	77.069991	FCRITB13
14.	Shioni	Bhandara	Maharashtra	20.191579	79.661286	FCRITB14
15.	Patur	Akola	Maharashtra	20.461537	76.943464	FCRITB15
16.	Jagnari slopes	Coimbatore	Tamil Nadu	11.323315	76.934989	FCRITB16
17.	Kalarayan Hills	Kallakurichi	Tamil Nadu	11.764162	76.415564	FCRITB17
18.	Pasighat	Eastsiang	Arunachal Pradesh	28.075837	95.325901	FCRITB18

Heritability and genetic advancement are pivotal metrics in unraveling the genetic intricacies of various agricultural traits. This study delves into the interplay between heritability and genetic advance as a percentage of mean, shedding light on the potential for effective selection strategies. The analysis reveals that traits exhibiting both high heritability and a high genetic advance as a percentage of the mean primarily operate under the influence of additive gene action. These findings signify the suitability of direct selection to enhance the performance of these traits, promising progress through selective breeding.

Conversely, traits characterized by moderate heritability and low genetic advance as a percentage of the mean are predominantly influenced by non-additive gene action. For such traits, direct selection may pose challenges, as a substantial portion of the variation is attributed to environmental factors. These environmental effects may arise from soil fertility disparities and other unpredictable variables, as suggested by Reddy et al. [49]. Researchers have proposed that traits governed by non-additive gene action may benefit more from management practices than direct selection for trait improvement. This perspective aligns with the recommendations to emphasizing the importance of tailored management approaches [50,51].

The findings of this investigation revealed all three kinds of genetic advances (low, moderate, and high) and uncovers biochemical traits with both high heritability and genetic advance as a percentage of the mean (>50). These high values indicate the prevalence of additive gene action for these specific traits, signifying the potential for effective trait enhancement through selective breeding.

Conversely, the study identifies traits (leaf length, leaf width, girth at breast height, and plant height) characterized by high heritability but low genetic advance as a percentage of the mean. Additionally, traits such as basal diameter, leaf area, and volume exhibit high heritability with moderate genetic advance as a percentage of the mean.

Hierarchical clustering, based on Ward's minimum variance cluster analysis, revealed phylogeographic patterns of genetic diversity. Length of the horizontal branches between clusters indicates that there is a high degree of dissimilarity between clusters. K-means clustering analysis demonstrated that trees from different geographic regions were grouped together in clusters. Interestingly, trees from the same geographical area were placed in different clusters, suggesting that geographical diversity did not necessarily correlate with genetic diversity and implying that it may have undergone divergent changes in various traits due to different selection pressures. This type of genetic diversity may arise from variations in adoption methods, selection criteria, natural selection pressures, and environmental factors [52]. This suggests that genetic drift has played a more significant role in generating diversity compared to geographic diversity [53]. The absence of any relationships between genetic diversity and geographical distribution in the current study is consistent with the findings of [54,55].

Furthermore, this clustering approach identified promising accessions with favorable traits, paving the way for the establishment of elite seedling nurseries and clonal seed nurseries for varietal and hybridization programs in the future.

The growth of a plant, as indicated by volume, basal diameter, and plant height is considered highly significant for improvement in the current study. Similarly, growth traits in black poplar are the most crucial based on principal component analysis [56]. In a study on the morphological characters of *P. deltoides* hybrid clones in a nursery, Ozel et al. [57] applied factor analysis, explaining 71.46% of the total variance with the first five components and captured 90% cumulative variability for the first two principal components to differentiate leaf characters of *Populus nigra* similar to the present study (0.948) [58]. Tunctaner [59] reported five principal components based on the study of fourteen traits in willow clones, a pattern also observed by Singh et al. [53] in Salix clones. The growth characters are attributed to distinct genetic constitution of the clones as highlighted in this study [60]. The promising clones selected for this study must undergo multi location trials to investigate the relationship between genotype and environment at various sites. This will allow for an analysis of the suitability of the clones and allow for the use of the clones for intra- and inter-specific control breeding (hybridization) aimed at producing more productive clones.

4. Materials and Methods

4.1. Genetic Material

A comprehensive and extensive survey of wild germplasm was conducted with the aim of identifying promising candidate plus trees (CPTs) of *T. bellerica*. This survey originated from five distinct states of the Indian subcontinent: Tamil Nadu, Maharashtra, Kerala, Karnataka, and Arunachal Pradesh, and examined the ecological impact on genetic diversity, growth, and eco-physiological traits (Figure 3, Table 5). These provinces were selected due to their inherent adaptability to the growing conditions suitable for *T. bellerica*. As all of the selected origins are distributed across the Indian subcontinent, they exhibit both commonalities and variations in their climatic origins. The selection process of CPTs involved utilizing the single-tree selection method, which relied on assessing the phenotypic traits with economic significance *viz.* total height, girth at breast height, bole height, and volume [50] (Figure S5, Table 6). Precautions were taken to ensure that the selected trees were free from pest and disease infestations and excluded isolated or poorly performing trees, commonly referred to as wolf trees. A total of 18 CPTs were collected from diverse locations between 10°54′ and 28°07′ E longitude, and 76°27′ and 95°32′ N latitude, across five states of the Indian subcontinent (Figures S6 and S7). Three kilograms of mature pods were harvested from each CPT by following a random sampling procedure.

These pods were collected from all four directions of the crown of each selected tree during the fruiting season between September and November in the year 2019. The gathering of potential CPTs was achieved through collaboration with officials from the respective forest departments while strictly adhering to required permissions and regulations.

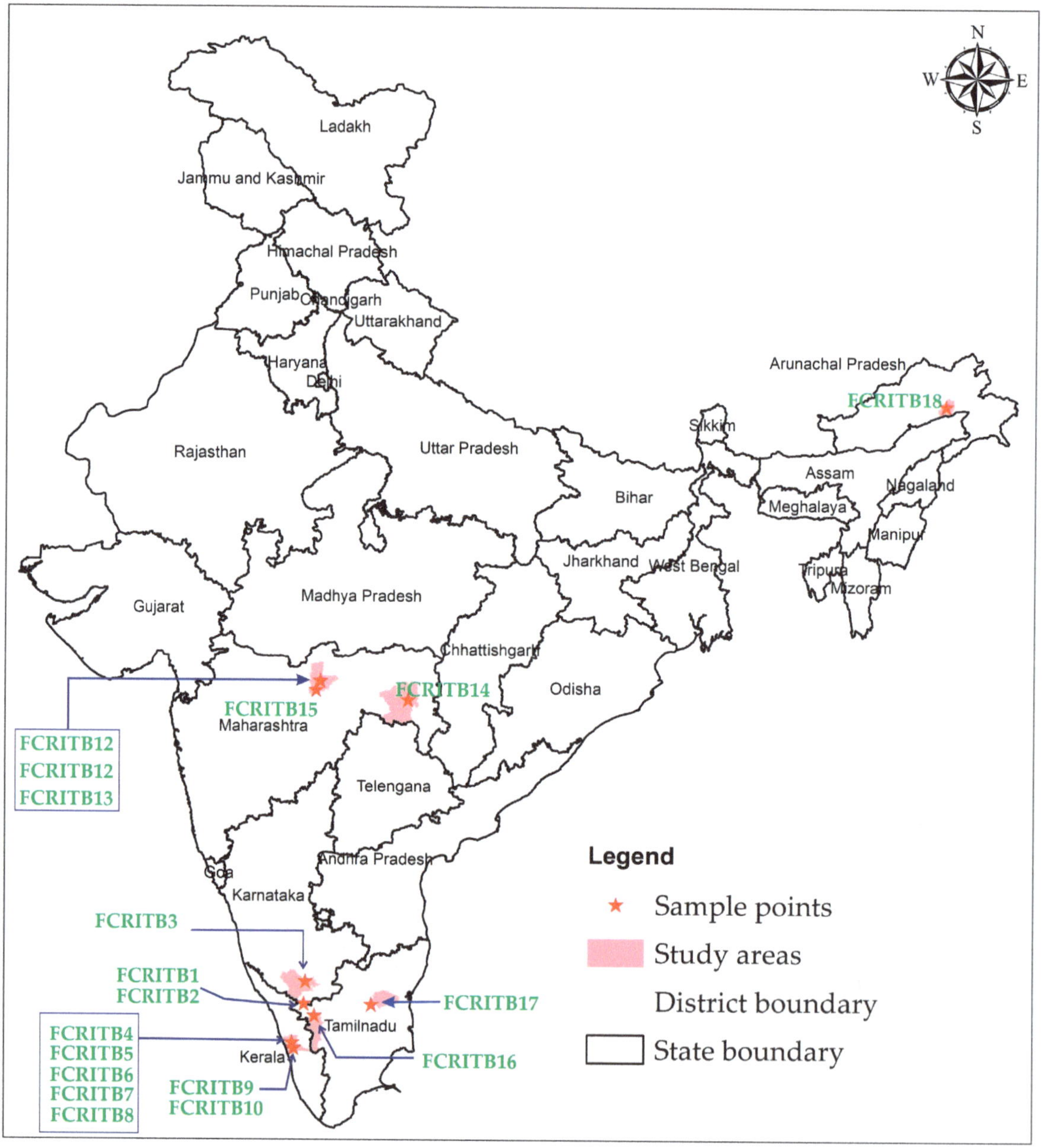

Figure 3. Map showing the progenies collection point.

Table 6. Morphometric attributes of selected Candidate Plus Trees of *T. bellerica*.

Accession Name	GBH (m)	Height (m)	Clear Bole Height (m)	Volume (m^3)
FCRITB01	3.7	17.0	8.2	731.00
FCRITB02	3.2	17.2	6.1	553.21
FCRITB03	0.90	12.0	5.7	30.53
FCRITB04	1.36	19.0	11.6	110.38
FCRITB05	1.45	18.0	9.8	118.87
FCRITB06	1.39	18.5	11.2	112.27
FCRITB07	1.90	17.3	10.4	196.16
FCRITB08	1.76	21.0	9.7	204.32
FCRITB09	2.4	19.3	10.7	349.17
FCRITB10	1.8	17.6	12.4	179.11
FCRITB11	2.8	9.2	3.6	226.55
FCRITB12	1.70	9.0	4.7	81.69
FCRITB13	1.40	11.0	3.3	67.71
FCRITB14	1.80	10.6	2.8	107.87
FCRITB15	1.40	10.2	4.7	62.79
FCRITB16	2.34	22.0	10.6	378.37
FCRITB17	2.30	23.0	7.9	204.37
FCRITB18	9.7	9.7	4.2	121.87

4.2. Study Site

After the collection process, the progenies were brought to the Forest College and Research Institute (FC&RI), TNAU, located in Mettupalayam, Tamil Nadu, India (geographical coordinate of 11.32° N latitude and 76.93° E longitude, 320 m MSL). Mettupalayam experiences a semi-arid climate characterized by a mean annual rainfall of 945 mm, along with an average of 73.6 rainy days per year. The annual temperature range varies from a minimum of 15.4 °C to a maximum of 34.9 °C. Typically, the lowest temperatures are recorded in January, while the highest temperatures occur in May each year. For the purpose of identifying elite progeny, a trial was initiated in the year 2020 at the FC&RI with three replications.

4.3. Progenies Planting

The plus trees' seeds were planted in raised beds, utilizing a mixture of red soil, sand, and farmyard manure (FYM) in a 2:1:1 ratio. These beds were consistently watered and meticulously tended to for a duration of two months. Following this period, the saplings with a collar region thickness exceeding 3–4 cm were carefully chosen and transplanted into polybags containing a blend of red soil, sand, and FYM in the same 2:1:1 ratio. Approximately one month after transplantation, these young seedlings were finally transferred and planted in the main field. No treatments or fertilizers were applied during the nursery stage. The establishment of the progeny evaluation trial in the field adhered to a randomized block design (RBD), with plants spaced at intervals of 4 × 4 m. Within each replication, four progenies per CPT were included for comprehensive evaluation. During the planting process, each seedling received additional nutrients in the form of 250 g of farmyard manure (FYM), 25 g of vermicompost and 5 g of di-ammonium phosphate (DAP). The subsequent data was acquired from the trees that were planted and observed at different time intervals.

4.4. Morphological

Data were meticulously recorded for all 18 progenies within each replication when the plants were 24 months old for the morphological traits. Field measurements were taken for each individual, including tree height (H) and basal diameter (BD). The plant's height was measured in meters (m) from the base of the stem to the tip using a measuring tape. The basal diameter of the trees at their base (in centimeters) was measured using a digital caliper from the Large SDN series. In cases where a tree had multiple basal stems, the diameters of all individual trunks were measured, and a single equivalent basal diameter (BD) value was calculated following the method outlined by Alvarez et al. [61].

4.5. Biochemical Parameters

Chlorophyll was extracted from fresh leaves using 80% acetone and 0.25 g leaf samples. The resulting extract was then measured spectrophotometrically at wavelengths of 475 nm, 645 nm, and 663 nm. The determination of total chlorophyll and carotenoid contents was carried out using established methodologies [62]. The Lowrey's method [63] was employed to evaluate the protein content of the leaves.

4.6. Genetic Estimates

Heritability, genetic advancement as a percentage of the mean, phenotypic, and genotypic coefficients of variation (PCV and GCV), were calculated for volume as well as growth traits, following the methodologies proposed by various researchers [64–66].

Broad-sense heritability in all the progenies was estimated by dividing the variance in measurements into two components: between-accessions and within-accessions [67].

4.7. Statistical Analysis

The initial dataset was created by calculating the averages for each trait across four CPTS within each replication and between replication in the experiment. These calculated means were then subjected to subsequent statistical and genetic analyses. Correlation between traits to reveal possible associations was calculated with raw data based on single plant estimates, using the Pearson correlation coefficient at $p \leq 0.05$. PCA was performed with progeny means to determine the relationships among progenies and to obtain an overview of correlation among traits. Various statistical analysis was conducted using the SPSS Windows software package (IBM SPSS version 26).

5. Conclusions

The ultimate objective of tree improvement is to enhance the growth and yield traits of tree species. These traits are intricate and are influenced by the interaction of various physiological and morphological characteristics. Therefore, solely relying on the performance of individual tree species for improvement might prove to be less effective. Hence, it can be concluded that for tree improvement of *T. bellerica* through the phenotypic selection process, the number of plus trees selected from a population should be sufficiently large in order to exploit the large intra-population genetic variation. Besides, significant differences were found between the features in the progeny study, which evaluated genetic correlations and variability in growth and physio-chemical parameters. For the majority of variables, estimates of broad-sense heritability were high, suggesting significant genetic control. Plant height, leaf area, and girth at breast height were found to be important characteristics for increasing *T. bellerica* volume through correlation studies. The study demonstrated the effectiveness of targeting girth at breast height and plant height for establishing elite seedling nurseries and clonal seed nurseries for future varietal and hybridization programs.

Supplementary Materials: The following supporting information can be downloaded at: https://www.mdpi.com/article/10.3390/plants13040470/s1, Figure S1: One way ANOVA plot showing the significant differences among the studied physiochemical properties amidst the eighteen progenies; Figure S2: Correlation map showing the relationship between 13 characters in the study; Figure S3:

PCA variance explained; Figure S4: Segregation of the 18 Terminalia progenies according to their growth, physiological and biochemical characteristics determined by principal component analysis; Figure S5: Selection of candidate plus; Figure S6: Walter-Leith diagram of the monthly rainfall and daily average temperature of Bandipur (Karnataka), Thrissur (Kerala), Kallakurichi (Tamil Nadu) and Jognari (Tamil Nadu); Figure S7: Walter-Leith diagram of the monthly rainfall and daily average temperature of Maharashtra, Mysuru (Karnataka), Pasighat (Arunachal Pradesh) and Vellanikkara (Kerala); Table S1: Component scores and loadings of *T. bellerica* traits.

Author Contributions: Conceptualization, S.U.K., K.T.P., K.S., S.V. and D.U.N.; methodology, S.U.K., K.T.P., K.S., S.V., D.U.N. and M.J.S.; software, S.U.K., K.S., S.V. and D.U.N.; validation, S.U.K., K.S., S.V., D.U.N., S.M.K., M.D., M.R. and V.G.; formal analysis, S.V and D.U.N.; investigation, S.U.K., K.T.P., K.S., S.V., D.U.N., S.M.K., M.D., M.R. and V.G.; resources, S.U.K., K.T.P., K.S., S.V., D.U.N., S.M.K., M.D., M.R. and V.G.; data curation, S.U.K., K.S., S.V., D.U.N., S.M.K., M.D., M.R. and V.G.; writing—original draft preparation, S.U.K., K.T.P., K.S., S.V. and D.U.N.; writing—review and editing, S.U.K., K.T.P., K.S., S.V., D.U.N., S.M.K., M.D., P.K., M.R. and V.G.; visualization, S.U.K., K.T.P., K.S., S.V., D.U.N., S.M.K., M.D., P.K., M.R. and V.G.; supervision, S.U.K., K.T.P., K.S., S.V., D.U.N., S.M.K., M.D., M.R. and V.G. All authors have read and agreed to the published version of the manuscript.

Funding: This research received no external funding.

Data Availability Statement: The data is available from corresponding author upon reasonable request.

Acknowledgments: The authors express their gratitude to the Forest College and Research Institute, Mettupalayam, Tamil Nadu, India. The contribution of their essential laboratory and instrumentation facilities has played a vital role in providing support for this study.

Conflicts of Interest: The authors declare no conflicts of interest.

References

1. Patil, U.H.; Gaikwad, D.K. Phytochemical evaluation and bactericidal potential of *Terminalia arjuna* stem bark. *Int. J. Pharm. Sci. Res.* **2011**, *2*, 614.
2. Hein, P.R.G.; Chaix, G. NIR spectral heritability: A promising tool for wood breeders? *J. Near Infrared Spectrosc.* **2014**, *22*, 141–147. [CrossRef]
3. Gray, L.K.; Rweyongeza, D.; Hamann, A.; John, S.; Thomas, B.R. Developing management strategies for tree improvement programs under climate change: Insights gained from long-term field trials with lodgepole pine. *For. Ecol. Manag.* **2016**, *377*, 128–138. [CrossRef]
4. Sæbø, A.; Borzan, Ž.; Ducatillion, C.; Hatzistathis, A.; Lagerstrom, T.; Supuka, J.; García-Valdecantos, J.L.; Rego, F.; Van Slycken, J. The selection of plant materials for street trees, park trees and urban woodland. In *Urban Forests and Trees: A Reference Book*; Springer: Berlin/Heidelberg, Germany, 2005; pp. 257–280.
5. Zobel, B.; Talbert, J. *Applied Forest Tree Improvement*; John Wiley & Sons: Hoboken, NJ, USA, 1984.
6. Schmidt, P.; Hartung, J.; Bennewitz, J.; Piepho, H.P. Heritability in plant breeding on a genotype-difference basis. *Genetics* **2019**, *212*, 991–1008. [CrossRef] [PubMed]
7. Kumar, M.S.; Parameswari, N.; Chin, C.F.; Baharum, Z.; Olalekan, K.K.; Aini, A.N. Selection and Screening of Superior Genotypes for Quality Planting Stock Based on Vegetative Growth Performance of Some Selected 12-Year-Old *Acacia* Species. *Open J. For.* **2016**, *6*, 217–229. [CrossRef]
8. ICES. Herring Assessment Working Group for the Area South of 62°N (HAWG). ICES Scientific Reports. 2023. Available online: https://doi.org/10.17895/ices.pub.22182034.v2 (accessed on 15 August 2023).
9. Kanna, S.U.; Krishnakumar, N.; Kather, M.M.A.; Jailani, K. Growth performance of *Ailanthus excelsa* through progeny test. *Pharma Innov. J.* **2019**, *8*, 204–210.
10. Krishnakumar, N.; Parthiban, K.T.; Jayamani, P.; Revathi, R.; Umeshkanna, S. Genetic variability of growth parameters among different progenies of *Santalum album* L. *J. Indian Soc. Coast. Agric. Res.* **2017**, *35*, 56–63.
11. Prakash, G.M. Genetic Evaluation and Wood Characterization Studies in *Neolamarckia cadamba* (Roxb.) Bosser. Ph.D. Thesis, Tamil Nadu Agricultural University, Coimbatore, India, 2017.
12. Mohanraj, K.; Umesh Kanna, S.; Parthiban, K.T.; Kumaran, K. Variability, Broad Sense Heritability, Genetic Advance of *Toona ciliata* M. Roem., Progenies. *Pharma Innov. J.* **2021**, *10*, 1247–1251.
13. Jawahar Vishnu, M.V.; Parthiban, K.T.; Umadevi, M.; Sudhagar, R.J.; Fernandaz, C.C.; Javed, T.; Alotaibi, S.S. Genetic evaluation of Jatropha backcross hybrid clones (BC4F1) for yield and oil quality. *Front. Genet.* **2022**, *13*, 953486. [CrossRef]
14. Migliore, J.; Lezine, A.M.; Hardy, O.J. The recent colonization history of the most widespread Podocarpus tree species in Afromontane forests. *Ann. Bot.* **2020**, *126*, 73–83. [CrossRef]
15. Dhivya, S.; Ashutosh, S.; Gowtham, I.; Baskar, V.; Harini, A.B.; Mukunthakumar, S.; Sathishkumar, R. Molecular identification and evolutionary relationships between the subspecies of Musa by DNA barcodes. *BMC Genom.* **2020**, *21*, 659. [CrossRef] [PubMed]

16. Donkpegan, A.S.; Doucet, J.L.; Migliore, J.; Duminil, J.; Dainou, K.; Pineiro, R.; Hardy, O.J. Evolution in African tropical trees displaying ploidy-habitat association: The genus Afzelia (Leguminosae). *Mol. Phylogenet. Evol.* **2017**, *107*, 270–281. [CrossRef] [PubMed]
17. Nadeem, M.A.; Nawaz, M.A.; Shahid, M.Q.; Doğan, Y.; Comertpay, G.; Yıldız, M.; Baloch, F.S. DNA molecular markers in plant breeding: Current status and recent advancements in genomic selection and genome editing. *Biotechnol. Biotechnol. Equip.* **2018**, *32*, 261–285. [CrossRef]
18. Ye, M.; Zhu, X.; Gao, P.; Jiang, L.; Wu, R. Identification of quantitative trait loci for altitude adaptation of tree leaf shape with *Populus szechuanica* in the Qinghai Tibetan Plateau. *Front. Plant Sci.* **2020**, *11*, 632. [CrossRef] [PubMed]
19. Lu, W.; Xiao, L.; Quan, M.; Wang, Q.; El-Kassaby, Y.A.; Du, Q.; Zhang, D. Linkage-linkage disequilibrium dissection of the epigenetic quantitative trait loci (epiQTLs) underlying growth and wood properties in *Populus*. *New Phytol.* **2020**, *225*, 1218–1233. [CrossRef] [PubMed]
20. Khan, I.; Shinwari, Z.K.; Zahra, N.B.; Jan, S.A.; Shinwari, S.; Najeebullah, S. DNA barcoding and molecular systematics of selected species of family Acanthaceae. *Pak. J. Bot.* **2020**, *52*, 205–212. [CrossRef] [PubMed]
21. Soltis, P.; Doyle, J.J. *Molecular Systematics of Plants II: DNA Sequencing*; Springer Science & Business Media: Berlin, Germany, 2012.
22. Bhandari, H.R.; Bhanu, A.N.; Srivastava, K.; Singh, M.N.; Shreya, H.A. Assessment of genetic diversity in crop plants-an overview. *Adv. Plants Agric. Res.* **2017**, *7*, 279–286.
23. Parthiban, K.T.; Thirunirai-Selvan, R.; Palanikumaran, B.; Krishnakumar, N. Variability and genetic diversity studies on *Neolamarckia cadamba* genetic resources. *J. Trop. For. Sci.* **2019**, *31*, 90–98. [CrossRef]
24. Parthiban, K.T.; Kanagaraj, N.; Palanikumaran, B.; Krishnakumar, N. Development of DUS descriptor for *casuarina* genetic resources. *Int. J. Genet.* **2018**, *10*, 333–338. [CrossRef]
25. Daneva, V.; Dhillon, R.S.; Johar, V. Plus tree selection and progeny testing of superior candidate plus trees (CPTs) of *Ailanthus excelsa*. *J. Pharmacogn. Phytochem.* **2018**, *7*, 543–545.
26. Krishnakumar, N.; Parthiban, K.T.; Kanna, S.U. Production, management and utilization technology for sandal wood (*Santalum album* L.). In *Forest Technol: Complete Value Chain Approach*; Scientific Publishers: New Delhi, India, 2017; pp. 372–383.
27. Sharma, A.; Bakshi, M. Growth and heritability estimates among clones of *Dalbergia sissoo* Roxb. in a clonal seed orchard. *For. Sci. Pract.* **2011**, *13*, 211. [CrossRef]
28. Sharma, J.P.; Singh, N.B.; Benal, V.; Gupta, D. Cultivation of shiitake mushroom on selected clones of willow (*Salix species*): A Case Study Under PPP Mode. In Proceedings of the 2nd International Conference on Bio-resource and Stress Management, Hyderabad, India, 7–10 January 2015; Volume 225.
29. Mohamed, M.N.; Parthiban, K.T.; Ravi, R.; Kumar, P. Provenance variation in growth and genetic potential of *Aquilaria malaccens* is under nursery condition. *Afr. J. Biotechnol.* **2015**, *14*, 2005–2013.
30. Meena, H.; Kumar, A.; Sharma, R.; Chauhan, S.K.; Bhargava, K.M. Genetic variation for growth and yield parameters in half-sib progenies of *Melia azedarach* (Linn.). *Turk. J. Agric. For.* **2014**, *38*, 531–539. [CrossRef]
31. Sangram, C.; Keerthika, A. Genetic variability and association studies among morphological traits of *Leucaena leucocephala* (Lam.) de Wit. genetic resources. *Res. J. Agric. For. Sci.* **2013**, *1*, 23–29.
32. Kundal, M.; Thakur, S.; Dhillon, G. Evaluation of growth performance of half sib progenies of *Toona ciliata* M. Roem under field conditions. *Genetika* **2020**, *52*, 651–660. [CrossRef]
33. Vijay, R.; Atul, R. Progeny performance of plus trees of *Toona ciliata* M. Roem. under nursery and field conditions. *Indian For.* **2009**, *135*, 92–98.
34. Selvan, R.T.; Parthiban, K.T. Clonal evaluation and genetic divergence studies in *Neolamarckia* cadamba roxb. *Electron. J. Plant Breed.* **2018**, *9*, 692–704. [CrossRef]
35. ThiruniraIselvan. Improvement and Utilization of *Neolamarckia cadamba* (Roxb.) Bosser Genetic Resources for Feed Quality and Utility. Ph.D. Thesis, Tamil Nadu Agricultural University, Coimbatore, India, 2017.
36. Mohamed, Z.; Abdelsalam, N.; Abdel Latif, K.; Abdelhady, R. Genetic Diversity of fig (*Ficus carica* L.) Based on Morphological Characters and Two-Way Hierarchical Cluster Analysis. *Alex. Sci. Exch. J.* **2017**, *38*, 168–174. [CrossRef]
37. Garima, G.; Handa, A.K.; Deepak, M. Variation in seed and seedling traits of *Pongamia pinnata*. *Indian For.* **2016**, *142*, 852–857.
38. Monclus, R.; Dreyer, E.; Delmotte, F.M.; Villar, M.; Delay, D.; Boudouresque, E.; Brignolas, F. Productivity, leaf traits and carbon isotope discrimination in 29 *Populus deltoids* × *P. nigra* clones. *New Phytol.* **2005**, *167*, 53–62. [CrossRef]
39. Mohit, G.; Neelu, G. Genetic variability and character association in Acacia catechu Willd. *Indian For.* **2006**, *132*, 785–794.
40. Mishra, S.; Bhatt, R. Varietal differences in photosynthetic pigments and biochemical constituents in *Leucaena leucocephala*. *Indian J. Plant Physiol.* **2004**, *9*, 86–89.
41. Zayed, M.Z.; Zaki, M.A.; Ahmad, F.B.; Ho, W.S.; Pang, S.L. Comparison of mimosine content and nutritive values of *Neolamarckia cadamba* and *Leucaena leucocephala* with medicago sativa as forage quality index. *Int. J. Sci. Technol. Res.* **2014**, *3*, 146–150.
42. Radhakrishnan, S. Genetic Divergence and DNA Based Molecular Characterization in *Albizia lebbeck* (L.) Benth. Ph.D. Thesis, Tamil Nadu Agricultural University, Coimbatore, India, 2001.
43. Ramachandra, N.G.; Nautiyal, S.; Negi, D.S.; Thapliyal, R.C. Seed source variation in chlorophyll contents of leaves of *Acacia catechu* Willd. under different water stress conditions. *Ann. For.* **1997**, *5*, 88–96.
44. Malik, A.B.; Makbdoom, M.I.; Hag, A. Investigations on the efficiency of exogenous synthetic growth regulators on fruit drop in mango (*Mangifera indica* L.). *Egypt. J. Hort.* **1993**, *20*, 1–14.

45. Cai, Y.C.; Ma, G.H.; Wang, Z.Q. Physiological characters of different geoprovenant populations of Siberian elm (*Ulmus pumila* L.). *Ningxia J. Agro-For. Sci. Technol.* **1990**, *2*, 20–23.
46. Rao, G.R.; Shanker, A.K.; Srinivas, I.; Korwar, G.R.; Venkateswarlu, B. Diversity and variability in seed characters and growth of *Pongamia pinnata* (L.) Pierre accessions. *Trees* **2011**, *25*, 725–734. [CrossRef]
47. Kadam, S.K. Evaluation of Full-Sib Progenies of Selected Clones of Poplar (*Populus deltoids* Bartr.). Ph.D. Thesis, Forest Research Institute, Dehradun, India, 2002.
48. Choudhary, P.; Singh, N.B.; Sharma, J.P.; Verma, A. Estimation of genetic parameters among intra and interspecific progenies of tree willows. *Indian For.* **2016**, *142*, 1157–1163.
49. Reddy, L.H.; Arias, J.L.; Nicolas, J.; Couvreur, P. Magnetic nanoparticles: Design and characterization, toxicity and biocompatibility, pharmaceutical and biomedical applications. *Chem. Rev.* **2012**, *112*, 5818–5878. [CrossRef]
50. Vashistha, R.; Dangi, A.K.; Kumar, A.; Chhabra, D.; Shukla, P. Futuristic biosensors for cardiac health care: An artificial intelligence approach. *3 Biotech* **2018**, *358*, 1–11. [CrossRef]
51. Wondimu, T.; Gizaw, A.; Tusiime, F.M.; Masao, C.A.; Abdi, A.A.; Gussarova, G.; Brochmann, C. Crossing barriers in an extremely fragmented system: Two case studies in the afro-alpine sky island flora. *Plant Syst. Evol.* **2014**, *300*, 415–430. [CrossRef]
52. Vivekananda, P.; Subramaninan, S. Genetic divergence in rainfed rice. *Oryza* **1993**, *39*, 60–62.
53. Singh, N.B.; Sharma, J.P.; Huse, S.A.; Thakur, I.K.; Gupta, R.K.; Sankhyan, H.P. Heritability, genetic gain, correlation and principal component analysis in introduced willow (*Salix* spp.) clones. *Indian For.* **2012**, *138*, 1100–1109.
54. Kaushik, N.; Kumar, S.; Kumar, K.; Beniwal, R.S.; Kaushik, N.; Roy, S. Genetic variability and association studies in pod and seed traits of *Pongamia pinnata* (L.) Pierre in Haryana, India. *Genet. Resour. Crop Evol.* **2007**, *54*, 1827–1832. [CrossRef]
55. Divakara, B.N.; Das, R. Variability and divergence in *Pongamia pinnata* for further use in tree improvement. *J. For. Res.* **2011**, *22*, 193–200. [CrossRef]
56. Isik, F.; Toplu, F. Variation in juvenile traits of natural black poplar (*Populus nigra* L.) clones from Turkey. *New For.* **2004**, *27*, 175–182. [CrossRef]
57. Ozel, H.B.; Ertekin, M.; Tunctaner, K. Genetic variationin growth traits and morphological characteristics ofeastern cottonwood (*Populus deltoides* Bartr.) hybridsat nursery stage. *Sci. Res. Essays* **2010**, *5*, 962–969.
58. Kajba, D.; Ballian, D.; Idžojtić, M.; Poljak, I. Leaf morphology variation of *Populus nigra* L. in natural populations along the rivers in Croatia and Bosnia and Herzegovina. *South-East Eur. For. Seefor* **2015**, *6*, 39–51. [CrossRef]
59. Tunctaner, K. Primary selection of willow clones for multi-purpose use in short rotation plantation. *Silvae Genet.* **2002**, *51*, 105–112.
60. Singh, N.B.; Sharma, J.P.; Choudhary, P.; Gupta, R.K. Genotype x environment interaction and growth stability of exotic tree willow (*Salix* spp.) clones. *Indian J. Genet.* **2014**, *74*, 222–228. [CrossRef]
61. Alvarez, J.A.; Villagra, P.E.; Villalba, R.; Debandi, G. Effects of the pruning intensity and tree size on multi-stemmed Prosopis flexuosa trees in the Central Monte, Argentina. *For. Ecol. Manag.* **2013**, *310*, 857–864. [CrossRef]
62. Yoshida, S.; Coronel, V. Nitrogen nutrition, leaf resistance, and leaf photosynthetic rate of the rice plant. *Soil Sci. Plant Nutr.* **1976**, *22*, 207–211. [CrossRef]
63. Lowry, O.H.; Rosenbough, H.I.; Fair, A.L.; Randall, R.I. Protein measurement with the Folin phenol reagent. *J. Biol. Chem.* **1951**, *193*, 265–275. [CrossRef] [PubMed]
64. Baenziger, S.P.; Gregory, S.; McMaster, R.; Wilhelm, W.W.; Weiss, A.; Hays, C.J. Putting genes into genetic coefficients. *Field Crops Res.* **2004**, *90*, 133–143. [CrossRef]
65. Pliura, A.; Zhang, S.Y.; MacKay, J.; Bousquet, J. Genotypic variation in wood density and growth traits of poplar hybrids at four clonal trials. *For. Ecol. Manag.* **2007**, *238*, 92–106. [CrossRef]
66. Yoshida, H.; Takeshi, H.; Katsura, K.; Tatsuhiko, S. A model explaining genotypic and environmental variation in leaf area development of rice based on biomass growth and leaf N accumulation. *Field Crops Res.* **2007**, *102*, 228–238. [CrossRef]
67. Falconer, D.S.; Mackay, T.F.C. *Introduction to Quantitative Genetics*, 4th ed.; Longmans Green: Harlow, UK, 1996.

plants

MDPI

Article

Nitrogen Addition and Heterotroph Exclusion Affected Plant Species Diversity–Biomass Relationship by Affecting Plant Functional Traits

Xile Xu [1], Luping Yang [1], Kai Shen [1], Huijuan Cao [1], Yishi Lin [1], Jinliang Liu [1,2] and Wenjuan Han [1,2,*]

1 College of Life and Environmental Science, Wenzhou University, Wenzhou 325035, China; 21451335046@stu.wzu.edu.cn (X.X.); 21461337123@stu.wzu.edu.cn (L.Y.); 21461337078@stu.wzu.edu.cn (K.S.); 22451335002@stu.wzu.edu.cn (H.C.); 22461337071@stu.wzu.edu.cn (Y.L.); jinliang.liu@foxmail.com (J.L.)

2 Zhejiang Provincial Key Laboratory for Water Environment and Marine Biological Resources Protection, Wenzhou University, Wenzhou 325035, China

* Correspondence: hanwj@wzu.edu.cn; Tel.: +86-0577-86689079

Abstract: (1) Background: Heterotrophs can affect plant biomass and alter species diversity–productivity relationships. However, these studies were conducted in systems with a low nitrogen (N) availability, and it is unclear how heterotroph removal affects the relationship between plant species diversity and productivity in different N habitats. (2) Methods: Three typical understory herbaceous plants were selected to assemble the plant species diversity (three plant species richness levels (1, 2, and 3) and seven plant species compositions), and the control, insecticide, fungicide, and all removal treatments were performed at each plant species diversity level in systems with or without N addition treatments. (3) Results: In systems without N addition, the insecticide treatment increased the plant aboveground biomass, total biomass, and leaf area, while the fungicide treatment reduced the plant belowground biomass, root length, and root tip number; the presence of *Bidens pilosa* increased the plant aboveground biomass. Similarly, the presence of *Bletilla striata* increased the plant belowground biomass and root diameter under each heterotroph removal treatment. In systems with N addition, all removal treatments reduced the plant belowground biomass and increased the plant leaf area; the presence of *B. pilosa* significantly increased the plant aboveground biomass, total biomass, and root length under each heterotroph removal treatment. The presence of *B. striata* significantly increased the plant belowground biomass and leaf area under insecticide and fungicide treatments. (4) Conclusions: Heterotroph removal alters the plant species diversity–biomass relationship by affecting the plant functional traits in systems with different N availabilities. The impact of biodiversity at different trophic levels on ecosystem functioning should be considered under the background of global change.

Keywords: nitrogen addition; heterotroph removal; species identity; productivity; functional traits

check for updates

Citation: Xu, X.; Yang, L.; Shen, K.; Cao, H.; Lin, Y.; Liu, J.; Han, W. Nitrogen Addition and Heterotroph Exclusion Affected Plant Species Diversity–Biomass Relationship by Affecting Plant Functional Traits. *Plants* **2024**, *13*, 258. https://doi.org/10.3390/plants13020258

Academic Editors: Xugao Wang, Zuoqiang Yuan and Chengjin Chu

Received: 9 November 2023
Revised: 3 January 2024
Accepted: 11 January 2024
Published: 16 January 2024

1. Introduction

Biodiversity is an important determinant of ecosystem function [1]. Plant species diversity (species richness and species identity) could enhance plant productivity through the selection effect and complementary effect [1–4]. However, most productivity measures did not account for the effects of heterotrophs on productivity [5–7]. Heterotrophs include herbivores, predators, scavengers, and pathogens. Previous studies showed that the removal of arthropods and foliar fungi increased plant biomass [8–10], while the removal of soil fungi increased the forb biomass in grassland systems [11]. Removing foliar fungi also increased the biomass of trees in forest systems [12]. Increasing the plant species diversity can increase the abundance of arthropods [13,14] or decrease the abundance of fungal pathogens [15], and the impact of heterotrophs on plant biomass may increase or decrease with an increasing plant species diversity. In addition, plant and microbial diversities may have complementary effects on nutrient cycling [16]; plant and herbivore diversities may have opposite effects on plant productivity [17]. Thus, considering the influence of heterotrophs on the plant species diversity–biomass relationship is necessary.

A few studies have concentrated on the impact of heterotroph removal on plant species diversity–biomass relationships [8,12]. In the grassland system, insecticide and fungicide treatments promoted the impact of plant diversity on productivity [8]. In forest systems, the positive relationship between tree species richness and productivity was eliminated when tree crowns were under a fungicide treatment [12]. However, all these studies were conducted in habitats with relatively low nitrogen (N) levels. The impact of heterotroph removal on the plant species diversity–biomass relationship in habitats with a high N level remains unclear.

Human activities such as industrial development and agricultural production have continuously increased atmospheric N deposition in the terrestrial ecosystem [18–21]. In habitats with a low N availability, N deposition could increase plant biomass [6,22,23]. Nevertheless, continuous N deposition could lead to N saturation, inhibited plant growth, and reduced plant biomass [24]. The increase in N availability in habitats may promote the growth of dominant plants, thereby increasing the selection effects [25]; it may also increase the complementary utilization of N by plants or promote interspecies interactions to enhance the complementary effect [7,26]. In addition, the increase in N availability in habitats may also alter the abundance of heterotrophs. For example, N addition reduced the number of soil microorganisms [27,28]. Thus, exploring the influence of heterotroph removal on plant species diversity–productivity relationships in high N habitats is necessary.

The functional traits of plant leaves and roots, such as the leaf area, root length, and root diameter, can reflect plants' adaptability to the environment, their self-regulation ability in complex habitats, and their essential characteristics and effective utilization of resources [29]. Previous research showed that N deposition promoted the growth of the aboveground biomass of plants and specific leaf area [30,31], but excessive N would decrease the specific root length and belowground biomass [32]. The presence of herbivorous insects reduced the plant leaf area [33]. There was a direct interaction between soil microbial communities and roots; fungi and rhizobia could affect the ability of roots to capture nutrients from the soil [8,34].

To test how N addition and heterotroph exclusion affect the effect of plant species diversity on plant biomass, we conducted a three-factor (N addition, plant species diversity, and heterotrophic removal) control experiment, selecting three typical understory herbaceous plants, *Perilla frutescens*, *Bletilla striata*, and *Bidens pilosa*, to assemble the plant species diversity, and heterotroph removal was performed at each plant species diversity level. N deposition was simulated by N addition (10 g N m^{-2} yr^{-1}). The plant above- and belowground biomasses and leaf and root functional traits of herbaceous plants were measured. We investigated the influence of heterotroph removal on plant biomass and functional traits in the system without/with N addition. We further investigated the effect of heterotroph removal on the plant diversity–biomass relationship in the system without/with N addition. We predicted that heterotroph exclusion and N addition may affect the plant species diversity–biomass relationship through the plant functional traits.

2. Results

2.1. Plant Biomass Responds to N Addition and Heterotroph Removal

N addition increased the plant biomass, with the plant aboveground, belowground, and total biomasses increased by 294.3%, 61.6%, and 178.5% on average, respectively (Figure 1). Under different heterotroph removal treatments, N addition improved the plant total and aboveground biomasses; under control and fungicide treatment groups, N addition also improved the plant belowground biomass (Figure 1).

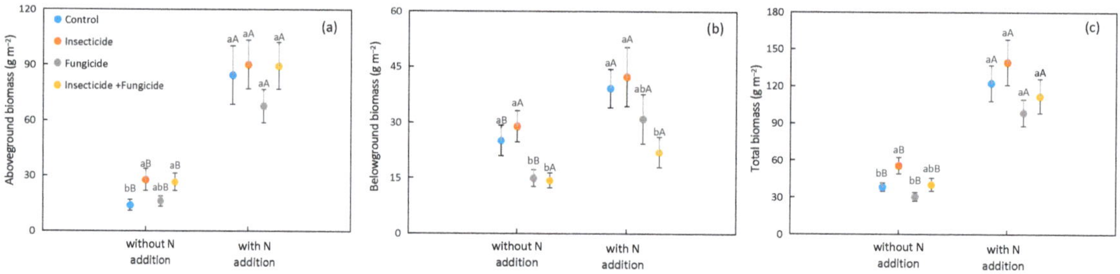

Figure 1. Difference in plant (**a**) aboveground, (**b**) belowground, (**c**) total biomass among heterotroph removal with or without N addition. Significant differences between systems without or with nitrogen addition were indicated in capital letters, and significant differences among heterotroph removal were indicated in lowercase letters. Each circle represents the average biomass of all species compositions under each heterotroph removal treatment. Blue: control; orange: insecticide; gray: fungicide; yellow: all removal.

In systems without N addition, the insecticide treatment increased the plant aboveground biomass by 98.9%, and all removal treatments increased the plant aboveground biomass by 90.3% relative to the control (Figure 1a); insecticide treatment also increased the plant total biomass by 45.9% relative to the control (Figure 1c). In systems with N addition, the study did not discover significant differences in the aboveground and total biomasses among various heterotroph removal treatments (Figure 1a,c). In systems with or without N addition, all removal treatments decreased the plant belowground biomass by 42.9% and 43.9% relative to the control, but insecticide treatment did not affect the plant belowground biomass (Figure 1b).

2.2. The Relationship between Plant Species Diversity and Plant Biomass

Plant species richness significantly improved the plant aboveground biomass, but plant belowground and total biomasses did not respond to plant species richness (Table S1). Plant species compositions also significantly affected the plant aboveground, belowground, and total biomasses (Table S1).

In systems without N addition, the aboveground biomass of the *B. pilosa* monoculture was significantly higher than that of the *P. frutescens* monoculture and *B. striata* monoculture (Figure 2a), and the presence of *B. pilosa* significantly increased the plant aboveground and belowground biomasses under each heterotroph removal treatment (Table 1). It is worth noting that the plant total biomass was not affected by plant species identity in the control treatment. However, the plant total biomass was improved with fungicide treatment by 110.2% and all removal treatments by 155.8% when *B. pilosa* was present (Table 1).

Table 1. Plant biomass (aboveground, belowground, and total) responses to species identity without N addition; *p* values are displayed in bold font when $p < 0.05$.

Source of Variation	Aboveground Biomass		Belowground Biomass		Total Biomass	
	p Value	Change	*p* Value	Change	*p* Value	Change
Control						
P. frutescens	0.223	ns	0.135	ns	0.061	ns
B. striata	0.357	ns	**<0.001**	↑ 381.67%	0.058	ns
B. pilosa	**<0.001**	↑ 343.78%	0137	ns	0.818	ns
Insecticide						
P. frutescens	0.121	ns	0.109	ns	**0.003**	↓ 49.72%
B. striata	0.942	ns	**<0.001**	↑ 356.69%	0.071	ns
B. pilosa	**<0.001**	↑ 300.77%	0.216	ns	0.189	ns

Table 1. *Cont.*

Source of Variation	Aboveground Biomass		Belowground Biomass		Total Biomass	
	p Value	Change	*p* Value	Change	*p* Value	Change
Fungicide						
P. frutescens	0.554	ns	0.152	ns	0.126	ns
B. striata	0.102	ns	**<0.001**	↑ 252.87%	0.645	ns
B. pilosa	**<0.001**	↑ 472.00%	1.000	ns	**0.001**	↑ 110.19%
Insecticide + Fungicide						
P. frutescens	0.567	ns	0.056	ns	0.144	ns
B. striata	0.397	ns	**0.003**	↑ 138.16%	0.765	ns
B. pilosa	**<0.001**	↑ 506.20%	0.635	ns	**<0.001**	↑ 155.78%

Notes: Values and arrows in 'Change' column show significant increase (↑) or decrease (↓) of the variables with the presence of a certain species compared to its absence, and 'ns' means no significant change.

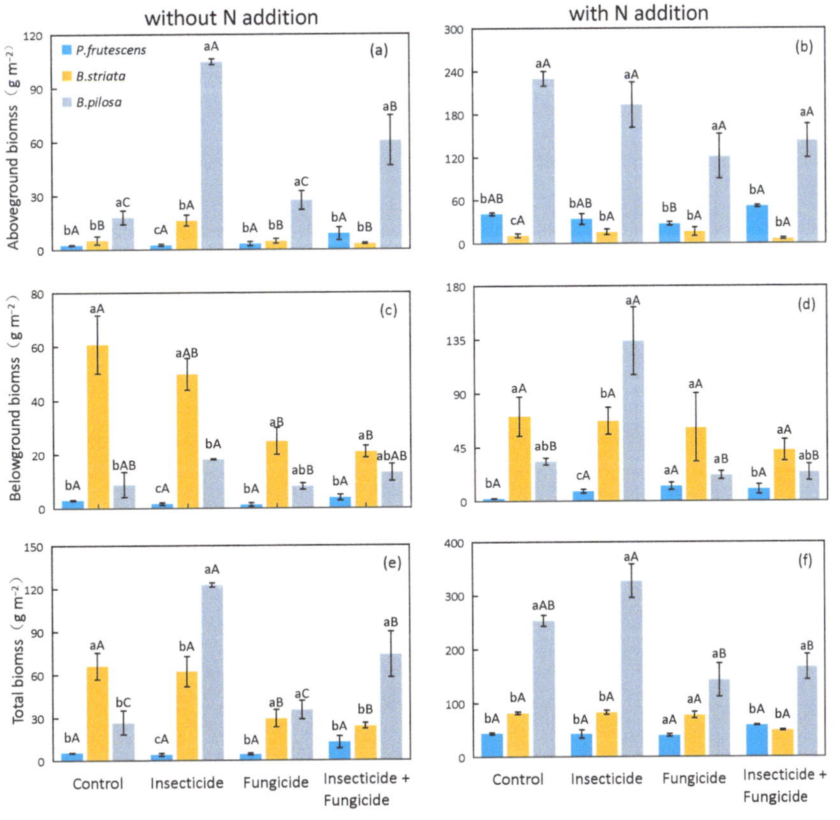

Figure 2. Difference in monoculture plant aboveground biomass (**a**), belowground biomass (**c**), and total biomass (**e**) in systems without N addition and plant aboveground biomass (**b**), belowground biomass (**d**), and total biomass (**f**) in systems with N addition among heterotroph exclusion with or without N addition. Significant differences between heterotroph removal groups were indicated in capital letters, and significant differences between plant species monocultures were indicated in lowercase letters. Blue bars: *P. frutescens* monoculture; yellow bars: *B. striata* monoculture; gray bars: *B. pilosa* monoculture.

In systems with N addition, the aboveground and total biomasses of the *B. pilosa* monoculture were significantly higher than those of the *P. frutescens* monoculture and *B.*

striata monoculture (Figure 2b,f). Plant aboveground and total biomasses were improved when *B. pilosa* was present under each heterotroph removal treatment (Table 2). Significantly, the presence of *B. striata* decreased the aboveground biomass by 58.1% under control treatment, while the presence of *B. striata* did not affect the plant aboveground biomass after heterotroph removal. The presence of *B. striata* also increased the plant belowground biomass by 170.9% and 174.2%, respectively, under control and fungicide treatments (Table 2).

Table 2. Plant biomass (aboveground, belowground, and total) responses to species identity with N addition; p values are displayed in bold font when $p < 0.05$.

Source of Variation	Aboveground Biomass		Belowground Biomass		Total Biomass	
	p Value	Change	*p* Value	Change	*p* Value	Change
Control						
P. frutescens	0.235	ns	0.487	ns	0.095	ns
B. striata	**0.016**	↓ 58.09%	**<0.001**	↑ 170.89%	0.357	ns
B. pilosa	**<0.001**	↑ 735.36%	**0.040**	↓ 46.65%	**0.002**	↑ 122.13%
Insecticide						
P. frutescens	0.247	ns	**0.002**	↓ 75.01%	**0.009**	↓ 52.30%
B. striata	0.111	ns	0.178	ns	0.114	ns
B. pilosa	**<0.001**	↑ 425.05%	0.516	ns	**<0.001**	↑ 220.75%
Fungicide						
P. frutescens	0.316	ns	0.134	ns	0.069	ns
B. striata	0.464	ns	**0.018**	↑ 174.19%	0.555	ns
B. pilosa	**<0.001**	↑ 270.44%	0.237	ns	**0.003**	↑ 85.24%
Insecticide + Fungicide						
P. frutescens	0.432	ns	**0.014**	↓ 60.88%	0.980	ns
B. striata	0.077	ns	0.236	ns	0.148	ns
B. pilosa	**<0.001**	↑ 355.43%	0.347	ns	**<0.001**	↑ 211.94%

Notes: Values and arrows in 'Change' column show significant increase (↑) or decrease (↓) of the variables with the presence of a certain species compared to its absence, and 'ns' means no significant change.

2.3. Functional Traits of Plant Leaves and Roots Respond to N Addition and Heterotroph Removal

N addition increased the plant leaf area, root length, and root tip number by 49.5%, 95.9%, and 53.0% on average, respectively (Figure 3a,b,d). N addition increased the leaf area under control and all removal treatments (Figure 3a); N addition also increased the root length and root tip number in the insecticide, fungicide, and all removal treatment groups (Figure 3b,d); N addition reduced the root diameter under fungicide treatment (Figure 3c).

In systems without N addition, the insecticide treatment increased the leaf area by 98.1% relative to the control (Figure 3a). Fungicide treatment decreased the root length by 46.2% relative to the control (Figure 3b). Insecticide treatment increased the root diameter by 19.5%, and fungicide treatment increased the root diameter by 20.0% relative to the all removal treatment (Figure 3c). The insecticide, fungicide, and all removal treatments decreased the root tip number by 30.0%, 41.7%, and 10.9% relative to the control, respectively (Figure 3d). In systems with N addition, the all removal treatment increased the leaf area by 54.7%, and fungicide treatment decreased the root diameter by 16.0% relative to the control (Figure 3a,c). There were no significant differences found in the plant root length and tip number among heterotroph removal treatments (Figure 3b,d).

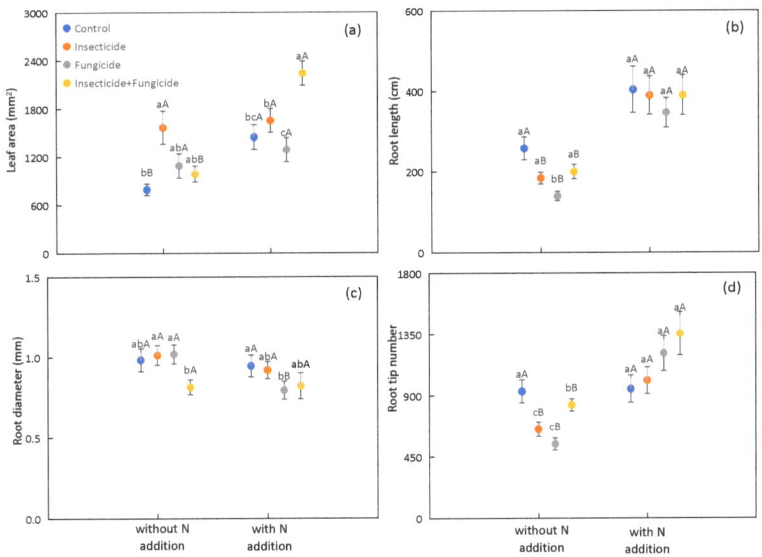

Figure 3. Difference in plant functional traits of plant leaf area (**a**), root length (**b**), root diameter (**c**), and root tip number (**d**) among heterotroph removal with or without N addition. Significant differences between systems without or with nitrogen addition were indicated in capital letters, and significant differences between heterotroph removals were indicated in lowercase letters. Each circle represents the average plant functional traits of all species compositions. Blue: control; orange: insecticide; gray: fungicide; yellow: insecticide and fungicide.

2.4. *The Relationship between Plant Species Diversity and Functional Traits of Plant Leaves and Roots*

Species richness significantly affected the plant leaf area and root tip number but did not affect the root length and diameter (Table S1). The plant leaf area decreased when the species richness increased to two and three, while the root tip number increased when species richness increased to three. Plant species compositions also significantly affected the plant leaf area, root length, root diameter, and root tip number (Table S1).

In systems without N addition, the leaf area of the *B. striata* monoculture was significantly higher than that of the *P. frutescens* monoculture under control treatment groups (Figure 4a). However, the root diameter of the *B. striata* monoculture was significantly higher than that of the *P. frutescens* monoculture under each heterotroph removal treatment (Figure 4e), and the plant leaf area and root diameter were improved when *B. striata* was present (Table 3). The responses of plant root length and root tip number to the species identity were various under different heterotroph removal treatments. The presence of *B. striata* reduced the root tip number and root length of plants under control and insecticide treatments (Table 3), but the plant species identity did not affect root length under all heterotroph removal treatments.

In systems with N addition, the root length of the *B. striata* monoculture was significantly lower than that of the *B. pilosa* monoculture (Figure 4d), and the presence of *B. striata* reduced the root length and root tip number under each heterotroph removal treatment. The presence of *B. pilosa* increased the plant root length under each heterotroph removal treatment (Table 4). The response of plant root length to the species identity remained unchanged after heterotroph removal treatment. The root diameter of the *B. striata* monoculture was significantly higher than that of the *P. frutescens* monoculture and *B. pilosa* monoculture (Figure 4f), and the presence of *B. striata* increased the root diameter under each heterotroph removal treatment (Table 4).

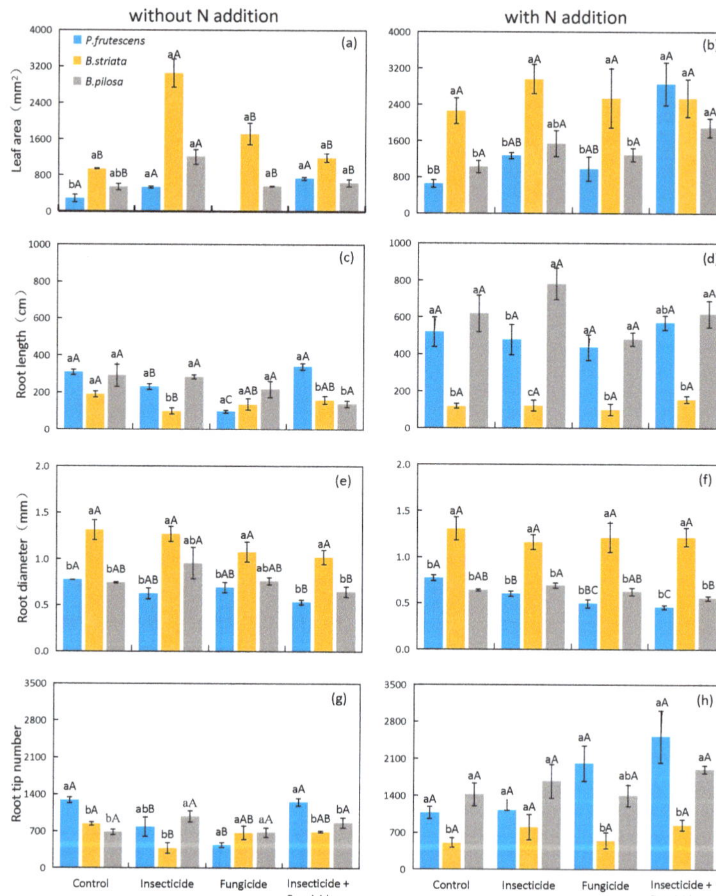

Figure 4. Difference in monoculture plant leaf area (**a**), root length (**c**), root diameter (**e**), root tip number (**g**) in systems without N addition and leaf area (**b**), root length (**d**), root diameter (**f**), root tip number (**h**) in systems with N addition among heterotroph exclusion with or without N addition. Significant differences between heterotroph removals were indicated in capital letters, and significant differences between plant species monocultures were indicated in lowercase letters. Blue bars: *P. frutescens* monoculture; yellow bars: *B. striata* monoculture; gray bars: *B. pilosa* monoculture.

Table 3. Functional traits of plant leaves and roots respond to species identity without N addition; p values are displayed in bold font when $p < 0.05$.

Source of Variation	Leaf Area		Root Length		Root Diameter		Root Tip Number	
	p Value	Change	*p* Value	Change	*p* Value	Change	*p* Value	Change
Control								
P. frutescens	0.136	ns	0.063	ns	0.208	ns	**0.007**	↑ 62.01%
B. striata	**<0.001**	↑ 95.19%	**0.002**	↓ 52.62%	**0.012**	↑ 51.69%	**0.016**	↓ 37.89%
B. pilosa	0.133	ns	0.341	ns	**0.037**	↓ 25.67%	0.138	ns
Insecticide								
P. frutescens	0.123	ns	0.311	ns	**0.035**	↓ 21.59%	0.480	ns
B. striata	**0.006**	↑ 96.78%	**<0.001**	↓ 40.80%	**<0.001**	↑ 53.61%	**0.010**	↓ 34.62%
B. pilosa	**0.004**	↓ 57.27%	0.235	ns	0.494	ns	0.589	ns

Table 3. *Cont.*

Source of Variation	Leaf Area		Root Length		Root Diameter		Root Tip Number	
	p Value	Change	p Value	Change	p Value	Change	p Value	Change
Fungicide								
P. frutescens	0.177	ns	0.180	ns	0.255	ns	**0.032**	↓ 25.80%
B. striata	**0.002**	↑ 129.53%	0.413	ns	**<0.001**	↑ 59.98%	0.432	ns
B. pilosa	0.368	ns	**0.036**	↑ 41.28%	0.619	ns	0.378	ns
Insecticide + Fungicide								
P. frutescens	0.744	ns	0.119	ns	0.268	ns	0.052	ns
B. striata	**0.002**	↑ 81.35%	0.135	ns	**<0.001**	↑ 51.78%	**0.039**	↓ 21.53%
B. pilosa	0.275	ns	0.876	ns	0.109	ns	0.445	ns

Notes: Values and arrows in 'Change' column show significant increase (↑) or decrease (↓) of the variables with the presence of a certain species compared to its absence, and 'ns' means no significant change.

Table 4. Functional traits of plant leaves and roots respond to species identity with N addition; p values are displayed in bold font when $p < 0.05$.

Source of Variation	Leaf Area		Root Length		Root Diameter		Root Tip Number	
	p Value	Change	p Value	Change	p Value	Change	p Value	Change
Control								
P. frutescens	0.111	ns	0.542	ns	0.693	ns	0.913	ns
B. striata	**0.025**	↑ 55.64%	**0.007**	↓ 53.30%	**<0.001**	↑ 73.57%	**<0.001**	↓ 48.59%
B. pilosa	0.762	ns	**<0.001**	↑ 225.29%	**<0.001**	↓ 36.05%	**0.001**	↑ 90.01%
Insecticide								
P. frutescens	**0.023**	↓ 34.55%	0.865	ns	0.077	ns	0.678	ns
B. striata	0.090	ns	**0.008**	↓ 64.61%	**<0.001**	↑ 67.27%	**<0.001**	↓ 41.13%
B. pilosa	0.664	ns	**0.018**	↑ 79.98%	0.900	ns	0.667	ns
Fungicide								
P. frutescens	0.180	ns	0.824	ns	0.406	ns	0.826	ns
B. striata	0.832	ns	**0.014**	↓ 36.42%	**<0.001**	↑ 62.31%	**0.048**	↓ 34.31%
B. pilosa	0.781	ns	**0.003**	↑ 94.93%	**0.046**	↓ 27.22%	0.239	ns
Insecticide + Fungicide								
P. frutescens	0.273	ns	0.432	ns	0.252	ns	0.622	ns
B. striata	0.810	ns	**<0.001**	↓ 64.46%	**<0.001**	↑ 90.17%	**<0.001**	↓ 57.71%
B. pilosa	**0.026**	↓ 26.22%	**0.015**	↑ 76.82%	0.164	ns	0.180	ns

Notes: Values and arrows in 'Change' column show significant increase (↑) or decrease (↓) of the variables with the presence of a certain species compared to its absence, and 'ns' means no significant change.

3. Discussion

3.1. The Effect of Heterotroph Removal on Plant Biomass

Previous work found that heterotroph removal can increase plant biomass, and the effects of different heterotroph removals on plant biomass were different, with the highest increase in insecticide treatment groups [8,12,35]. In systems without N addition, the insecticide treatment increased the plant aboveground biomass by 99.0% and total biomass by 54.0%, relative to the control (Figure 1a). Meanwhile, the insecticide treatment increased the leaf area by 98.1% relative to the control (Figure 3a). According to the plant survival strategy, the plant biomass increased with the increase in plant leaf area [36]. We also found a positive correlation between the plant total biomass and leaf area (Figure 5). Insecticide treatment may increase plant biomass by increasing the plant leaf area. Another reason may be that arthropods have a negative impact on biomass [11]; thereby, insecticide treatment accumulates the biomass removed by herbivorous insects. Unlike previous research results, the plant biomass usually increases after fungicide treatment [11]; the plant aboveground and total biomasses did not increase, and even the plant belowground biomass decreased after fungicide treatment in this study (Figure 1b). Fungicide could remove some pathogens. However, some symbiotic bacteria in the soil that are beneficial for plant growth were affected by the fungicide treatment [9]. In addition, the plant root length and root tip number for fungicide treatment groups were lower than those under the control treatment (Figure 3b,d). These results indicated that heterotroph removal may influence the plant biomass by affecting the plant functional traits.

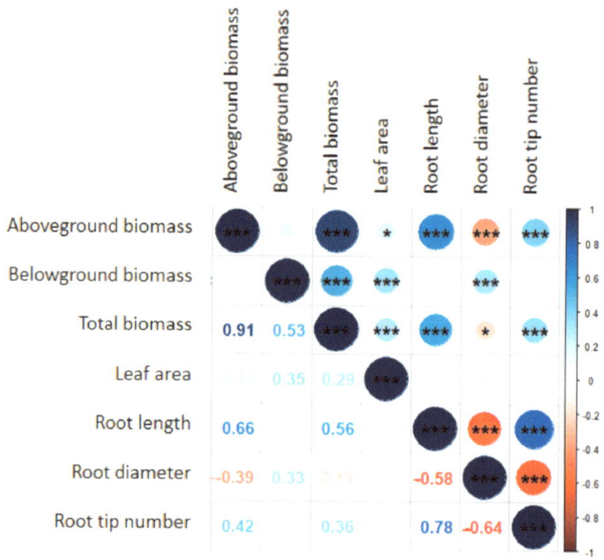

Figure 5. Correlation analysis between various parameters. * represents a significant relationship ($p < 0.05$), *** represents a significant relationship ($p < 0.001$).

In systems with N addition, heterotroph removal did not affect the plant aboveground and total biomasses (Figure 1a,c). This result differs from those in systems without N addition (Figure 1a,c). N addition altered the effect of heterotroph removal on plant biomass. The possible reason may be that N addition provided sufficient environmental resources, increased plant N absorption, promoted photosynthesis, and increased plant biomass production [37], thereby reducing the effect of heterotroph removal on the plant biomass. The root system is the organ in which plants absorb nutrients from the soil, and roots can affect the plant biomass by affecting the soil nutrient turnover, nutrient utilization efficiency, and mycorrhizal infection. Heterotroph removal may affect the ecosystem function through plant functional traits [29,38,39]. In this study, there were no significant differences in the plant root length and diameter among heterotroph removal treatments in systems with N addition (Figure 3b,c). In addition, N addition may reduce the soil microbial community [40,41], further weakening the influence of heterotroph removal on plant biomass. These results indicated that the effect of heterotroph removal on plant biomass depends on the habitat N availability.

3.2. The Effect of Heterotroph Removal on Plant Species Richness–Biomass Relationship

Most research showed that plant productivity increased with the increase in plant species richness, and heterotroph removal altered the plant species diversity effect [4,8,12,42]. In the grassland system, removing the insecticide and fungicide treatments promoted the effect of plant species diversity on productivity [8]. However, in forest systems, a fungicide treatment eliminated the positive relationship between tree species richness and productivity [12]. In this study, the plant biomass did not respond to species richness under each heterotroph removal treatment in systems without N addition (Figure S1a,c,e). The reason may be that only three lower levels of richness were set in this experiment (1, 2, and 3), while most experiments were set to high levels of richness [9,14,43].

In systems with N addition, the impact of the plant species richness on plant aboveground and total biomasses was positive under all removal treatments (Figure S1b,f). High plant species richness improved the plant biomass through enhancing nutrient utilization [44], and N addition may promote this effect. Meanwhile, insecticide and fungicide treatments removed arthropods, leaf fungi, and soil fungi that could reduce plant biomass, causing a significant increase in the plant aboveground and total biomasses when species richness was three (Figure S1b,f). However, plant species richness had a negative impact on the plant belowground biomass under an insecticide treatment (Figure S3d). The reason may be that the abundance of arthropods increased with an increasing plant species richness [45], which may consume more plant leaf area. There was a positive correlation between the plant leaf area and belowground biomass (Figure 5), ultimately leading to a decrease in the plant belowground biomass with an increasing species richness under an insecticide treatment.

3.3. The Effect of Heterotroph Removal on the Effect of Plant Species Identity on Plant Biomass

Plant species identity is an important part of plant species diversity [1], and many studies have shown that the plant species identity affects plant biomass [42,46]. In systems without N addition, the presence of *B. striata* increased the plant belowground biomass under each heterotroph removal treatment (Table 1); the presence of *B. striata* also increased the plant leaf area and root diameter (Table 3). There was a significant positive correlation between the leaf area, root diameter, and plant belowground biomass (Figure 5). These results suggested that plant species identity affected the plant biomass by influencing the plant functional traits.

In systems with N addition, the presence of *B. striata* increased the plant belowground biomass in the control group, while this effect was dismissed after insecticide and fungicide treatments (Table 2). Meanwhile, the change pattern in the plant leaf area of *B. striata* was consistent with that of the plant belowground biomass (Table 4). Moreover, there was a significant positive correlation between the plant leaf area and plant belowground biomass (Figure 5), indicating that heterotroph removal altered the effect of species identity on the plant biomass by influencing the plant leaf area. In addition, N addition may change the interaction between plant species [26]. In this study, under insecticide or fungicide treatments, the relative yield of *B. striata* in systems with N addition was higher than that in systems without N addition (Figure S2). We also found that the selection effect became increasingly important in systems with N addition, resulting in an increase in the net biodiversity effect (Figure S3). Interestingly, the presence of *B. striata* decreased the plant aboveground biomass in the control group but did not affect the plant aboveground biomass after heterotroph removal treatment (Table 4). This study also found that the presence of *B. striata* increased the plant root diameter but decreased the plant root length and root tip number under each heterotroph removal treatment (Table 4). We observed a significant positive correlation between the plant root length, root tip number, and plant aboveground biomass, and a negative correlation between the root diameter and plant aboveground (Figure 5). These results also indicated that heterotroph removal changed the plant species identity's effect on plant biomass by affecting the plant functional traits.

4. Materials and Methods

4.1. Experimental Design

The experiment was set up in a greenhouse at Wenzhou University in Wenzhou, Zhejiang Province, China (120°42′4″ E, 27°55′46″ N). The climate was a subtropical monsoon climate. The greenhouse has a transparent plastic roof, shielding the experiment from rainwater while maintaining temperature and humidity levels. A shading net was installed above the plastic roof to simulate the light environment under the forest. A three-factor control experiment was conducted (Figure 6): (1) species diversity: based on the functional trait, three local common understory herbaceous plants were chosen: *Perilla frutescens* (L.) *Britt*, *Bletilla striata*, and *Bidens pilosa* L. (Table S2) for plant species diversity configuration (all seven plant species compositions); (2) heterotroph removal treatments: control, insecticide, fungicide, and both insecticide and fungicide treatments; (3) N addition: N deposition was simulated by N addition, using without N addition as the control. There are four repetitions for each treatment. In total, 224 pots (30 cm diameter and 20 cm height) were constructed.

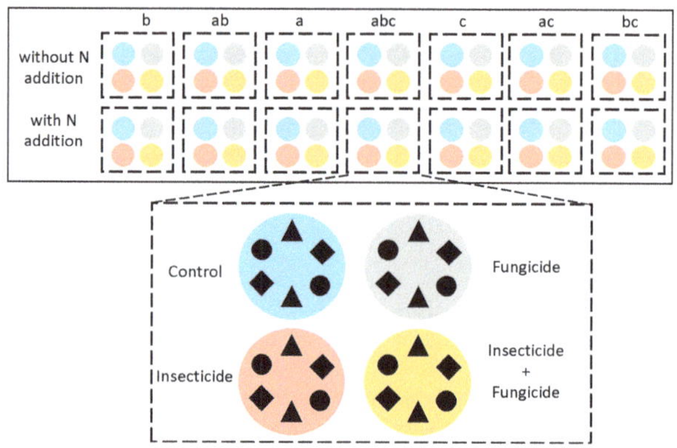

Figure 6. Experimental design (one block). The letters above the uppermost boxes represent plant species compositions, and species compositions within each group are randomly arranged. Colors represent heterotroph removal treatments: blue: control; orange: insecticide; gray: fungicide; yellow: insecticide and fungicide. The three-species treatment is depicted here, with different shapes representing different species. Six plants are planted in each experimental system uniformly.

In April 2022, the seedlings of the plants were transplanted into pots, with six individuals planted in each pot. From the end of April to the beginning of September, heterotroph removal treatment and N addition treatment were conducted once a month. According to Seabloom et al. (2017), insects were removed by spraying an insecticide water emulsion (0.03% permethrin), fungi were removed by spraying fungicide (30% carbendazim), and the control group added an equal volume of water. Based on the environmental wet N deposition rate in Zhejiang Province (2.69 g N m^{-2} yr^{-1}), we added ammonium nitrate (NH_4NO_3) solution every month to simulate high N conditions, with an average amount added each time (10 g N m^{-2} yr^{-1}), and we added water as a control group.

Pesticides may impact plant growth even in the system without heterotrophs. Therefore, we designed a laboratory to test the impact of insecticides and fungicides on the plant biomass. The soil was homogenized and subjected to high-pressure steam treatment. Three plant species were treated with four heterotroph removal treatments, and each treatment had four replicates, totaling 48 pots with one individual in each pot. The heterotroph removal treatment was applied once a month, and the application amount was the same as the field experiment. Plants were allowed to grow for a total of ten weeks. After ten weeks, harvest each plant and divide it into aboveground and belowground biomasses. Results showed that heterotroph removal treatments do not affect plant biomass in the indoor experiment without consumers (Figure S4).

4.2. Sample Collection and Calculation

Plants were harvested at the end of the plant growth period. After washing harvested plants, three complete leaves and three roots were taken from each plant in each pot. After scanning with a scanner (EPSON GT-X980, Hangzhou, China), the images were analyzed and processed using the Wanshen leaf processing system (version 2018; www.Wseen.com) to obtain the leaf area, root length, root diameter, and number of root tips. Divide the plants into aboveground and belowground parts, dry at 105 °C for 20 min, then dry at 65 °C for 48 h to obtain each species' aboveground and belowground biomass for each pot.

The net effects, complementary effects, and selection effects were calculated according to Loreau and Hector's calculation method [47]. The net effect refers to the difference between the observed yield (actual yield) and the expected yield (weighted average of individual yield corresponding to species in the mixture based on planting proportion) of the mixture. Complementary effects are measured from changes in the relative yield of species. The selection effect is measured by subtracting the complementary effect from the net effect.

1. Plant functional traits of leaves and roots:

To evaluate the response of plant functional traits to N addition and heterotroph removal treatments at the community level, the functional traits' community weighted mean (CWM) was calculated:

$$\mathrm{CWM} = \sum_{i=1}^{s} \mathrm{E_{ic}} \times \mathrm{B_{ic}} \tag{1}$$

where $\mathrm{E_{ic}}$ represents species i's plant functional traits in composition c, $\mathrm{B_{ic}}$ represents species i's biomass in composition c (when calculating plant leaf area's CWM, the $\mathrm{B_{ic}}$ referred to the proportion of species i's aboveground biomass in composition c's aboveground biomass; when calculating plant root traits' CWM, the $\mathrm{B_{ic}}$ referred to the proportion of plant species i's belowground biomass in composition c's belowground biomass of composition c), and s referred to species' amount in composition c.

2. Relative yield:

To evaluate whether N addition alter the competitiveness of specific species in the mixture, relative yield (RY) of the plant was calculated:

$$RY_i = \frac{O_i}{E_i} \tag{2}$$

where O_i represents the aboveground biomass of species i per plant in the mixture, while E_i represents the aboveground biomass of species i per plant in the monoculture. If $RY_i > 1$, species i is the dominant species in the mixture.

4.3. Statistical Analysis

The influence of species diversity (species compositions and species richness), N addition, and heterotroph removal treatment on the plant biomass (aboveground, belowground, and total) and functional traits of plant leaves and roots was determined using a three-way ANOVA. The effects of plant species richness on plant biomass under each heterotroph removal treatment were tested using linear regression analysis. The difference in the above parameters between systems with and without N addition under the same heterotroph removal treatment was tested using an independent sample *t*-test. The effect of heterotroph removal treatment or plant species compositions on the above parameters under the same N habitat level was determined using a one-way ANOVA. If there were significant differences, the Tukey method was conducted. The effect of the plant species identity (the presence of certain species) on each parameter was determined using an independent sample *t*-test. The difference between the zero and net effects, complementary effects, and selection effects was examined using a single sample *t*-test. The correlations of various parameters were verified using Pearson's correlation analysis. Before analysis, the data were ln-transformed to satisfy the equality of variance (Levene's test) and assumptions of normality (Kolmogorov–Smirnov test). If the data after conversion still did not fulfill the assumptions, a nonparametric Kruskal–Wallis test was employed. All statistical analyses were conducted using the R4.1.1 program. All data were delivered as the mean $\pm$ standard error, and the statistical significance level was set as $\alpha = 0.05$.

5. Conclusions

This study investigated the relationship between the plant species diversity and biomass response to heterotroph removal in systems with and without N addition. Our research findings indicate that heterotroph removal affected plant biomass by influencing the plant leaf area in both systems with or without N addition, and altered the effect of the plant species richness–plant biomass relationship by influencing the plant leaf area in systems with N addition but not in systems without N addition. Heterotroph removal also affected the effect of species identity on the plant biomass by influencing the plant functional traits in both systems with or without N addition. Therefore, it is recommended that in the background of global N deposition, the impact of other trophic level organisms on ecosystem functioning cannot be ignored when analyzing the species diversity–ecosystem functions relationship. In terms of ecosystem management, biodiversity at different trophic levels should be protected. In the future, more high plant species diversity experiments with long-term research are needed to determine the impact of heterotrophs on biodiversity–ecosystem function relationships in high-N habitats.

Supplementary Materials: The following supporting information can be downloaded at: https://www.mdpi.com/article/10.3390/plants13020258/s1, Table S1: Three-way ANOVA Table (a) Effects of nitrogen addition, species compositions, heterotrophs removal treatment, and (b) Effects of nitrogen addition, species richness, and heterotrophs removal treatment on above-, below-ground, total biomass and functional traits of plant leaves and roots; Table S2: The plant functional traits of *Perilla frutescens*, *Bletilla striata* and *Bidens pilosa*; Figure S1. Linear regression of species richness on plant aboveground biomass (a), belowground biomass (c), and total biomass (e) in systems without N addition and plant aboveground biomass (b), belowground biomass (d), and total biomass (f) in systems with N addition under different heterotrophs removal treatment. Figure S2. The relative yield of *B. pilosa* under different heterotroph treatments and N availability. Significant differences between heterotroph removal were indicated in capital letters, and significant differences between N availability were indicated in lowercase letters. Blue circle: without N addition; yellow, with N addition. Figure S3. Net effect of aboveground biomass (a), belowground biomass (b), and total biomass (c); complementary effect of aboveground biomass (d), belowground biomass (e), and total biomass (f); selection effect of aboveground biomass (g), belowground biomass (h), and total biomass (j) under different heterotroph removal treatments. * Represents a significant effect under this treatment. Blue bar: control; orange, insecticide; gray, fungicide; yellow, insecticide and fungicide. Figure S4. Aboveground (a), belowground (b), and total biomass (c) of plant monoculture under different heterotrophs treatments. Same lowercase letters indicate no difference between het-erotrophs removal. Blue bar: control; orange, insecticide; gray, fungicide; yellow, insecticide and fungicide.

Author Contributions: Conceptualization, W.H.; funding acquisition, W.H.; investigation, X.X., L.Y., K.S., H.C. and Y.L.; writing—original draft, X.X.; writing—review and editing, X.X., L.Y., K.S., H.C., Y.L., J.L. and W.H. All authors have read and agreed to the published version of the manuscript.

Funding: This research was funded by the National Natural Science Foundation of China (31901213) and Zhejiang Provincial Natural Science Foundation (LY22C030003).

Data Availability Statement: Data are available on request from the authors. The data are not publicly available due to [Some data are acquired from other collaborators].

Acknowledgments: We thank Yanru Huang, Wenhao Wu, and Shengtao Zhou for their assistance in the field and laboratory work.

Conflicts of Interest: The authors declare no conflicts of interest.

References

1. Tilman, D.; Isbell, F.; Cowles, M.J. Biodiversity and ecosystem functioning. *Annu. Rev. Ecol. Evol. Syst.* **2014**, *45*, 471–493. [CrossRef]
2. Tilman, D.; Knops, J.; Wedin, D.; Reich, P.; Ritchie, M.; Siemann, E. The influence of functional diversity and compositions on ecosystem processes. *Science* **1997**, *277*, 1300–1302. [CrossRef]
3. Tilman, D.; Reich, P.B.; Knops, J.; Wedin, D.; Mielke, T.; Lehman, C. Diversity and productivity in a long-term grassland experiment. *Science* **2001**, *294*, 843–845. [CrossRef]
4. van der Plas, F. Biodiversity and ecosystem functioning in naturally assembled communities. *Biol. Rev.* **2019**, *94*, 1220–1245. [CrossRef]

5. He, J.S.; Bazzaz, F.A.; Schmid, B. Interactive effects of diversity nutrients and elevated CO_2 on experimental plant communities. *Oikos* **2002**, *97*, 337–348. [CrossRef]
6. Isbell, F.; Reich, P.B.; Tilman, D.; Hobbie, S.E.; Polasky, S.; Binder, S. Nutrient enrichment biodiversity loss and consequent declines in ecosystem productivity. *Proc. Natl. Acad. Sci. USA* **2013**, *110*, 11911–11916. [CrossRef]
7. Roscher, C.; Schmid, B.; Kolle, O.; Schulze, E.D. Complementarity among four highly productive grassland species depends on resource availability. *Oecologia* **2016**, *181*, 571–582. [CrossRef]
8. Seabloom, E.W.; Kinkel, L.; Borer, E.T.; Hautier, Y.; Montgomery, R.A.; Tilman, D. Food webs obscure the strength of plant diversity effects on primary productivity. *Ecol. Lett.* **2017**, *20*, 505–512. [CrossRef]
9. Zaret, M.M.; Kuhs, M.A.; Anderson, J.C.; Seabloom, E.W.; Borer, E.T.; Kinkel, L.L. Seasonal shifts from plant diversity to consumer control of grassland productivity. *Ecol. Lett.* **2022**, *25*, 1215–1224. [CrossRef]
10. Zaret, M.M.; Kinkel, L.L.; Borer, E.T.; Seabloom, E.W. Soil nutrients cause threefold increase in pathogen and herbivore impacts on grassland plant biomass. *J. Ecol.* **2023**, *111*, 1629–1640. [CrossRef]
11. Seabloom, E.W.; Borer, E.T.; Kinkel, L.L. No evidence for trade-offs in plant responses to consumer food web manipulations. *Ecology* **2018**, *99*, 1953–1963. [CrossRef] [PubMed]
12. Huang, Y.Y.; Schuldt, A.; Hönig, L.; Yang, B.; Liu, X.J.; Bruelheide, H.; Ma, K.P.; Schmid, B.; Niklaus, P.A. Effects of enemy exclusion on biodiversity–productivity relationships in a subtropical forest experiment. *J. Ecol.* **2022**, *110*, 2167–2178. [CrossRef]
13. Borer, E.T.; Seabloom, E.W.; Tilman, D. Plant diversity controls arthropod biomass and temporal stability. *Ecol. Lett.* **2012**, *15*, 1457–1464. [CrossRef]
14. Porazinska, D.L.; Farrer, E.C.; Spasojevic, M.J.; de Mesquita, C.P.B.; Sartwell, S.A.; Smith, J.G.; White, C.T.; King, A.J.; Suding, K.N.; Schmidt, S.K. Plant diversity and density predict belowground diversity and function in an early successional alpine ecosystem. *Ecology* **2018**, *99*, 1942–1952. [CrossRef] [PubMed]
15. Keesing, F.; Holt, R.D.; Ostfeld, R.S. Effects of species diversity on disease risk. *Ecol. Lett.* **2006**, *9*, 485–498. [CrossRef]
16. Jing, X.; Sanders, N.J.; Shi, Y.; Chu, H.Y.; Classen, A.T.; Zhao, K.; Chen, L.T.; Shi, Y.; Jiang, Y.X.; He, J.S. The links between ecosystem multifunctionality and above- and belowground biodiversity are mediated by climate. *Nat. Commun.* **2015**, *6*, 8159. [CrossRef]
17. Duffy, J.E.; Cardinale, B.J.; Mcintyre, P.B.; Thebault, E.; Loreau, M. The functional role of biodiversity in ecosystems: Incorporating trophic complexity. *Ecol. Lett.* **2007**, *10*, 522–538. [CrossRef]
18. Galloway, J.N.; Dentener, F.J.; Capone, D.G.; Boyer, E.W.; Howarth, R.W.; Seitzinger, S.P.; Asner, G.P.; Cleveland, C.C.; Green, P.A.; Holland, E.A.; et al. Nitrogen cycles: Past, present, and future. *Biogeochemistry* **2004**, *70*, 153–226. [CrossRef]
19. Ackerman, D.; Millet, D.B.; Chen, X. Global estimates of inorganic nitrogen deposition across four decades. *Glob. Biogeochem. Cycles* **2019**, *33*, 100–107. [CrossRef]
20. Wilcots, M.E.; Schroeder, K.M.; Delancey, L.C.; Kjaer, S.J.; Hobbie, S.E.; Seabloom, E.W.; Borer, E.T. Realistic rates of nitrogen addition increase carbon flux rates but do not change soil carbon stocks in a temperate grassland. *Glob. Chang. Biol.* **2022**, *28*, 4819–4831. [CrossRef]
21. Ke, Y.G.; Yu, Q.; Wang, H.Q.; Zhao, Y.; Jia, X.T.; Yang, Y.D.; Zhang, Y.L.; Zhou, W.; Wu, H.H.; Xu, C.; et al. The potential bias of nitrogen deposition effects on primary productivity and biodiversity. *Glob. Chang. Biol.* **2022**, *29*, 1054–1061. [CrossRef] [PubMed]
22. Bai, Y.F.; Wu, J.G.; Clark, C.M.; Naeem, S.; Pan, Q.M.; Huang, J.H.; Zhang, L.X.; Han, X.G. Tradeoffs and thresholds in the effects of nitrogen addition on biodiversity and ecosystem functionsing: Evidence from inner Mongolia Grasslands. *Glob. Chang. Biol.* **2010**, *16*, 358–372. [CrossRef]
23. Seabloom, E.W.; Adler, P.B.; Alberti, J.; Biederman, L.; Buckley, Y.M.; Cadotte, M.W.; Collins, S.L.; Dee, L.; Fay, P.A.; Firn, J.; et al. Increasing effects of chronic nutrient enrichment on plant diversity loss and ecosystem productivity over time. *Ecology* **2021**, *102*, e03218. [CrossRef] [PubMed]
24. Tian, D.H.; Wang, H.; Sun, J.; Niu, S.L. Global evidence on nitrogen saturation of terrestrial ecosystem net primary productivity. *Environ. Res. Lett.* **2016**, *11*, 024012. [CrossRef]
25. Fridley, J.D. Resource availability dominates and alters the relationship between species diversity and ecosystem productivity in experimental plant communities. *Oecologia* **2002**, *132*, 271–277. [CrossRef] [PubMed]
26. Grime, J.P. Evidence for the existence of three primary strategies in plants and its relevance to ecological and evolutionary theory. *Am. Nat.* **1977**, *111*, 1221–1226. [CrossRef]
27. Frey, S.D.; Knorr, M.; Parrent, J.L.; Simpson, R.T. Chronic nitrogenenrichment affects the structure and function of the soil microbialcommunity in temperate hardwood and pine forests. *Forest Ecol. Manag.* **2004**, *196*, 159–171. [CrossRef]
28. Zhang, Y.J.; Jin, Y.H.; Xu, J.W.; He, H.S.; Tao, Y.; Yang, Z.P.; Bai, Y.Y. Effects of exogenous N and endogenous nutrients on alpine tundra litter decomposition in an area of high nitrogen deposition. *Sci. Total Environ.* **2021**, *805*, 150388. [CrossRef]
29. Liu, C.C.; Li, Y.; Yan, P.; He, N.P. How to Improve the Predictions of Plant Functional Traits on ecosystem functioning? *Front. Plant Sci.* **2021**, *12*, 622260. [CrossRef]
30. Tatarko, A.R.; Knops, J.M.H. Nitrogen addition and ecosystem functioning: Both species abundances and traits alter community structure and function. *Ecosphere* **2018**, *9*, e02087. [CrossRef]
31. Yan, P.; He, N.P.; Yu, K.L.; Xu, L.; Van Meerbeek, K. Integrating multiple plant functional traits to predict ecosystem productivity. *Commun. Biol.* **2023**, *6*, 239. [CrossRef] [PubMed]

32. Gao, W.; Chen, D.; Hu, X.; Fang, X.; Li, Q.; Huang, Q.; Sun, F.; Zhou, J.; Bai, Y.; Zhang, J.; et al. Nitrogen deposition drives the intricate changes of fine root traits. *Glob. Ecol. Conserv.* **2023**, *43*, e02443. [CrossRef]
33. Price, P.W. The plant vigor hypothesis and herbivore attack. *Oikos* **1991**, *62*, 244–251. [CrossRef]
34. Bardgett, R.D.; Mommer, L.; De Vries, F.T. Going belowground: Root traits as drivers of ecosystem processes. *Trends Ecol. Evol.* **2014**, *29*, 692–699. [CrossRef] [PubMed]
35. Bruno, J.F.; Boyer, K.E.; Duffy, J.E.; Lee, S.C. Relative and interactive effects of plant and grazer richness in a benthic marine community. *Ecology* **2008**, *89*, 2518–2528. [CrossRef] [PubMed]
36. Reich, P.B.; Walters, M.B.; Ellsworth, D.S. From tropics to tundra: Global convergence in plant functioning. *Proc. Natl. Acad. Sci. USA* **1997**, *94*, 13730–13734. [CrossRef]
37. Lawlor, D.W.; Lemaire, G.; Gastal, F. Nitrogen, plant growth and crop yield. In *Plant Nitrogen*; Lea, P.J., Morot-Gaudry, J.F., Eds.; Springer: Berlin/Heidelberg, Germany, 2001; pp. 343–367.
38. Diaz, S.; Cabido, M. Vive la différence: Plant functional diversity matters to ecosystem processes. *Trends Ecol. Evol.* **2001**, *16*, 646–655. [CrossRef]
39. Faucon, M.P.; Houben, D.; Lambers, H. Plant functional traits: Soil and ecosystem services. *Trends Plant Sci.* **2017**, *22*, 385–394. [CrossRef]
40. Zhou, Z.H.; Wang, C.K.; Zheng, M.H.; Jiang, L.F.; Luo, Y.Q. Patterns and mechanisms of response by soil microbial communities to nitrogen addition. *Soil Biol. Biochem.* **2017**, *115*, 433–441. [CrossRef]
41. Averill, C.; Waring, B. Nitrogen limitation of decomposition and decay: How can it occur? *Glob. Chang. Biol.* **2018**, *24*, 1417–1427. [CrossRef]
42. Han, W.J.; Chang, J.; Jiang, H.; Niu, S.D.; Liu, Y.; Xu, J.M.; Wu, J.Z.; Ge, Y. Plant species diversity affects plant nutrient pools by affecting plant biomass and nutrient concentrations in high-nitrogen ecosystems. *Basic Appl. Ecol.* **2021**, *56*, 213–225. [CrossRef]
43. Tilman, D. Biodiversity: Population Versus Ecosystem Stability. *Ecology* **1996**, *77*, 350–363. [CrossRef]
44. Knops, J.M.H.; Bradley, K.L.; Wedin, D.A. Mechanisms of plant species impacts on ecosystem nitrogen cycling. *Ecol. Lett.* **2002**, *5*, 454–466. [CrossRef]
45. Haddad, N.M.; Tilman, D.; Haarstad, J.; Ritchie, M.; Knops, J.M.H. Contrasting effects of plant richness and composition on insect communities: A field experiment. *Am. Nat.* **2001**, *158*, 17–35. [CrossRef]
46. Luo, B.; Du, Y.Y.; Han, W.J.; Geng, Y.; Wang, Q.; Duan, Y.Y.; Ren, Y.; Liu, D.; Chang, J.; Ge, Y. Reduce health damage cost of greenhouse gas and ammonia emissions by assembling plant diversity in floating constructed wetlands treating wastewater—ScienceDirect. *J. Clean. Prod.* **2020**, *244*, 118927. [CrossRef]
47. Loreau, M.; Hector, A. Partitioning selection and complementarity in biodiversity experiments. *Nature* **2001**, *412*, 72–76. [CrossRef]

plants

MDPI

Article

Relationships between Phyllosphere Bacterial Communities and Leaf Functional Traits in a Temperate Forest

Zuoqiang Yuan [1,2], Ji Ye [2], Fei Lin [2,*], Xing Wang [2,3,*], Teng Yang [4], Boyuan Bi [1], Zikun Mao [2,5], Shuai Fang [2,5], Xugao Wang [2,5], Zhanqing Hao [1] and Arshad Ali [6]

[1] Shaanxi Key Laboratory of Qinling Ecological Intelligent Monitoring and Protection, School of Ecology and Environment, Northwestern Polytechnical University, Xi'an 710072, China; zqyuan@nwpu.edu.cn (Z.Y.); bby0710@nwpu.edu.cn (B.B.)

[2] CAS Key Laboratory of Forest Ecology and Management, Institute of Applied Ecology, Chinese Academy of Sciences, Shenyang 110016, China; yeji1011@163.com (J.Y.); wangxg@iae.ac.cn (X.W.)

[3] Plant Ecology and Nature Conservation, Wageningen University & Research, 6708 PB Wageningen, The Netherlands

[4] State Key Laboratory of Soil and Sustainable Agriculture, Institute of Soil Science, Chinese Academy of Sciences, East Beijing Road 71, Nanjing 210008, China; tyang@issas.ac.cn

[5] Key Laboratory of Terrestrial Ecosystem Carbon Neutrality, Institute of Applied Ecology, Chinese Academy of Sciences, Shenyang 110016, China

[6] Forest Ecology Research Group, College of Life Sciences, Hebei University, Baoding 071002, China; arshadforester@gmail.com

* Correspondence: linfei@iae.ac.cn (F.L.); xingwang.wang@wur.nl (X.W.)

Abstract: As a vital component of biodiversity, phyllosphere bacteria in forest canopy play a critical role in maintaining plant health and influencing the global biogeochemical cycle. There is limited research on the community structure of phyllosphere bacteria in natural forests, which creates a gap in our understanding of whether and/or how phyllosphere bacteria are connected to leaf traits of their host. In this study, we investigated the bacterial diversity and composition of the canopy leaves of six dominant tree species in deciduous broad-leaved forests in northeastern China, using high-throughput sequencing. We then compare the differences in phyllosphere bacterial community structure and functional genes of dominant tree species. Fourteen key leaf functional traits of their host trees were also measured according to standard protocols to investigate the relationships between bacterial community composition and leaf functional traits. Our result suggested that tree species with closer evolutionary distances had similar phyllosphere microbial alpha diversity. The dominant phyla of phyllosphere bacteria were Proteobacteria, Actinobacteria, and Firmicutes. For these six tree species, the functional genes of phyllosphere bacteria were mainly involved in amino acid metabolism and carbohydrate metabolism processes. The redundancy and envfit analysis results showed that the functional traits relating to plant nutrient acquisition and resistance to diseases and pests (such as leaf area, isotope carbon content, and copper content) were the main factors influencing the community structure of phyllosphere bacteria. This study highlights the key role of plant interspecific genetic relationships and plant attributes in shaping phyllosphere bacterial diversity.

Keywords: phyllosphere bacteria; plant phylogeny; functional traits; interspecific variation; community structure

check for updates

Citation: Yuan, Z.; Ye, J.; Lin, F.; Wang, X.; Yang, T.; Bi, B.; Mao, Z.; Fang, S.; Wang, X.; Hao, Z.; et al. Relationships between Phyllosphere Bacterial Communities and Leaf Functional Traits in a Temperate Forest. *Plants* **2023**, *12*, 3854. https://doi.org/10.3390/plants12223854

Academic Editor: Oleg Chertov

Received: 15 September 2023
Revised: 9 November 2023
Accepted: 11 November 2023
Published: 15 November 2023

1. Introduction

The concept of phyllosphere was first introduced by a British Pathologist (named F.T. Last) in 1955 and is defined as the outer surface environment of leaves with complex microbial communities [1]. However, scientists have recently found a further connection between plant leaves' outer surface microbial communities and their inner parts. For example, microbes can enter the leaves from the outer surface through the stomata and veins of plants [2]. Recent studies suggest that the surface area of all leaves may be twice as large as the global land surface area, making it a significant microbial habitat on Earth [3]. Most studies have traditionally focused on the surface of leaves, but in recent years, researchers have also begun to include the leaf margin (i.e., above-ground plant parts). This area is now recognized as providing a suitable living environment for certain microbes, according to studies such as those by Farre-Armengol et al. [4] and Stone Bram and Jackson Colin [5].

Phyllsophere microbes are essential in maintaining plant health and ecosystem functions [6–8]. Previous studies have found that the phyllosphere microorganism *Pseudomonas lurida* has an antagonistic effect on pathogenic microbes to maintain plant host health [9]. Further, the phyllosphere microorganism *Staphylococcus* can improve the adaptability of host plants to adversity due to the production of indole-3-acetic acid (IAA) [10]. As for factors affecting the composition of the phyllosphere microbial community, most studies argue that the type of host plants is important [11–16]. For example, bacterial community structure on leaves was highly correlated with host evolutionary relatedness and suites of plant functional traits related to host ecological strategies for resource uptake and growth/mortality tradeoffs, whereas the abundance of several bacterial taxa was correlated with host growth, mortality, and function [11–16].

Furthermore, plant functional traits could regulate microbial communities due to the lack of nutrients on the leaf surface [13,17–19]. Phyllosphere microbes rely on the host plant to obtain their required nutrients [20]. Therefore, the physiological characteristics of the host plant can determine whether the phyllosphere microbes can successfully colonize or not [13]. Studies have reported that leaf nitrogen and phosphorus elements strongly influence phyllosphere bacterial communities, which depend on leaf traits [19,21]. Moreover, aluminum and copper in the leaves of tropical forests in Panama are related to the function of phyllosphere bacteria, which is associated with the ability of these two elements to resist pests and diseases [17]. Nonetheless, in contrast to our understanding of rhizosphere microbes, the role of phyllosphere microbes within the environment, as well as the underlying interaction mechanisms between phyllosphere microbes and plants, remain largely uncharted.

The forest canopy is the main interface where forests interact with the external environment, and it is responsible for supporting a significant number of species on Earth [22–24]. The conservation and sustainable use of forest canopy biodiversity in the face of climate change is a crucial topic in ecological research and has been widely studied [22,25]. In this context, the broad-leaved Korean pine mixed forest, as the typical vegetation in temperate forests, is famous for its complex structure and unique species composition, in comparison to regions at the same latitude worldwide [26]. Although there are many studies on the interactions among animals, plants, and soil microbes in the study region [27–29], there is still weak knowledge of canopy phyllosphere bacteria which restricts our understanding of microbial composition and its impact on forest community structure and function in natural ecosystems [7]. This study aims to determine the main the community composition and and the driving factors of canopy phyllosphere bacteria in the temperate broad-leaved Korean pine mixed forest of Changbai Mountain in northeastern China. To do so, we collected canopy leaves of six dominant tree species to answer the following three questions: (1) what are the similarities and differences in the alpha diversity and community composition of phyllosphere bacteria in different tree species? (2) are there unique biomarkers for each tree species? and (3) are there potential associations between phyllosphere bacterial communities and leaf functional traits? We hypothesize that: (1) tree species with closer evolutionary distance should have similar alpha phyllosphere microbial diversity; (2) tree species do not have unique dominate phyla, as microbes generally communicate between long-term coexisting tree species; and (3) functional traits regulate microbial community composition.

2. Results

2.1. Phyllosphere Bacterial Diversity and Composition

The phyllosphere bacterial richness exhibited significant variations among different tree species, with the highest richness observed in *Q. mongolica*, while *P. koraiensis* had notably lower bacterial richness compared to the other tree species. When comparing the phyllosphere bacterial diversity between *Q. mongolica* and *U. pumila*, no significant differences were detected. However, both tree species exhibited significant dissimilarities in bacterial diversity compared to the remaining tree species, except for *F. mandshurica* (Figure 1). Notably, bacterial richness exhibits significant differences among tree species with closer phylogenetic relationships, such as *A. mono* and *T. amurensis*, as well as *Q. mongolica* and *U. japonica*. In contrast, significant differences in bacterial richness were observed between host plants with more distantly related evolutionary backgrounds, for instance, *F. mandshurica* and *P. koraiensis*.

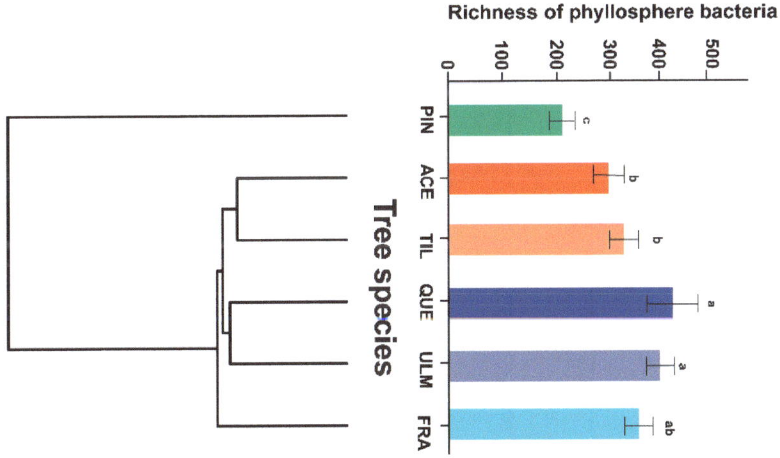

Figure 1. The richness of phyllosphere bacteria arranged according to tree's phylogenetic relationships in a temperate mixed forest of Changbai Mountain. If the branches between two tree species are shorter, it signifies that they are closer in terms of evolution. Conversely, longer branches indicate a more distant genetic relationship. Letters "a", "b", and "c" represent distinct groups identified through ANOVA. Groups sharing the same letter signify no significant difference ($p > 0.05$), whereas groups with different letters indicate a significant difference ($p < 0.05$). Abbreviations of tree species: ACE: *Acer mono*; FRA: *Fraxinus mandshurica*; PIN: *Pinus koraiensis*; QUE: *Quercus mongolica*; TIL: *Tilia amurensis*; ULM: *Ulmus japonica*.

A total of 11 phyla, 46 classes, 147 orders, 401 families, and 1077 genera were identified from the leaves of the 6 tree species studied. Proteobacteria (67.7%), Actinobacteria (27.8%), and Firmicutes (1.0%) are dominant at the phylum level. At the class level, the relative abundance of bacteria on the leaves was ranked as follows: Gammaproteobacteria (34.4%), Alphaproteobacteria (26.6%), and Actinobacteria (23.2%) (see Figure 2 for a graphical representation). These findings highlight significant variations in the composition of bacteria among different tree species. Additionally, at the phylum level, the relative abundance of Proteobacteria was highest on the leaves of *F. mandshurica* (80.6%) and lowest on *T. amurensis* (46.2%). The relative abundance of Actinobacteria was highest on *T. amurensis* (40.9%) and lowest on *F. mandshurica* (14.2%). Moreover, the relative abundance of Firmicutes was highest on *A. mono* (4.2%) and lowest on *F. mandshurica* (0.1%).

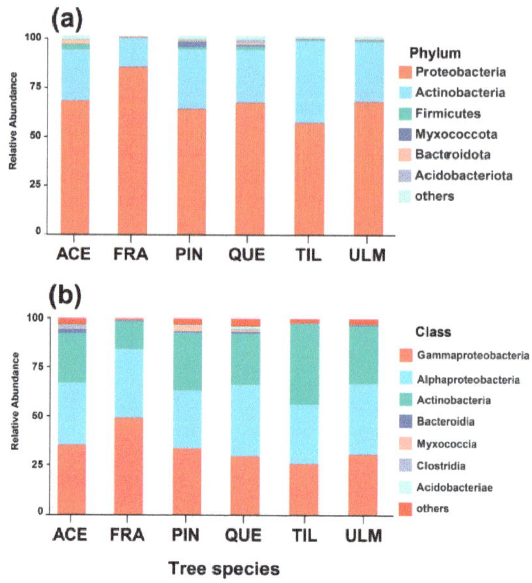

Figure 2. Histogram of relative abundance of phyllosphere bacteria of dominant tree species in a temperate mixed forest of Changbai Mountain at the level of phylum (**a**) and class (**b**) level. Abbreviations of tree species: ACE: *Acer mono;* FRA: *Fraxinus mandshurica;* PIN: *Pinus koraiensis;* QUE: *Quercus mongolica;* TIL: *Tilia amurensis;* ULM: *Ulmus japonica.*

2.2. Differences in Phyllosphere Bacterial Community among Tree Species

ANOSIM analysis and NMDS showed that there are significant differences in microbial community composition among six tree species (Figure 3; Table 1). Based on the LEfSe analysis, the primary biomarkers for *F. mandshurica* were Myxococcaceae, *Aureimonas,* and Oxalobacteraceae. For *P. koraiensis,* the main biomarkers were Actinobacteria, Myxococcia, Myxococcales, *Beijerinckia,* and *Methylobacterium.* *Q. mongolica* had Rhizobiales and *Pseudomonas* as the primary biomarkers, while *T. amurensis* had Micrococcales, Microbacteriaceae, *Curtobacterium, Amnibacterium,* and P3OB-42 as the main biomarkers.

Table 1. Analysis results of bacteria ANOSIM in leaves of dominant species of a temperate mixed forest in Changbai Mountain.

Comparison among Tree Species	R	p
Global	0.300	<0.001
ULM vs. ACE	0.216	0.025
ULM vs. TIL	0.228	0.023
ULM vs. PIN	0.504	0.010
ULM vs. FRA	0.636	0.008
ULM vs. QUE	0.796	0.008
ACE vs. TIL	0.260	0.016
ACE vs. PIN	0.480	0.020
ACE vs. FRA	0.356	0.016
ACE vs. QUE	0.312	0.025
TIL vs. PIN	0.524	0.009
TIL vs. FRA	0.792	0.011
TIL vs. QUE	0.696	0.010
PIN vs. FRA	0.540	0.011
PIN vs. QUE	0.488	0.007
FRA vs. QUE	0.660	0.011

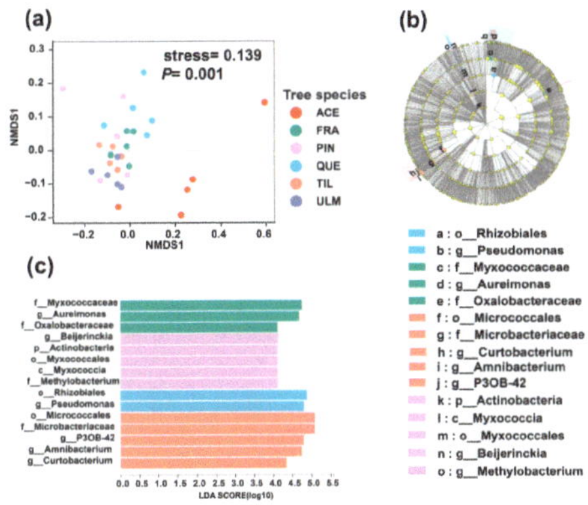

Figure 3. NMDS visualization diagram (**a**); cladogram of taxa which were differentially expressed based on Linear discriminant analysis Effect Size (LEfSe) (**b**); and Linear discriminant analysis (LDA) scores indicating the effect size of each differentially expressed taxon (**c**) of phyllosphere bacteria community structure of dominant tree species in a temperate mixed forest of Changbai Mountain. The stress value measures the goodness of fit of the NMDS solution to the original data. A lower stress value (closer to 0) indicates a better fit of the NMDS solution to the data, suggesting that the reduced-dimensional representation preserves the underlying structure of the original data. Abbreviations of tree species: ACE: *Acer mono*; FRA: *Fraxinus mandshurica*; PIN: *Pinus koraiensis*; QUE: *Quercus mongolica*; TIL: *Tilia amurensis*; ULM: *Ulmus japonica*.

2.3. Correlations between Phyllosphere Bacterial Community and Leaf Functional Traits

Our study examined leaf functional traits across six plant species. Leaf area (LA), leaf thickness (SLA), leaf dry matter content (LDMC), leaf nitrogen content (LNC15), leaf carbon content (LCC), leaf calcium content (LCaC), leaf aluminum content (LAlC), leaf copper content (LCuC), leaf zinc content (LZnC), and leaf stomatal area (LSA) significant differences were observed among the plant species. Specifically, *Q. mongolica* exhibited the highest leaf area, measuring 65.368 cm^2, while *P. koraiensis* displayed the lowest with a mere 0.730 cm^2. *A. mono* had the highest carbon isotope composition, with a value of −27.789%., while *T. amurensis* had the lowest at −29.681%. *U. japonica* showed the highest leaf copper content at 5.111 g/kg, whereas *F. mandshurica* displayed the lowest content at 4.118 g/kg (Table 2).

Table 2. Leaf functional traits of six tree species ANOSIM in leaves of dominant species of the temperate mixed forest in Changbai Mountain. Asterisks indicate significant differences between tree species (* = $p < 0.05$, ** = $p < 0.01$, *** = $p < 0.001$). Abbreviations of tree species: ACE: *Acer mono*; FRA: *Fraxinus mandshurica*; PIN: *Pinus koraiensis*; QUE: *Quercus mongolica*; TIL: *Tilia amurensis*; ULM: *Ulmus japonica*.

	ACE (Mean ± SE)	FRA (Mean ± SE)	PIN (Mean ± SE)	QUE (Mean ± SE)	TIL (Mean ± SE)	ULM (Mean ± SE)
LA (cm^2) ***	24.52 ± 1.69	36.08 ± 2.45	0.73 ± 0.035	65.36 ± 10.88	38.26 ± 3.86	17.02 ± 0.53
SLA (cm^2/g) **	190.05 ± 30.52	260.03 ± 36.44	84.84 ± 11.57	310.555 ± 14.59	315.12 ± 20.76	76.56 ± 12.01
LDMC (g/g) **	0.092 ± 0.014	0.15 ± 0.04	0.02 ± 0.00	0.30 ± 0.04	0.17 ± 0.02	0.10 ± 0.01
LCC_{13} (d13C/12C) **	−30.58 ± 0.30	−27.78 ± 0.53	−30.51 ± 0.478	−29.47 ± 0.24	−29.68 ± 0.52	−29.32 ± 0.30
LNC_{15} (%) *	−2.39 ± 0.77	0.18 ± 0.67	0.48 ± 0.40	−0.20 ± 0.19	1.01 ± 0.49	−1.55 ± 0.44
LCC (%)	41.73 ± 0.09	42.34 ± 0.38	46.53 ± 0.70	43.04 ± 0.15	43.87 ± 0.38	39.30 ± 0.28
LNC (%)	1.96 ± 0.08	1.90 ± 0.26	1.604 ± 0.12	2.32 ± 0.08	2.09 ± 0.16	1.92 ± 0.11
LPC (g/kg)	1.87 ± 0.15	1.94 ± 0.49	2.10 ± 0.10	2.68 ± 0.26	3.03 ± 0.30	2.11 ± 0.11
LKC (g/kg)	0.07 ± 0.01	0.08 ± 0.00	0.04 ± 0.01	0.01 ± 0.00	0.08 ± 0.01	0.07 ± 0.01
LCaC (g/100 g) **	0.24 ± 0.01	0.25 ± 0.03	0.07 ± 0.02	0.139 ± 0.01	0.17 ± 0.02	0.30 ± 0.02
LAlC (g/kg) ***	0.63 ± 0.07	0.46 ± 0.07	1.63 ± 0.31	0.65 ± 0.14	0.45 ± 0.08	0.45 ± 0.06
LCuC (g/kg) **	4.90 ± 0.62	4.12 ± 0.52	4.77 ± 0.14	4.90 ± 0.62	3.44 ± 0.158	5.11 ± 0.50
LZnC (mg/kg) ***	63.99 ± 7.39	25.56 ± 1.15	64.49 ± 5.77	41.42 ± 8.66	30.34 ± 9.19	33.21 ± 5.33
LSA (10^{-9} m^2) ***	7.37 ± 0.02	17.06 ± 0.05	12.96 ± 0.10	24.24 ± 0.07	7.90 ± 0.07	18.74 ± 0.09

The redundancy and envfit analysis results showed that the functional traits relating to plant nutrient acquisition and resistance to diseases and pests (such as leaf area, isotope carbon content, and copper content) were the main factors influencing the community structure of phyllosphere bacteria (Figure 4).

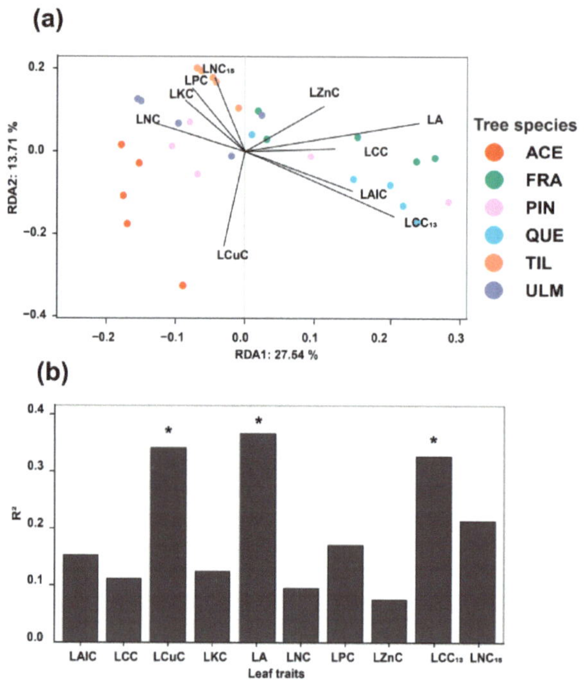

Figure 4. Redundancy analysis result (RDA) of phyllosphere bacterial community and leaf functional traits (**a**); and the explanatory power of each predictive factor (**b**). See Table 3 for the symbolic meanings of each factor in the figure; The asterisk (*) in the figure indicates that the factor has a significant influence on the microbial community ($p < 0.05$). Abbreviations of tree species: ACE: *Acer mono*; FRA: *Fraxinus mandshurica*; PIN: *Pinus koraiensis*; QUE: *Quercus mongolica*; TIL: *Tilia amurensis*; ULM: *Ulmus japonica*.

2.4. Functional Genes of Phyllosphere Bacteria

For the analysis of bacterial functions, the functions related to metabolism were the richest in our dataset, accounting for 76% of all functional annotation sequences (Figure 4a). The main metabolic processes were amino acid metabolism (9.1%), carbohydrate metabolism (8.4%), and energy metabolism (4.2%). Environmental and genetic information processing also had a high relative abundance, mainly membrane transport (3.9%), signal transduction (3.1%), and translation (2.1%) (Figure 5a). The variance test among tree species with secondary functions showed that there were significant differences in relative abundance among tree species with six metabolic processes including amino acid metabolism, membrane transport, xenobiotics biodegradation and metabolism, biosynthesis of other secondary metabolites, and glycan biosynthesis and metabolism. *P. koraiensis*'s amino acid metabolism and membrane transport were significantly higher than other tree species (Figure 5b).

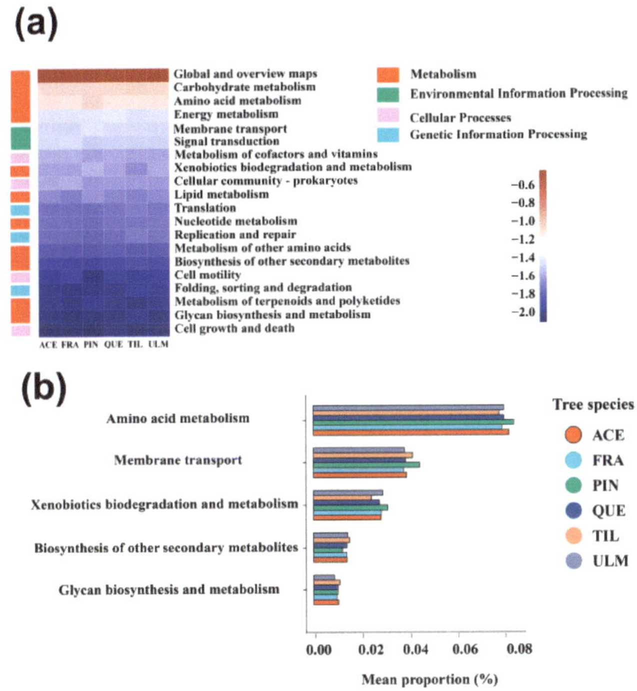

Figure 5. The function genes of phyllosphere bacteria of dominant tree species in a temperate mixed forest of Changbai Mountain. A heatmap illustrates the relative abundance of functions (**a**), and specifically in panel (**b**), only functions with statistical significance ($p < 0.05$) are presented in the comparison between tree species.Abbreviations of tree species: ACE: *Acer mono*; FRA: *Fraxinus mandshurica*; PIN: *Pinus koraiensis*; QUE: *Quercus mongolica*; TIL: *Tilia amurensis*; ULM: *Ulmus japonica*.

3. Discussion

We found that tree species with closer evolutionary distances had more similar phyllosphere microbial diversity, which could be attributed to the assembly of host-associated microbiomes in long-term evolutionary processes [30,31]. Compared to tree species with smaller evolutionary distances, tree species with a long evolutionary distance may have generated greater selection pressure on the species pool of phyllosphere microbes, resulting in more significant differences in phyllosphere microbial diversity [32]. This finding is in line with the strong plant–bacterial interactions on leaves found in a neotropical forest in Panama [13], where most of the dominant bacterial taxa in the phyllosphere have significant evolutionary association with host tree species. These evolutionary associations are likely to be related to phylogenetic variation in host traits, given the interaction between microbial community variance explained by host traits and taxonomy [12].

Our results show that Proteobacteria, Actinobacteria, and Firmicutes were the dominant phyla observed in our samples. The dominant position of Proteobacteria is consistent with many other studies on the bacterial community composition of plant leaves [32–36]. This observation is attributed to the relatively faster replication rate of Proteobacteria compared to other phyla. The ability of Proteobacteria to replicate more rapidly may result in their higher abundance and dominance within the leaf microbiome, as they can colonize and proliferate more efficiently in the leaf environment [37]. Further, Proteobacteria have diverse metabolisms and perform important plant functions, such as nitrogen fixation, nitrification, methylation, and oxygen-free photosynthesis [38–41]. Actinobacteria and Firmicutes are typically associated with arid environments [40–42], which may explain their adaptability to thrive on leaf surfaces, often exposed to dry air and UV radiation [43–45].

Additionally, we found significant biomarkers for all tree species except *U.japonica* and *A. mono*. It is possible that the composition of the microbial community reflects how the relative microbial abundance responds to changes in the environment, according to studies by Liu et al. [30] and Maestre et al. [31]. This means that differences in microhabitats on the surface of leaves could be a factor in the unique microbial flora of different tree species. However, because *U. japonica* and *Acero mono* are usually in the secondary forest layer [46], they may be less affected by external environmental changes and overlap with the canopy layer and the primary forest layer, so there is no unique microbial flora [47]. Specifically, LEfse analysis shows that the *P. koraiensis* biomarker is Actinobacteria, primarily saprophytic bacteria, which produces enzymes that can degrade cellulose. One of the biomarkers for *P. koraiensis* is *Beijerinckia*, a free-living, nitrogen-fixing aerobic microorganism known for its abundant nitrogenase enzymes, which facilitate efficient nitrogen reduction [48]. Additionally, *Methylobacterium*, another biomarker for Korean pine, is recognized for its utilization of C1 compounds released by plants [49]. This shows that *P. koraiensis* may be essential in regulating forest climate and carbon and nitrogen cycles. Under drought stress, *Pseudomonas*, a biomarker of *Q. mongolica*, can produce extracellular polysaccharides to protect bacteria from water threats. Prior research has shown that the inoculation of *Pseudomonas* sp. into host plants can lead to a significant increase in the levels of proline, amino acids, and soluble sugars, ultimately bolstering the plants' drought resistance [50]. *Q. mongolica*, a tree species renowned for its drought resistance in the Changbai Mountain area [51], may also derive potential benefits from the presence of Pseudomonas. The advantages that the presence of drought-resistant bacteria can bring to *Q. mongolica* include enhancing its ability to withstand water stress, facilitating the accumulation of osmoprotectants, and promoting overall resilience in challenging arid conditions [52]. Thus, the interaction with *Pseudomonas* could further contribute to the drought resistance observed in *Q. mongolica*. The natural products found in this family are expected to become an important source of drugs in the future [53]. This kind of bacteria is the biomarker of *U. pumila*, so the application value of *U. pumila* leaves should be further explored.

The redundancy analysis confirmed that certain traits are linked explicitly to phyllosphere microbial community composition. The carbon isotope content, leaf area, and copper in the leaves of plants have significant effects on the bacterial community in the canopy. Previous studies have suggested that leaf environmental conditions such as element concentration, resource availability, and defensive compounds can act as ecological filters and affect the microbial community composition on leaves [54]. Specifically, leaf area and leaf carbon isotope content mainly represent the resource acquisition ability of plants. Therefore, our findings show that the characteristics of the phyllosphere microbial composition of tree species are closely related to their nutrient acquisition ability [55]. In addition, leaf copper content is also essential to the composition of the phyllosphere bacterial community because they can also be used as a defensive compound for plants [40,56].

The abundance of functional genes affects the transformation of ecological processes by affecting microbial processes. The results of functional prediction support the critical role of carbohydrate and amino acid metabolism in the phyllosphere bacteria [57–59]. For example, an experimental study on the colonization of phyllosphere bacteria [58] showed that carbohydrate metabolism is produced during plant photosynthesis and is then consumed by other organisms, aiding the process of regulating the metabolic formation, decomposition, and mutual transformation of microbes [60]. Amino acid metabolism can help bacteria absorb amino acids, which is beneficial to the survival and reproduction of phyllosphere bacteria [61,62]. Membrane transporters are also reported as an important part of the functional library of epiphytic microbes, which can maximize the ability to monopolize other restricted resources [63]. The richness of signal transduction pathways involves rapid sensing and response to environmental changes, which will eventually be consistent with the high variability of humidity, light, and temperature conditions in the microbial habitat [17]. The variance analysis showed that amino acid metabolism and membrane transport of *P. koraiensis* was significantly higher than those of other tree species, which may imply that the phyllosphere bacteria of conifer species, *P. koraiensis*, had developed a stronger ability to absorb nutrients and transport energy due to their lower leaf area (i.e., insufficient nutrient acquisition), compared with broadleaf species. Generally speaking, phyllosphere bacteria in the broad-leaved Korean pine mixed forest can absorb more carbohydrates and amino acids from the leaves through these functional genes, which improves the diversity of phyllosphere bacteria. These potential biological functions lay the foundation for the interaction between plants and microbes. Phyllophere microbes may participate in many life processes of plants through these potential functions and help plants grow in nonideal conditions.

4. Materials and Methods

4.1. Study Site

The experimental plot is located in the broad-leaved Korean pine mixed forest (42°21′01″ N, 128°42′51″ E) of the National Nature Reserve in Changbai Mountain in northeastern China. The elevation ranges between 830 and 850 m within our study area. The study site has a typical temperate continental mountain climate, with an average annual temperature of 3.6 °C and average annual precipitation of 700 mm [64,65]. The zonal soil in this area is mountainous dark brown forest soil. The broad-leaved Korean pine mixed forest is rich in species composition and has a clear vertical structure. The dominant tree species include *Pinus koraiensis*, *Tilia amurensis*, *Quercus mongolica*, *Acer mono*, *Fraxinus mandshurica*, and *Ulmus japonica* [66].

4.2. Leaf Sampling

In August 2020, 6 dominant tree species (*P. koraiensis*, *T. amurensis*, *Q. mongolica*, *A. mono*, *F. mandshuric*, *U. japonica*) in broad-leaved Korean pine mixed forest were sampled by selecting 5 individual trees per species, with a total of 30 trees (i.e., 30 samples). The interval between each tree was greater than 20 m. We use averruncator to cut leaves from all four directions of the target tree species. To avoid the influence of leaf age and disease, we selected 20 healthy and undamaged leaves at the top of each branch, mixed them, and stored them at 4 °C until they could be returned to the laboratory to be kept at −80 °C.

4.3. Measurements and Calculations of Leaf Functional Traits

Fourteen common leaf morphological and chemical traits related to plant life history and nutrient and water use efficiency were measured based on 5 to 10 leaves [67]. Leaf area (LA) was calculated using a portable scanner (Canon LiDE 110, Tokyo, Japan) and Image Pro Plus 6.0 software (Media Cybernetics, Silver Spring, MD, USA). Leaf dry matter content (LDMC) was determined after drying in a constant mass oven for 48 h at 65 °C. Specific leaf area (SLA) was calculated as the ratio of fresh leaf area to leaf dry matter content. The dried leaf samples were then ground to a fine powder using a ball mill (RETSCH, GmbH, Haan, Germany). Leaf carbon (LCC) and nitrogen (LNC) contents were measured using an elemental analyzer (Vario EL III, Elementar, Hanau, Germany). Leaf carbon isotope content (LCC13) and nitrogen isotope content (LNC15) were analyzed using a stable isotope analyzer (CMCRDS system, Picarro, CA, USA). Other chemical elements, including phosphorus (LPC), potassium (LKC), aluminum (LAlC), copper (LCuC), calcium (LCaC), and zinc (LZnC), were measured using an ICP Optima 8000 (Perkin-Elmer, Waltham, MA, USA). The size of the leaf stomatal area was measured using a nail polish blotting method [68]. The main functions of each of these traits are listed in Table 3.

Table 3. Plant traits and their corresponding functions used in this study.

Abbreviations	Functional Traits	Functions
LA	Leaf Area	Resource allocation capacity [69]
SLA	Specific Leaf Area	Resource allocation capacity [69]
LDMC	Leaf Dry Matter Content	Resource allocation capacity [70]
LCC_{13}	Leaf Stable Carbon 13 Content	Water utilization efficiency [70]
LNC_{15}	Leaf Stable Nitrogen 15 Content	Resource utilization efficiency [71]
LCC	Leaf Carbon Content	Plant photosynthesis [72]
LNC	Leaf Nitrogen Content	Plant photosynthesis [72]
LPC	Leaf Phosphorus Content	Plant photosynthesis [72]
LKC	Leaf Potassium Content	Plant photosynthesis [72]
LCaC	Leaf Calcium Content	Plant metabolism [73]
LAlC	Leaf Aluminum Content	Plant metabolism [73]
LCuC	Leaf Copper Content	Plant metabolism; resistance to diseases, pests [74]
LZnC	Leaf Zinc Content	Plant metabolism; resistance to diseases, pests [74]
LSA	Leaf Stomatal Area	Transpiration, photosynthesis [75]

4.4. Leaf DNA Extraction and Sequencing

For the extraction of phyllosphere bacteria, we ground 5–10 fresh leaves, weighed 5 g of them into sterile tubes, and added 10 mL of 0.1 M potassium phosphate buffer (pH = 8.0) to each gram of sample [12,14]. The samples underwent ultrasonic cleaning twice for 1 min and centrifuged for 10 s. The potassium phosphate buffers from the two washes were combined and filtered through a 0.2 μm sterile filter (Supor EAV, Pall Corporation, Ann Arbor, MI, USA). The filters were chopped, and DNA was extracted using the FastDNA® SPIN Kit (Qbiogene, Irvine, CA, USA). DNA extracts were checked on a 1% agarose gel, and DNA concentration and purity were determined using a NanoDrop-2000 (Thermo Scientific, Wilmington, NC, USA). Sequencing was performed by Shanghai Meiji Biomedical Technology Co., Ltd. on the Illumina NovaSeq PE250 platform. The amplification primers were 799F (AACMGGATTAGATACCCKG) and 1193R (ACGTCATCCCCACCTTCC). Raw reads were saved to the NCBI database (SUB9956652).

4.5. Data Analyses

Raw gene sequencing reads were filtered using QIIME (version 1.7). Low-quality sequences (length < 200 bp, ambiguous bases > 0, average base quality score < 25) were removed. Samples were differentiated based on barcodes and primers, and sequence orientation was adjusted. UPARSE version 7.1 [76] was used to cluster sequences into operational taxonomic units (OTUs) with 97% similarity and to identify and delete chimeric sequences. The classification of the representative sequences of each OTU was species annotated by the RDP classifier [77] according to the Silva v138 bacterial database alignment. The alignment threshold was set to 80%, and 656,940 high-quality sequences were obtained. To avoid the influence of sequencing depth on subsequent analysis, all samples were diluted to the same sequencing depth (21,898 sequences per sample), resulting in 5231 OTUs.

We used the R package V.PhyloMaker [78] to construct a phylogenetic tree for six trees species. Based on the results of OTU cluster analysis, α-diversity (i.e., species richness) was calculated. ANOVA (Analysis of Variance) was used to explore whether α-diversity and leaf functional traits was significantly different among tree species. The Bray–Curtis matrix between samples was calculated at the OTU level and visualized by NMDS (Non-Metric Multidimensional Scaling) to depict differences in the bacterial community in a two-dimensional space. Significance was evaluated by ANOSIM (Analysis of Similarities) [79]. LEfSe analysis (Linear discriminant analysis Effect Size) was used to identify biomarkers causing differences in microbial community structure between groups using a Linear discriminant analysis (LDA) score threshold of 4.0 and an alpha level of 0.05 at the genus level and above on the Hutlab Galaxy website application (http://huttenhowe.sph.harvard.edu/galaxy/

10 June 2023) [80]. The study utilized RDA (Redundancy Analysis) to investigate the relationship between the bacterial community in the canopy phyllosphere and leaf functional traits. The analysis also utilized the 'envfit' method to determine the explanatory power of each factor [77]. Finally, the functional annotation of PICRUSt2 predictions was obtained based on the Kyoto Encyclopedia of Genes and Genomes (KEGG) database [81]. Except for LEfSe analysis, other statistical analyses were conducted using R software (version 4.3.2).

5. Conclusions

Our findings suggest that tree species with a close phylogenetic relationship tend to exhibit similar alpha phyllosphere bacterial diversity. Additionally, we observed multiple biomarkers for all tree species except *Ulmus japonica* and *Quercus mongolica*. The biomarkers of *Pinus koraiensis* may suggest that it is essential for forest climate as well as carbon and nitrogen cycles, and the biomarkers of *Quercus mongolica* may contribute to the drought resistance of the host. Our study also shows that the leaf traits of host tree species, such as leaf area, leaf carbon isotope content, and leaf copper and zinc contents could regulate the composition of the phyllosphere bacterial community. The results of functional prediction show that the main functional genes are carbohydrate metabolism and membrane transport, which can improve resource utilization capacity. The abundance of these two functions of *Pinus koraiensis* is higher than that of other tree species, which may be an evolutionary strategy of microbes in *Pinus koraiensis* under the condition of low leaf area. The results provide basic data for revealing the canopy biodiversity composition, structure, and driving factors of temperate forests and provide a useful reference for the study of plant-microbes interaction under global changes.

Author Contributions: Z.Y., X.W. (Xing Wang), and F.L. conceived the idea; Z.Y., J.Y., F.L., X.W. (Xing Wang), T.Y., B.B., Z.M., S.F., X.W. (Xugao Wang), and Z.H. performed the research and collected the data; X.W. (Xing Wang), Z.Y. and A.A. analysed the data; Z.Y. and X.W. (Xing Wang) wrote the first draft of the manuscript, and all the authors contributed substantially to the revisions. All authors have read and agreed to the published version of the manuscript.

Funding: This work was supported by the Strategic Priority Research Program of the Chinese Academy of Sciences (XDA 23080301 & XDB 31030000), the National Natural Science Foundation of China (32171581, 31670632), Natural Science Basic Research Program of Shaanxi (22JHQ036), and the Fundamental Research Funds for the Central Universities. AA is currently supported by Hebei University (Special Project No. 521100221033).

Institutional Review Board Statement: This article does not contain any studies with human participants or animals performed by any of the authors.

Informed Consent Statement: All authors and their institution consent their participation in the present work and the final manuscript publication.

Data Availability Statement: Microbe data are available from the NCBI Sequence Read ArchivAeArchive (SRA) database (Accession Number: PRJNA824213), other data that support the findings of this study are available from the corresponding author upon reasonable request.

Acknowledgments: We thank numerous works in Changbai Mountain Nature Reserve who contributed to the field survey.

Conflicts of Interest: We confirm that this manuscript has not been published elsewhere and is not under consideration by another journal. We declare no conflict of interest associated with this publication.

References

1. Last, F. Seasonal incidence of Sporobolomyces on cereal leaves. *Trans. Br. Mycol. Soc.* **1955**, *38*, 221–239. [CrossRef]
2. Zhu, Y.G.; Xiong, C.; Wei, Z.; Chen, Q.L.; Ma, B.; Zhou, S.Y.D.; Tan, J.; Zhang, L.M.; Cui, H.L.; Duan, G.L. Impacts of global change on phyllosphere microbiome. *New Phytol.* **2021**, *234*, 1977–1986. [CrossRef] [PubMed]
3. Vorholt, J.A. Microbial life in the phyllosphere. *Nat. Rev. Microbiol.* **2012**, *10*, 828–840. [CrossRef] [PubMed]

4. Farre-Armengol, G.; Filella, I.; Llusia, J.; Penuelas, J. Bidirectional Interaction between Phyllospheric Microbiotas and Plant Volatile Emissions. *Trends Plant Sci.* **2016**, *21*, 854–860. [CrossRef]
5. Stone Bram, W.G.; Jackson Colin, R. Biogeographic patterns between bacterial phyllosphere communities of the southern magnolia (*Magnolia grandiflora*) in a small forest. *Microb. Ecol.* **2016**, *71*, 954–961. [CrossRef]
6. Chen, T.; Nomura, K.; Wang, X.; Sohrabi, R.; Xu, J.; Yao, L.; Paasch, B.C.; Ma, L.; Kremer, J.; Cheng, Y.; et al. A plant genetic network for preventing dysbiosis in the phyllosphere. *Nature* **2020**, *580*, 653–657. [CrossRef]
7. Laforest-Lapointe, I.; Paquette, A.; Messier, C.; Kembel, S.W. Leaf bacterial diversity mediates plant diversity and ecosystem function relationships. *Nature* **2017**, *546*, 145–147. [CrossRef]
8. Liu, H.; Brettell, L.E.; Singh, B. Linking the phyllosphere microbiome to plant health. *Trends Plant Sci.* **2020**, *25*, 841–844. [CrossRef]
9. Lee, D.W.; Hong, J.-S.; Kim, S.-H.; Kim, J.W.; Kim, B.S. First Report of *Pseudomonas* lurida Causing Bacterial Leaf Spot on *Miscanthus sinensis*. *J. Phytopathol.* **2014**, *162*, 195–200. [CrossRef]
10. Arun, D.K.; Sabarinathan, K.G.; Gomathy, M.; Kannan, R.; Balachandar, D. Mitigation of drought stress in rice crop with plant growth-promoting abiotic stress-tolerant rice phyllosphere bacteria. *J. Basic Microbiol.* **2020**, *60*, 768–786. [CrossRef]
11. Yao, H.; Sun, X.; He, C.; Maitra, P.; Li, X.-C.; Guo, L.-D. Phyllosphere epiphytic and endophytic fungal community and network structures differ in a tropical mangrove ecosystem. *Microbiome* **2019**, *7*, 57. [CrossRef] [PubMed]
12. Yao, H.; Sun, X.; He, C.; Li, X.-C.; Guo, L.-D. Host identity is more important in structuring bacterial epiphytes than endophytes in a tropical mangrove forest. *FEMS Microbiol. Ecol.* **2020**, *96*, fiaa038. [CrossRef] [PubMed]
13. Kembel, S.W.; O'Connor, T.K.; Arnold, H.K.; Hubbell, S.P.; Wright, S.J.; Green, J.L. Relationships between phyllosphere bacterial communities and plant functional traits in a neotropical forest. *Proc. Natl. Acad. Sci. USA* **2014**, *111*, 13715–13720. [CrossRef] [PubMed]
14. Laforest-Lapointe, I.; Messier, C.; Kembel, S.W. Host species identity, site and time drive temperate tree phyllosphere bacterial community structure. *Microbiome* **2016**, *4*, e2367. [CrossRef] [PubMed]
15. Leveau, J.H.J. A brief from the leaf: Latest research to inform our understanding of the phyllosphere microbiome. *Curr. Opin. Microbiol.* **2019**, *49*, 41–49. [CrossRef]
16. Wang, M.; Sun, B. Unlocking the connection: Aging as a lens to examine the effects of climate warming. *Innov. Life* **2023**, *1*, 100003. [CrossRef]
17. Lajoie, G.; Maglione, R.; Kembel, S.W. Adaptive matching between phyllosphere bacteria and their tree hosts in a neotropical forest. *Microbiome* **2020**, *8*, 70. [CrossRef]
18. Hunter, P.J.; Hand, P.; Pink, D.; Whipps, J.M.; Bending, G.D. Both leaf properties and microbe-microbe interactions influence within-species variation in bacterial population diversity and structure in the lettuce (*Lactuca* species) phyllosphere. *Appl. Environ. Microbiol.* **2010**, *76*, 8117–8125. [CrossRef]
19. Liu, L.; Lu, L.; Li, H.; Meng, Z.; Dong, T.; Peng, C.; Xu, X. Divergence of phyllosphere microbial communities between females and males of the Dioecious *Populus cathayana*. *Mol. Plant-Microbe Interact.* **2021**, *34*, 351–361. [CrossRef]
20. Remus-Emsermann, M.N.; Schlechter, R.O. Phyllosphere microbiology: At the interface between microbial individuals and the plant host. *New Phytol.* **2018**, *218*, 1327–1333. [CrossRef]
21. Kembel Steven, W.; Mueller Rebecca, C. Plant traits and taxonomy drive host associations in tropical phyllosphere fungal communities. *Botany* **2014**, *92*, 303–311. [CrossRef]
22. Nakamura, A.; Kitching, R.L.; Cao, M.; Creedy, T.J.; Fayle, T.M.; Freiberg, M.; Hewitt, C.N.; Itioka, T.; Koh, L.P.; Ma, K.; et al. Forests and their canopies: Achievements and horizons in canopy science. *Trends Ecol. Evol.* **2017**, *32*, 438–451. [CrossRef] [PubMed]
23. Wilkes, P.; Jones, S.D.; Suarez, L.; Mellor, A.; Woodgate, W.; Soto-Berelov, M.; Haywood, A.; Skidmore, A.K. Mapping forest canopy height across large areas by upscaling ALS estimates with freely available satellite data. *Remote Sens.* **2015**, *7*, 12563–12587. [CrossRef]
24. Ozanne, C.M.; Anhuf, D.; Boulter, S.L.; Keller, M.; Kitching, R.L.; Körner, C.; Meinzer, F.C.; Mitchell, A.; Nakashizuka, T.; Dias, P.S. Biodiversity meets the atmosphere: A global view of forest canopies. *Science* **2003**, *301*, 183–186. [CrossRef] [PubMed]
25. Oliveira, B.F.; Scheffers, B.R. Vertical stratification influences global patterns of biodiversity. *Ecography* **2019**, *42*, 249. [CrossRef]
26. Anderson-Teixeira, K.J.; Davies, S.J.; Bennett, A.C.; Gonzalez-Akre, E.B.; Muller-Landau, H.C.; Joseph Wright, S.; Abu Salim, K.; Almeyda Zambrano, A.M.; Alonso, A.; Baltzer, J.L. CTFS-Forest GEO: A worldwide network monitoring forests in an era of global change. *Glob. Chang. Biol.* **2015**, *21*, 528–549. [CrossRef]
27. Bai, F.; Sang, W.; Axmacher, J.C. Forest vegetation responses to climate and environmental change: A case study from Changbai Mountain, NE China. *For. Ecol. Manag.* **2011**, *262*, 2052–2060. [CrossRef]
28. Liu, J.; Meng, Z.; Liu, X.; Zhang, X.-H. Microbial assembly, interaction, functioning, activity and diversification: A review derived from community compositional data. *Mar. Life Sci. Technol.* **2019**, *1*, 112–128. [CrossRef]
29. Sun, X.; Deharveng, L.; Bedos, A.; Chang, L.; Scheu, S.; Wu, D. Changes in diversity and body size of Onychiurinae (Collembola: Onychiuridae) along an altitudinal gradient in Changbai Mountain, China. *Soil. Ecol. Lett.* **2020**, *2*, 230–239. [CrossRef]
30. Liu, N.; Hu, H.; Ma, W.; Deng, Y.; Dimitrov, D.; Wang, Q.; Shrestha, N.; Su, X.; Feng, K.; Liu, Y.; et al. Relationships between soil microbial diversities across an aridity gradient in temperate grasslands. *Microb. Ecol.* **2022**, *85*, 1013–1027. [CrossRef]

31. Maestre, F.T.; Delgado-Baquerizo, M.; Jeffries, T.C.; Eldridge, D.J.; Ochoa, V.; Gozalo, B.; Quero, J.L.; García-Gómez, M.; Gallardo, A.; Ulrich, W.; et al. Increasing aridity reduces soil microbial diversity and abundance in global drylands. *Proc. Natl. Acad. Sci. USA* **2015**, *112*, 15684–15689. [CrossRef]
32. Redford Amanda, J.; Bowers Robert, M.; Knight, R.; Linhart, Y.; Fierer, N. The ecology of the phyllosphere: Geographic and phylogenetic variability in the distribution of bacteria on tree leaves. *Environ. Microbiol.* **2010**, *12*, 2885–2893. [CrossRef]
33. Espenshade, J.; Thijs, S.; Gawronski, S.; Boye, H.; Weyens, N.; Vangronsveld, J. Influence of urbanization on epiphytic bacterial communities of the *platanus* × *hispanica* tree leaves in a biennial study. *Front. Microbiol.* **2019**, *10*, 35–45. [CrossRef] [PubMed]
34. Gandolfi, I.; Canedoli, C.; Imperato, V.; Tagliaferri, I.; Gkorezis, P.; Vangronsveld, J.; Padoa Schioppa, E.; Papacchini, M.; Bestetti, G.; Franzetti, A. Diversity and hydrocarbon-degrading potential of epiphytic microbial communities on *Platanus* × *acerifolia* leaves in an urban area. *Environ. Pollut.* **2017**, *220*, 650–658. [CrossRef] [PubMed]
35. Smets, W.; Wuyts, K.; Oerlemans, E.; Wuyts, S.; Denys, S.; Samson, R.; Lebeer, S. Impact of urban land use on the bacterial phyllosphere of ivy (*Hedera* sp.). *Atmos. Environ.* **2016**, *147*, 376–383. [CrossRef]
36. Lymperopoulou, D.S.; Adams, R.I.; Lindow, S.E. Contribution of vegetation to the microbial composition of nearby outdoor air. *Appl. Environ. Microbiol.* **2016**, *82*, 3822–3833. [CrossRef]
37. Leff, J.W.; Del Tredici, P.; Friedman, W.E.; Fierer, N. Spatial structuring of bacterial communities within individual Ginkgo biloba trees. *Environ. Microbiol.* **2015**, *17*, 2352–2361. [CrossRef]
38. Watanabe, K.; Kohzu, A.; Suda, W.; Yamamura, S.; Takamatsu, T.; Takenaka, A.; Koshikawa, M.K.; Hayashi, S.; Watanabe, M. Microbial nitrification in throughfall of a Japanese cedar associated with archaea from the tree canopy. *Springerplus* **2016**, *5*, 1596. [CrossRef]
39. Atamna-Ismaeel, N.; Finkel, O.; Glaser, F.; von Mering, C.; Vorholt, J.A.; Koblizek, M.; Belkin, S.; Beja, O. Bacterial anoxygenic photosynthesis on plant leaf surfaces. *Environ. Microbiol. Rep.* **2012**, *4*, 209–216. [CrossRef]
40. Mina, D.; Pereira, J.A.; Lino-Neto, T.; Baptista, P. Impact of plant genotype and plant habitat in shaping bacterial pathobiome: A comparative study in olive tree. *Sci. Rep.* **2020**, *10*, 3475. [CrossRef]
41. Coleman-Derr, D.; Desgarennes, D.; Fonseca-Garcia, C.; Gross, S.; Clingenpeel, S.; Woyke, T.; North, G.; Visel, A.; Partida-Martinez, L.P.; Tringe, S.G. Plant compartment and biogeography affect microbiome composition in cultivated and native *Agave* species. *New Phytol.* **2016**, *209*, 798–811. [CrossRef]
42. Mina, D.; Pereira, J.A.; Lino-Neto, T.; Baptista, P. Epiphytic and endophytic bacteria on olive tree phyllosphere: Exploring tissue and cultivar effect. *Microb. Ecol.* **2020**, *80*, 145–157. [CrossRef] [PubMed]
43. Ruibal, C.; Gueidan, C.; Selbmann, L.; Gorbushina, A.A.; Crous, P.W.; Groenewald, J.Z.; Muggia, L.; Grube, M.; Isola, D.; Schoch, C.L.; et al. Phylogeny of rock-inhabiting fungi related to Dothideomycetes. *Stud. Mycol.* **2009**, *61*, 123–133. [CrossRef] [PubMed]
44. Schoch, C.L.; Shoemaker, R.A.; Seifert, K.A.; Hambleton, S.; Spatafora, J.W.; Crous, P.W. A multigene phylogeny of the Dothideomycetes using four nuclear loci. *Mycologia* **2006**, *98*, 1041–1052. [CrossRef] [PubMed]
45. Chen, Q.-L.; Cai, L.; Wang, H.-C.; Cai, L.-T.; Goodwin, P.; Ma, J.; Wang, F.; Li, Z. Fungal composition and diversity of the tobacco leaf phyllosphere during curing of leaves. *Front. Microbiol.* **2019**, *11*, 554051. [CrossRef] [PubMed]
46. Takeda, Y.; Nakanishi, S.; Choe, D. Phytosociological study on natural summer-green forests in Korea. *Ecol. Res.* **1994**, *9*, 21–32. [CrossRef]
47. Mills, G. An urban canopy-layer climate model. *Theor. Appl. Climatol.* **1997**, *57*, 229–244. [CrossRef]
48. Rico, L.; Ogaya, R.; Terradas, J.; Penuelas, J. Community structures of N-2-fixing bacteria associated with the phyllosphere of a Holm oak forest and their response to drought. *Plant Biol.* **2014**, *16*, 586–593. [CrossRef]
49. Abanda-Nkpwatt, D.; Muesch, M.; Tschiersch, J.; Boettner, M.; Schwab, W. Molecular interaction between Methylobacterium extorquens and seedlings: Growth promotion, methanol consumption, and localization of the methanol emission site. *J. Exp. Bot.* **2006**, *57*, 4025–4032. [CrossRef]
50. Kang, S.-M.; Radhakrishnan, R.; Khan, A.L.; Kim, M.-J.; Park, J.-M.; Kim, B.-R.; Shin, D.-H.; Lee, I.-J. Gibberellin secreting rhizobacterium, *Pseudomonas putida* H-2-3 modulates the hormonal and stress physiology of soybean to improve the plant growth under saline and drought conditions. *Plant Physiol. Biochem.* **2014**, *84*, 115–124. [CrossRef]
51. He, F.; Yang, B.; Wang, H.; Yan, Q.; Cao, Y.; He, X. Changes in composition and diversity of fungal communities along Quercus mongolica forests developments in Northeast China. *Appl. Soil. Ecol.* **2016**, *100*, 162–171. [CrossRef]
52. Khan, N.; Ali, S.; Tariq, H.; Latif, S.; Yasmin, H.; Mehmood, A.; Shahid, M.A. Water conservation and plant survival strategies of rhizobacteria under drought stress. *Agronomy* **2020**, *10*, 1683. [CrossRef]
53. Bull, C.T.; Shetty, K.G.; Subbarao, K.V. Interactions between myxobacteria, plant pathogenic fungi, and biocontrol agents. *Plant Dis.* **2002**, *86*, 889–896. [CrossRef] [PubMed]
54. Meyer Katrin, M.; Leveau Johan, H.J. Microbiology of the phyllosphere: A playground for testing ecological concepts. *Oecologia* **2012**, *168*, 621–629. [CrossRef]
55. Lindow, S.E.; Brandl, M.T. Microbiology of the phyllosphere. *Appl. Environ. Microbiol.* **2003**, *69*, 1875–1883. [CrossRef]
56. Kaczmarczyk, A.; Hochstrasser, R.; Vorholt, J.A.; Francez-Charlot, A. Complex two-component signaling regulates the general stress response in Alphaproteobacteria. *Proc. Natl. Acad. Sci. USA* **2014**, *111*, E5196–E5204. [CrossRef]
57. Ryffel, F.; Helfrich, E.J.N.; Kiefer, P.; Peyriga, L.; Portais, J.-C.; Piel, J.; Vorholt, J.A. Metabolic footprint of epiphytic bacteria on *Arabidopsis thaliana* leaves. *ISME J.* **2016**, *10*, 632–643. [CrossRef]

58. Mueller, D.B.; Schubert, O.T.; Roest, H.; Aebersold, R.; Vorholt, J.A. Systems-level Proteomics of Two Ubiquitous Leaf Commensals Reveals Complementary Adaptive Traits for Phyllosphere Colonization. *Mol. Cell. Proteom.* **2016**, *15*, 3256–3269. [CrossRef]
59. Yadav, R.K.P.; Papatheodorou, E.M.; Karamanoli, K.; Constantinidou, H.-I.A.; Vokou, D. Abundance and diversity of the phyllosphere bacterial communities of Mediterranean perennial plants that differ in leaf chemistry. *Chemoecology* **2008**, *18*, 217–226. [CrossRef]
60. Nehls, U.; Mikolajewski, S.; Magel, E.; Hampp, R. Carbohydrate metabolism in ectomycorrhizas: Gene expression, monosaccharide transport and metabolic control. *New Phytol.* **2001**, *150*, 533–541. [CrossRef]
61. Neis, E.P.; Dejong, C.H.; Rensen, S.S. The role of microbial amino acid metabolism in host metabolism. *Nutrients* **2015**, *7*, 2930–2946. [CrossRef] [PubMed]
62. Näsholm, T.; Kielland, K.; Ganeteg, U. Uptake of organic nitrogen by plants. *New Phytol.* **2009**, *182*, 31–48. [CrossRef] [PubMed]
63. Delmotte, N.; Knief, C.; Chaffron, S.; Innerebner, G.; Roschitzki, B.; Schlapbach, R.; von Mering, C.; Vorholt Julia, A. Community proteogenomics reveals insights into the physiology of phyllosphere bacteria. *Proc. Natl. Acad. Sci. USA* **2009**, *106*, 16428–16433. [CrossRef] [PubMed]
64. Shen, C.; Xiong, J.; Zhang, H.; Feng, Y.; Lin, X.; Li, X.; Liang, W.; Chu, H. Soil pH drives the spatial distribution of bacterial communities along elevation on Changbai Mountain. *Soil. Biol. Biochem.* **2013**, *57*, 204–211. [CrossRef]
65. Yuan, Z.; Ali, A.; Ruiz-Benito, P.; Jucker, T.; Mori, A.; Wang, S.; Zhang, X.; Li, H.; Hao, Z.; Wang, X.; et al. Above- and below-ground biodiversity jointly regulate temperate forest multifunctionality along a local-scale environmental gradient. *J. Ecol.* **2020**, *108*, 2012–2024. [CrossRef]
66. Shao, G.; Schall, P.; Weishampel, J.F. Dynamic simulations of mixed broadleaved-Pinus koraiensis forests in the Changbaishan Biosphere Reserve of China. *For. Ecol. Manag.* **1994**, *70*, 169–181. [CrossRef]
67. Perez-Harguindeguy, N.; Diaz, S.; Garnier, E.; Lavorel, S.; Poorter, H.; Jaureguiberry, P.; Bret-Harte, M.S.; Cornwell, W.K.; Craine, J.M.; Gurvich, D.E.; et al. New handbook for standardised measurement of plant functional traits worldwide. *Aust. J. Bot.* **2013**, *61*, 167–234. [CrossRef]
68. Zhao, W.L.; Chen, Y.J.; Brodribb, T.J.; Cao, K.F. Weak co-ordination between vein and stomatal densities in 105 angiosperm tree species along altitudinal gradients in Southwest China. *Funct. Plant Biol.* **2016**, *43*, 1126–1133. [CrossRef]
69. Wright, I.; Reich, P.; Westoby, M.; Ackerly, D.; Baruch, Z.; Bongers, F.; Cavender-Bares, J.; Cornelissen, J.; Diemer, M.; Flexas, J.; et al. The world-wide leaf economics spectrum. *Nature* **2004**, *21*, 821–827. [CrossRef]
70. Ripullone, F.; Lauteri, M.; Grassi, G.; Amato, M.; Borghetti, M. Variation in nitrogen supply changes water-use efficiency of Pseudotsuga menziesii and Populus x euroamericana; a comparison of three approaches to determine water-use efficiency. *Tree Physiol.* **2004**, *24*, 671–679. [CrossRef]
71. Saitoh, T.; Seiwa, K.; Nishiwaki, A. Effects of resource heterogeneity on nitrogen translocation within clonal fragments of Sasa palmata: An isotopic (15N) assessment. *Ann. Bot.* **2006**, *98*, 657–663. [CrossRef] [PubMed]
72. Longstreth, D.J.; Nobel, P.S. Nutrient Influences on Leaf Photosynthesis: Effects of nitrogen, phosphorus, and potassium for *Gossypium hirsutum* L. *Plant Physiol.* **1980**, *65*, 541–543. [CrossRef] [PubMed]
73. Bollard, E.; Butler, G. Mineral nutrition of plants. *Annu. Rev. Plant Physiol.* **1966**, *17*, 77–112. [CrossRef]
74. Pollard, A.J.; Powell, K.D.; Harper, F.A.; Smith, J.A.C. The genetic basis of metal hyperaccumulation in plants. *Crit. Rev. Plant Sci.* **2002**, *21*, 539–566. [CrossRef]
75. Sharkey, T.D. Transpiration-induced changes in the photosynthetic capacity of leaves. *Planta* **1984**, *160*, 143–150. [CrossRef]
76. Edgar, R.C. UPARSE: Highly accurate OTU sequences from microbial amplicon reads. *Nat. Methods* **2013**, *10*, 996–998. [CrossRef]
77. Wang, Q.; Garrity, G.M.; Tiedje, J.M.; Cole, J.R. Naive Bayesian classifier for rapid assignment of rRNA sequences into the new bacterial taxonomy. *Appl. Environ. Microbiol.* **2007**, *73*, 5261–5267. [CrossRef]
78. Jin, Y.; Qian, H.V. PhyloMaker: An R package that can generate very large phylogenies for vascular plants. *Ecography* **2019**, *42*, 1353–1359. [CrossRef]
79. Dixon, P. Vegan, a package of R functions for community ecology. *J. Veg. Sci.* **2003**, *14*, 927–930. [CrossRef]
80. Segata, N.; Izard, J.; Waldron, L.; Gevers, D.; Miropolsky, L.; Garrett, W.S.; Huttenhower, C. Metagenomic biomarker discovery and explanation. *Genome Biol.* **2011**, *12*, R60. [CrossRef]
81. Douglas, G.M.; Maffei, V.J.; Zaneveld, J.R.; Yurgel, S.N.; Brown, J.R.; Taylor, C.M.; Huttenhower, C.; Langille, M.G. PICRUSt2 for prediction of metagenome functions. *Nat. Biotechnol.* **2020**, *38*, 685–688. [CrossRef] [PubMed]

plants

MDPI

Article

Analysis of the Relative Importance of Stand Structure and Site Conditions for the Productivity, Species Diversity, and Carbon Sequestration of *Cunninghamia lanceolata* and *Phoebe bournei* Mixed Forest

Yiru Wang [1,2,3], Zhaohua Liu [1,2,3], Tao Tang [1,2,3,*] and Jiping Li [1,2,3,*]

1 Faculty of Forestry, Central South University of Forestry and Technology, Changsha 410004, China
2 Key Laboratory of State Forestry Administration on Forest Resources Management and Monitoring in Southern Area, Changsha 410004, China
3 Research Centre of Forest Remote Sensing and Information Engineering, Central South University of Forestry and Technology, Changsha 410004, China
* Correspondence: tangtao945348782@163.com (T.T.); lijiping@vip.163.com (J.L.); Tel.: +86-17752789218 (T.T.); +86-13787319186 (J.L.)

Abstract: Forest stand structure (the characteristics and interrelationships of live trees) and site conditions (the physical and environmental characteristics of a specific location) have been linked to forest regeneration, nutrient cycling, wildlife habitat, and climate regulation. While the effects of stand structure (i.e., spatial and non-spatial) and site conditions on the single function of *Cunninghamia lanceolata* and *Phoebe bournei* (*CLPB*) mixed forest have been studied in previous studies, the relative importance of stand structure and site conditions in terms of productivity, species diversity, and carbon sequestration remains unresolved. In this study, a structural equation model (SEM) was adopted to analyze the relative importance of stand structure and site conditions for the forest productivity, species diversity, and carbon sequestration of *CLPB* mixed forest in Jindong Forestry in Hunan Province. Our research demonstrates that site conditions have a greater influence on forest functions than stand structure, and that non-spatial structures have a greater overall impact on forest functions than spatial structures. Specifically, the intensity of the influence of site conditions and non-spatial structure on functions is greatest for productivity, followed by carbon sequestration and species diversity. In contrast, the intensity of the influence of spatial structure on functions is greatest for carbon sequestration, followed by species diversity and productivity. These findings provide valuable insights for the management of *CLPB* mixed forest in Jindong Forestry and have significant reference value for the close-to-natural forest management (CTNFM) of pure *Cunninghamia lanceolata* forests.

Keywords: stand structure; site conditions; forest functions; structural equation model; *Cunninghamia lanceolata; Phoebe bournei*

Citation: Wang, Y.; Liu, Z.; Tang, T.; Li, J. Analysis of the Relative Importance of Stand Structure and Site Conditions for the Productivity, Species Diversity, and Carbon Sequestration of *Cunninghamia lanceolata* and *Phoebe bournei* Mixed Forest. *Plants* **2023**, *12*, 1633. https://doi.org/10.3390/plants12081633

Academic Editors: Xugao Wang, Zuoqiang Yuan and Chengjin Chu

Received: 3 March 2023
Revised: 27 March 2023
Accepted: 7 April 2023
Published: 12 April 2023

1. Introduction

Forest ecosystems play a vital role in timber production, biodiversity preservation, and carbon sequestration [1–3]. Forests can be separated into mixed forests and pure forests according to the number and volume of tree species. Compared to pure forests, mixed forests contain a greater diversity of tree species and a more complex forest structure, allowing them to serve a greater range of forest ecological functions [4,5]. In this context, forest managers are increasingly accepting of close-to-natural forest management (CTNFM) practices. CTNFM, a production system based on the principles of multifunctional forest management, advocates for mixed uneven forests as an ecologically more stable alternative to uniform monocultures [6–8]. Therefore, transforming pure plantations into mixed forests by CTNFM is an effective way to improve ecological functions, which is of vital significance for the implementation of China's forest quality precision enhancement project.

The pure Chinese fir (*Cunninghamia lanceolata*) forests account for the largest proportion of the area of plantations in southern China. Transforming pure Chinese fir forests into mixed forests through CTNFM is an important challenge, aimed at improving the quality of Chinese fir forests. In recent years, a mixed forest management model of *Cunninghamia lanceolata* and *Phoebe bournei* (*CLPB*) has been recognized as the most successful CTNFM method for pure Chinese fir forests. In this model, the rare and valuable broad-leaved species *Phoebe bournei* has been replanted artificially in pure *Cunninghamia lanceolata* plantations to establish *CLPB* mixed forests, which can effectively promote the growth of both trees and improve the ecological function of pure *Cunninghamia lanceolata* plantations [9,10]. In addition, the State Forestry Administration of China has pushed the *CLPB* mixed forest management model as one of the most successful methods for enhancing the quality of Chinese fir forests in southern China.

The stand structure of forests, which encompasses the characteristics and interrelationships of live trees, is a fundamental attribute of forest ecosystems [11]. A well-designed stand structure has important implications for forest ecosystems, including the provision of wildlife habitat, carbon storage for climate regulation, and effective forest regeneration. By optimizing the structure of the stand, the maximum functional benefits of the forest ecological system can be realized [12–17]. The forest stand structure can be subdivided into spatial and non-spatial structures. Spatial structure pertains to the arrangement and interrelationships of living tree features and is typically measured using indicators such as mixing degree, angular scale, and size ratio. Non-spatial structure, on the other hand, characterizes the quality of individual trees and is often gauged by factors such as DBH, tree height, and density [18–21]. Nevertheless, even within a single tree species, growth rates can fluctuate significantly across different stands due to varying forest site conditions, which encompass the physical, chemical, and biological attributes of a specific location such as soil characteristics, topography, and vegetation. Site conditions may thus be one of the important factors that affect forest functions [22,23].

Jindong Forest Farm is located in the south of Hunan Province, which is the main planting area of *Cunninghamia lanceolata* in the Province. In recent years, with the increasing attention paid by the Chinese government to the ecological functions of forests, the forest management objectives of Jindong Forest Farm have shifted from producing wood to enhancing the ecological functions of forests. Hence, converting a vast tract of pure Chinese fir plantation into a mixed forest with enhanced biological functions is one of the greatest challenges facing forest management at Jindong Forest Farm. Through repeated studies on the model of mixing Chinese fir with other needles and broadleaves, Jindong Forest Farm discovered *CLPB* as the most effective CTNFM model for Chinese fir forest. It has thus become a national demonstration site for the *CLPB* mixed forest model [24,25]. Although previous studies have explored the effects of stand structure and site conditions on the single forest function of *CLPB* mixed forest in Jindong Forest Farm [26–30], the relative importance of stand structure and site conditions in terms of productivity, species diversity, and carbon sequestration remain unresolved.

The aim of this study was to explore the relative importance of the stand structure and site conditions of *CLPB* mixed forest to productivity, species diversity and carbon sequestration. To achieve this goal, we have adopted a structural equation model (SEM) approach, which can provide insights into the relative importance of various aspects of stand structure and site condition on forest functions [31–34]. The following hypotheses were tested: (1) site condition has a greater influence on forest functions (productivity, species diversity and carbon sequestration) than stand structure. (2) The total impact intensity of spatial and non-spatial structures on forest functions is equal.

2. Results

2.1. Response between Observed Variables

According to the correlation heat map (Figure 1), there is a very significant correlation between the slope position, dominant tree breast diameter and dominant tree height. The slope position has a significant correlation with DBH, and the dominant tree breast diameter is significantly related to the dominant tree height. The DBH is significantly negatively correlated with the stand density. The negative correlation between stand density and DBH is mainly due to the growth factors of the forest trees themselves and competition. Within a certain range, stand density will promote the growth of DBH, and beyond this range, the growth of DBH is inhibited. The DBH, tree height, stand density and stand productivity demonstrate a very significant correlation. Higher tree height and DBH usually mean higher biomass accumulation and higher productivity. In addition, higher density may also have an effect on productivity, but its effect may vary depending on environmental conditions and stand type.

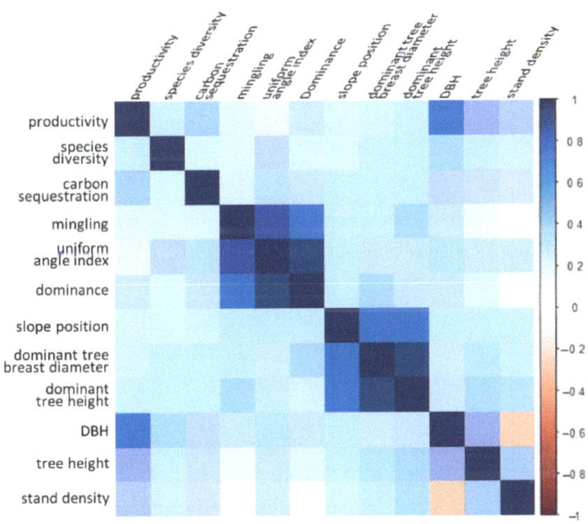

Figure 1. Correlation heat map between structure variables and function variables.

The dominance, mingling and uniform angle index are very significantly correlated. The influence of stand structure indicators on the carbon sequestration function is mainly through aspects such as biomass and tree age. For example, higher tree height and DBH usually imply higher biomass accumulation and higher carbon storage capacity. Due to the conversion relationship between carbon sequestration and productivity, there is a highly significant correlation between them. Forest species diversity is influenced by many factors, and our analysis found a significant correlation between forest species diversity and DBH. Among the possible reasons, higher tree height and DBH may provide more micro-environmental variation in habitat and thus support the presence of more species.

It should be noted that the correlation between stand structure, stand environment and forest productivity, species diversity and carbon sequestration functions is complex and may be influenced by many other factors, such as climate and soil. Therefore, more comprehensive studies and analyses are needed to more accurately assess the correlations between them.

2.2. Path Diagram and Standardized Coefficients in the SEM Analysis

The test results (Table 1) showed that the RMSEA parameter value of the initial model was 0.055, indicating that the model did not adequately fit the observed data, so further modification of the initial model was required. We improved the previous hypothesis, supplemented the relationship between forest species diversity and carbon sequestration, and reconstructed the path map (Figure 2). After calculating the fitting index of the model and comparing this with the detection standards, the Chi-square degrees of freedom ratio of the optimal model was 1.889, which between 1 and 3, indicating a good fit. RMSEA was less than 0.05, which met the evaluation criteria. It is assumed that the optimal model is adapted to the observation data, and that the test indicators of each adaptation statistic have reached the evaluation standard, indicating that the hypothesis model is well adapted to the data [35].

Table 1. Fitting parameters of structural equation model.

Statistics	Fitting Index	Evaluation Standard	A Priori Model	Optimal Model
Absolute fit statistics	χ^2/df	Between 1–3 means the model fits well	1.889	1.889
	GFI	>0.90	0.949	0.935
	RMSEA	<0.05	0.055	0.038
	NCP	The smaller the better	39.995	28.367
Value-added adaptation statistics	NFI	Between 0–1, the closer to 1, the better the model fit	0.948	0.954
	RFI		0.931	0.933
	IFI		0.975	0.978
	TLI		0.962	0.967
	CFI		0.974	0.978
Minimal adaptation statistics	PGFI	>0.5, The higher the value, the better	0.548	0.550
	PNFI	>0.5, The higher the value, the better	0.649	0.651

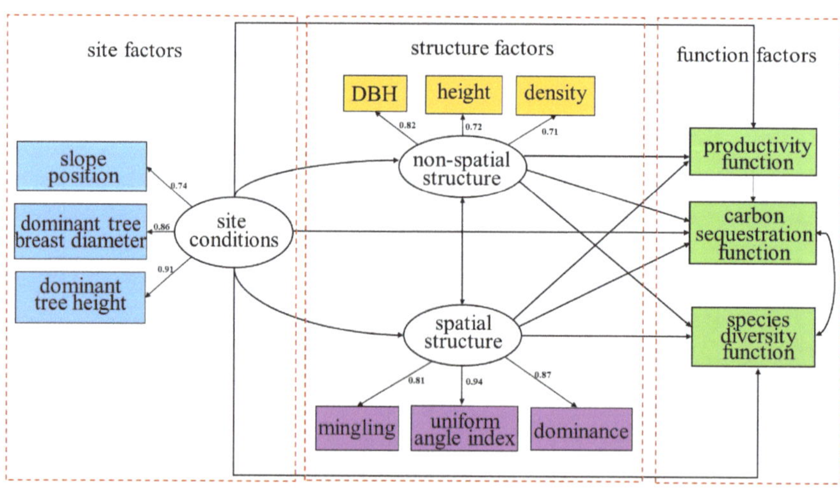

Figure 2. The optimal SEM between stand structures, site conditions and forest functions. The slope position, dominant tree breast diameter and dominant tree height are the explicit variables of site conditions. The DBH, height and density are the explicit variables of non-spatial structure. The mingling, uniform angle index and dominance are the explicit variables of spatial structure.

The SEM showed that site conditions had strongly positive effects on the forest productivity function, carbon sequestration function and species diversity function, and the standardized total effects were 1.221, 0.850 and 0.413, respectively (Figure 3). Site conditions also had an indirect influence on productivity, carbon sequestration and species diversity, and these indirect effects were 0.490, 0.321 and 0.102. Comparing stand structure with site conditions, the site conditions had the largest total impact on forest multi-functionality; the coefficient is 2.484, indicating that the forest functions are mainly affected by site conditions.

Site conditions had direct and negative effects on spatial structure (−0.409) and direct and positive effects on non-spatial structure (0.593).

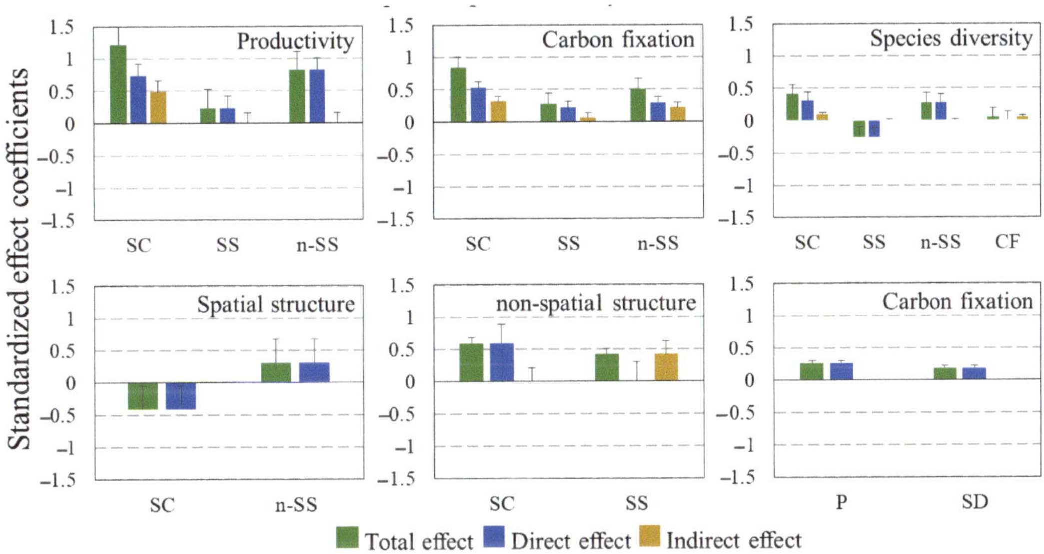

Figure 3. Standardized total, direct and indirect effects of site conditions, spatial structure and non-spatial structure on forest functions. (SC is site conditions, SS is spatial structure, n-SS is non-spatial structure, CF is carbon fixation, P is productivity, SD is species diversity.)

There are interactions between spatial structure and non-spatial structure. Spatial structure had indirect and positive effects on non-spatial structure (0.417), while non-spatial structure had direct and positive effects on spatial structure (0.309). The total impact of spatial structure on forest functions (0.267) is less than that of non-spatial structure (1.615). Spatial structure had direct and positive effects on productivity (0.234) and carbon sequestration (0.220), direct and negative effects on species diversity (−0.249), and indirect and positive effects on carbon sequestration (0.062). Non-spatial structure had direct and positive effects on productivity (0.826), species diversity (0.282), and carbon sequestration (0.290), and indirect and positive effects on carbon sequestration (0.217).

There are also interactions between the multiple forest functions. Forest productivity had direct and positive effects on carbon sequestration (0.263). Plant species diversity had a direct impact on carbon sequestration (0.182). Meanwhile, forest carbon sequestration had an indirect impact on species diversity (0.053).

2.3. Multi-Factors Analysis of Stand Structures, Site Conditions and Forest Functions

Factors of stand structure and site condition can be quantitatively analyzed using some indicators that can be directly measured, and the degree of response is different. The total effect coefficients of the site conditions on the forest non-spatial structure and on the spatial structure are 0.593 and −0.409, respectively, that is, when the site conditions change by 1, the forest non-spatial structure changes by 0.593 and the forest spatial structure changes by −0.409. The non-spatial structure of forest stands has a correlation coefficient of 0.82 for the breast diameter and 0.72 for the tree height, indicating that changes in the non-spatial structure of the forest stands are more likely to affect the DBH [36].

We found that in the current phase, forest productivity was most closely related to site conditions, not tree size (that is, non-spatial structure, DBH and height). At present, the mixed forest of *CLPB* in this study is in the stage of middle-aged and near mature forest, and the influence of tree size (DBH and tree height) on growth is weaker than that of site conditions. That is, the fertile soil and the slope position of the trees played a significant role in the growth of the forest.

Through the analysis of the SEM, we found that plant species diversity had a direct impact on carbon sequestration (0.182). Meanwhile, forest carbon sequestration had an indirect impact on species diversity (0.053). There was a direct relationship between carbon sequestration and productivity, Moreover, carbon sequestration was weakly correlated with species diversity, while species diversity and productivity were not found to be direct related [37,38]. The relationship between species diversity and productivity is affected by species richness, stand type and environmental heterogeneity, and these influencing factors do not exist independently but rather interact with each other [39–41]. Therefore, when studying the relationship between stand functions, it is worth considering the factors that may affect the relationship between them.

3. Discussion

The first hypothesis was strongly supported by our results, which showed that site condition has a greater influence on forest functions of *CLPB* mixed forest than stand structure. Figure 1 indicates that forest functions are directly influenced by site conditions; the intensity of the influence in greatest for productivity, followed by carbon sequestration and species diversity. These findings suggest that environmental factors have a more substantial direct impact on productivity than species diversity and carbon sequestration, indicating that productivity exhibits a more pronounced response to site conditions, which is consistent with Liu's [42] conclusions on the effects of environmental and stand structure factors on productivity. Improvements in site conditions were found to have a more significant impact on forest functions [43].

Our research demonstrates that non-spatial structures have a greater overall impact on forest functions than spatial structures, which was not exactly the same as our second hypothesis. The non-spatial structure has a direct positive effect on forest functions; the intensity of the influence is greatest for productivity, followed by carbon sequestration and species diversity. Forest spatial structure also has a direct effect on forest functions; the intensity of the influence is greatest for carbon sequestration, followed by species diversity and productivity. In addition to the direct effects, stand structure also has an indirect effect on carbon sequestration; the non-spatial structure has a stronger effect than the spatial structure. The indirect impact of the stand structure on productivity and species diversity is zero, and thus, the direct impact coefficient is equal to the total impact coefficient.

The DBH responds more to the non-spatial structure compared to tree height. Geir et al. also found in their study on the relationship between stand density and DBH and tree height that the response of DBH to changes in stand density is greater than that for tree height, which is consistent with the findings of our study [44]. Among the three quantitative indicators of spatial structure, the uniform angle index has the largest response to spatial structure, followed by dominance and then mingling. This is because the mixed forest of *CLPB* in our study has low tree species diversity. Most studies have reported a positive relationship between tree growth and the degree of mixing. Zhang discovered that in stands with greater tree species diversity, the effect of mingling on the stand growth would be enhanced, and the importance of mingling on the spatial structure would increase [45].

4. Materials and Methods

4.1. Study Site

Jindong Forest Farm is located in the southern part of Qiyang County, Hunan Province, China (Figure 4). It is located in the middle and upper reaches of the Xiangjiang River Basin, with dense mountains and steep slopes, with an average slope of 34° and 95.2% of grade IV and above (26° or above). The highest altitude is 1435 m and the lowest is 108 m. The soil of the forest farms is mainly yellow-red and yellow. The thickness of the soil layer is generally more than 60 cm, the gravel content is about 20% to 30%, and the average soil organic matter content is more than 2%, with the highest value reaching 11%.

It belongs to the subtropical southeast monsoon humid climate zone, with an average annual temperature of 18 °C, an extreme maximum temperature of 41 °C and an extreme minimum temperature of −8 °C. The average annual effective sunshine duration is 1617 h, the average annual precipitation is 1600–1890 mm, and the average annual evaporation is 1225 mm. The relative humidity is 75–82%, the annual frost-free period is 265–349 days, and the vegetation has 281–301 natural days [33]. There are 972 species belonging to 135 families in the forest farm. According to the survey, there are over 1500 species of higher plants belonging to more than 200 families. There are 654 species of woody plants in 98 families. At present, *Ginkgo biloba*, *Taxus chinensis* and others are first-class plants under state protection. The second-class protected plants are *Cinnamomum bodinieri*, *Pseudotsuga sinensis*, *Fokienia hodginsii*, *Phoebe bournei*, *Eucommia ulmoides*, and others. There are more than 190 species of terrestrial vertebrates in this area, of which 31 species are in the category of National Key Protected Animals, such as *Syrmaticus ellioti*, *Moschus berezovskii*, *Neofelis nebulosa*, and others.

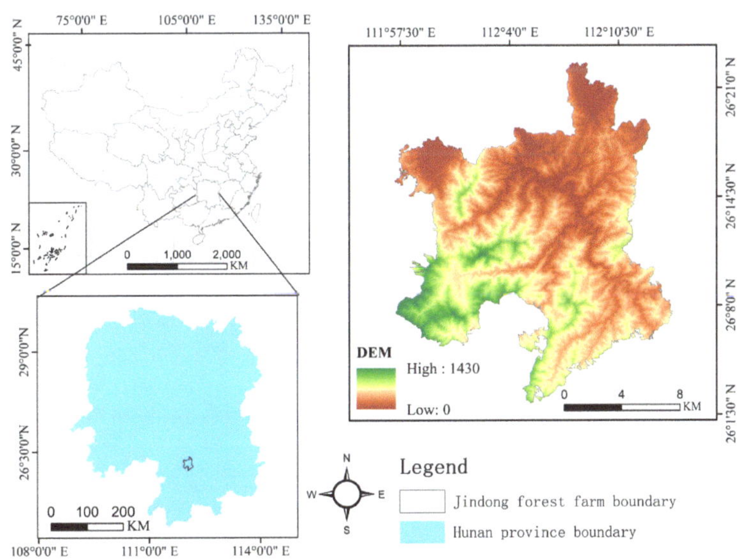

Figure 4. Location of the study site.

4.2. Data Collection

The field survey was conducted from July to August every year from 2015 to 2019. The sample plots were randomly selected from mixed forests of *CLPB* with the same stand phase and age, and were representative (Figure 5). A total of 40 plots (20 × 30 m) were monitored for five consecutive periods. Each sample plot was divided into 6 survey units (10 × 10 m), giving a total of 240 data points. According to the growth and dispersion characteristics of undergrowth vegetation, representative shrub quadrats (5 × 5 m) were set up in the upper, middle and lower sample plots of each plot, and one (1 × 1 m) herb quadrat was set in each shrub quadrat. A total of 120 shrub quadrats and 120 herbaceous quadrats were established in this study (Figure 6). Three soil profiles with a width of 0.8~1.0 m and a depth of 0.6~0.8 m were evenly set in each standard land area, and four samples from layers 0~15 cm, 15~30 cm, 30~45 cm and 45~60 cm deep were obtained with a ring knife, each with a volume of 100 cm^3. Three soil samples were taken from each layer to determine their physical properties.

Figure 5. Our sample plots were replanted with *Phoebe bournei* in the understory of *Cunninghamia lanceolata* trees to create a mixed forest of *CLPB*, changing the stand structure to form a heterogeneous, complex, mixed forest.

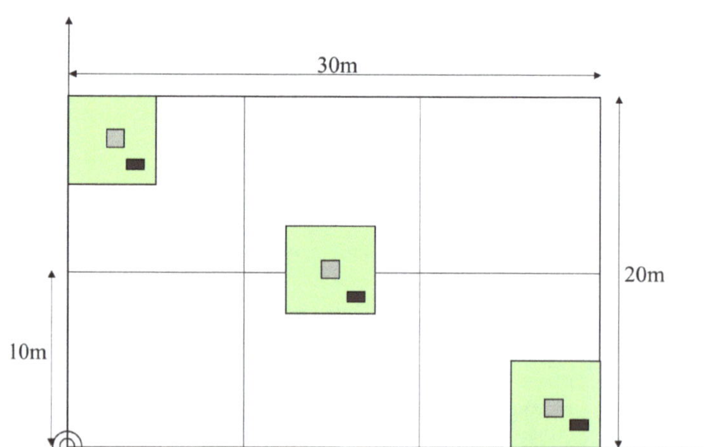

Figure 6. Survey units for trees, shrubs, herbs, and soils.

All trees in the plots with a diameter at breast height (DBH) larger than 5 cm were investigated. Starting from the lower-left corner of each plot, trees in the plot were numbered according to an "S" shape. For each tree, the tree species, tree height (m), DBH (cm) and position were recorded. The species, number, height (m), and coverage of shrubs (%) and herbs in each plot were recorded. The surveyed information of the plots also included altitude (m), terrain, slope (°), position, aspect, and soil type.

4.3. Data Processing

After collecting preliminary data for the forest and land samples, we proceeded to analyze this data to obtain multifunctional index values for the forest stands.

4.3.1. Productivity Measurement

In this study, the index of the stand's productivity was denoted by the live wood stock. The storage capacity of living trees was calculated by the binary volume formula (Table 2) and converted into the storage amount per unit area.

Table 2. Two-dimensional volume table of main trees.

Species	Formula	a	b	c
Cunninghamia lanceolata	$V = a \times D^b \times H^c$	0.000058777042	1.969983	0.896462
Other conifer	$V = a \times D^b \times H^c$	0.000062341803	1.855150	0.956825
One type of hardwood broad-leaf	$V = a \times D^b \times H^c$	0.000068563400	1.933221	0.867885
Second type of hardwood broad-leaf	$V = a \times D^b \times H^c$	0.000050479055	1.88452	0.990765
Soft broad-leaf	$V = a \times D^b \times H^c$	0.000041028005	1.80063	1.130599

Note: D is DBH (cm), H is tree height (m). *Phoebe bournei* is a type of hardwood broad-leaf tree.

4.3.2. Carbon Sequestration of Vegetation and Soil

Forest carbon comprises vegetation and soil carbon reserves. In this study, vegetation carbon was subdivided into tree-layer carbon, undergrowth-layer, shrub-layer and herb-layer carbon and litter-layer carbon. The tree-layer carbon was calculated using the forest biomass allometric growth equation [46–50], while we used the biomass method to estimate forest carbon at all levels (Table 3).

Table 3. Biomass allometric growth equation of different tree species in the tree layer.

Species	Biomass Equation	R^2
Cunninghamia lanceolata	$W_{trunk} = 37.9323D^{2.598}$	0.975
	$W_{branch} = 1.6255D^{2.0074}$	0.764
	$W_{leaf} = 5.2619D^{2.1515}$	0.788
	$\lg W_{root} = -1.995 + 2.45411\lg D$	0.962
Hard broad-leaf class	$W_{trunk} = 0.065D^{2.548}$	0.972
	$W_{branch} = 0.025D^{2.390}$	0.91
	$W_{leaf} = 0.036D^{1.818}$	0.876
	$W_{root} = 0.027D^{2.394}$	0.922
Soft broad-leaf class	$W_{trunk} = 0.080D^{2.348}$	0.995
	$W_{branch} = 0.027D^{1.762}$	0.975
	$W_{leaf} = 0.027D^{1.371}$	0.954
	$W_{root} = 0.027D^{2.165}$	0.873

Note: D is DBH (cm). See references [46–50].

Biomass was measured in the shrub, herb and litter layers using the quadrat all-harvest method. Fresh weights were weighed immediately after harvesting. After returning to the laboratory, the dry biomass was baked at 105 °C for 6–8 h to a constant weight, and the dry biomass was calculated and converted into biomass per unit area. The conversion methods of carbon storage and biomass are as follows:

$$C = BC_C \quad (1)$$

where C represents carbon storage (t); B stands for forest biomass (t); and C_c represents the carbon content, where the carbon conversion coefficient is 0.5 for coniferous forest and 0.45 for broadleaved forest.

Soil carbon storage is represented by the product of soil bulk density, soil thickness and soil organic carbon content. The calculation formula is as follows:

$$TOC = \theta \cdot D \cdot C \tag{2}$$

where *TOC* represents soil carbon storage (t/hm^2), θ represents soil bulk density (g/cm^3), *D* represents soil thickness (cm), and *C* represents soil organic carbon content (%).

4.3.3. Species Diversity

The relative abundance, relative cover (relative significance) and relative frequency of understory vegetation were calculated to represent the important values (IV) of different shrub and herb species in the plots. The species diversity index was calculated based on the species' importance values. In this paper, we use the Patrick richness index (*D*), Shannon–Wiener diversity index (*H*), Simpson dominance index (*H′*) and Pielou uniformity index (*J*) to comprehensively evaluate the species diversity of a community, which is calculated as follows:

$$\text{IV} = (\text{Relative frequency} + \text{relative coverage} + \text{relative abundance})/3 \times 100\% \tag{3}$$

$$H = -\sum_{i=1}^{s} p_i ln p_i \tag{4}$$

$$J = \left(-\sum_{i=1}^{s} p_i ln p_i\right)/lnS \tag{5}$$

$$H' = 1 - \sum_{i=1}^{s} p_i^2 \tag{6}$$

$$D = S \tag{7}$$

where p_i is the ratio of the number of individuals of the *i* species to the total number of individuals, and *S* refers to the total number of species in the sample.

This study uses survey units (10 × 10 m) as data points, so all data refer to the average of one survey unit. All statistical analyses were performed using SPSS 20.0 for Windows, and the SEM was constructed with Amos 22.0 software.

4.4. Structural Equation Model

The structural equation model consists of a measurement model and a structural model [24,51,52]. SEM can analyze the causal relationships between multiple variables in a system and clearly determine the relative importance of each relationship. The model can test the relationship between the explicit variables, latent variables, and error variables of the data, and then obtain the total, direct and indirect effects of the independent variables on the dependent variables [53–55].

Based on the field survey as well as the theoretical and experimental studies conducted in these forest ecosystems [56–59], we established an initial SEM to assess the relative importance of structure variables (spatial structure and non-spatial structure) and site characteristics as the drivers of the response variables to forest functions. In the SEM, three latent variables were established, namely site conditions (determined by observed variables of slope position, dominant tree breast diameter and dominant tree height), spatial structure (determined by observed variables of dominance, mingling and uniform angle index), and non-spatial structure (determined by observed variables of DBH, tree height, and stand density). To test whether these indicators related to the functions of the forest area, Pearson correlation coefficients were calculated before establishing the SEM. The indicators showing significant relationship effects were selected for constructing the SEM. We built an initial model based on the a priori hypothesis (Figure 7), and the model was tested using the goodness of fit index (GFI), comparative fit index (CFI), and root mean square error of approximation (RMSEA) [60,61]. The optimal model should have the lowest RMSEA value < 0.05, and the highest GFI value and CFI value > 0.90.

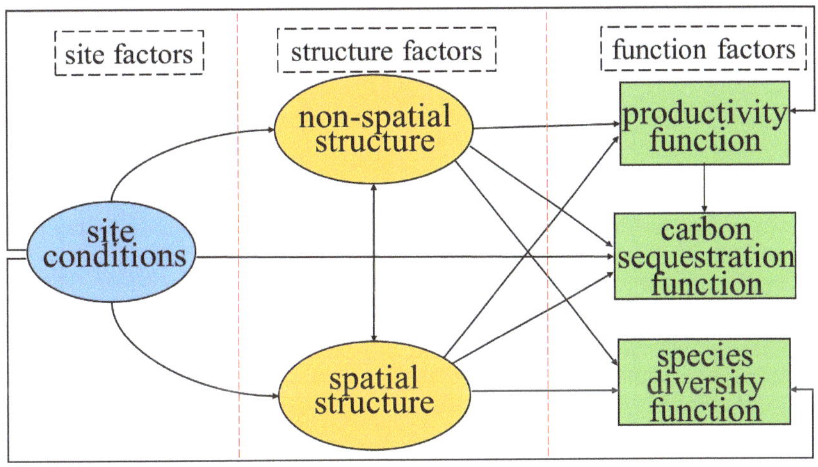

Figure 7. The a priori model based on the hypothesis relating stand structures, site conditions and stand functions. Structure factors include non-spatial structure and spatial structure; function factors include productivity, carbon sequestration and species diversity functions.

5. Conclusions

In this study, we used the SEM to quantify the relative importance of the stand structure and site conditions of *CLPB* to productivity, species diversity and carbon sequestration in Jindong Forest Farm, Hunan Province. Using data from 40 plots, with 240 survey units, we found that site conditions have a greater impact on forest function than stand structure. In addition, compared to spatial structures, non-spatial structures have a greater overall impact on forest functions. The results demonstrated that the close-to-natural forest management practices should be prioritized for pure Chinese fir forests with better site conditions, which can improve the ecological function of pure Chinese fir forests more effectively. In addition, adjusting the stand structure of Chinese fir forests can substantially improve the ecological performance.

Author Contributions: Conceptualization, Y.W., Z.L. and T.T.; methodology, Y.W. and Z.L.; software, Y.W.; validation, T.T. and J.L.; formal analysis, Y.W., Z.L. and T.T.; investigation, Y.W., Z.L., T.T. and J.L.; resources, Y.W., T.T. and J.L.; data curation, Y.W. and T.T.; writing—original draft preparation, Y.W. and Z.L.; writing—review and editing,, T.T. and J.L.; visualization, Y.W. and T.T.; supervision, J.L. and T.T.; project administration, J.L. and T.T.; funding acquisition, J.L. and T.T. All authors have read and agreed to the published version of the manuscript.

Funding: This research was funded by the Scientific Research Project of Education Department of Hunan Province (Project number: 22B0252), the National Forest Management Science and Technology Support Research Project: Research and Demonstration of Multifunctional Forest Management Operation Method in Jindong Forest Farm, Hunan Province (Project number: 169201531-6).

Data Availability Statement: The data from the sample plots in this study are available on request from the corresponding author. Those data are not publicly available due to privacy and confidentiality.

Acknowledgments: We are indebted to Yong Lv for his great support and help in improving our work. We also appreciate Yu Sun, Pengcheng Liao, Xue Wang, Xia Wei, Qian Tang, Danmei Wang, and Yifan Pang for their assistance in the fieldwork. We acknowledge the Jindong Forest Farm in Hunan Province for their support during our research.

Conflicts of Interest: The authors declare no conflict of interest.

References

1. Aubinet, M.; Chermanne, B.; Vandenhaute, M.; Longdoz, B.; Yernaux, M.; Laitat, E. Long term carbon dioxide exchange above a mixed forest in the Belgian Ardennes. *Agric. For. Meteorol.* **2001**, *108*, 293–315. [CrossRef]
2. Lagergren, F.; Eklundh, L.; Grelle, A.; Lundblad, M.; Molder, M.; Lankreijer, H.; Lindroth, A. Net primary production and light use efficiency in a mixed coniferous forest in Sweden. *Plant Cell Environ.* **2005**, *28*, 412–423. [CrossRef]
3. Lu, Y. From Normal Forest to Close-to-nature Forest: Multi-functional Forestry and Its Practice at National, Regional and Forest Management Unit Levels in Germany. *World For. Res.* **2010**, *23*, 1–11. [CrossRef]
4. Song, G.T. Comparison of species diversity between *Larix gmelini* pure forest and *Larix gmelini-betula platyphylla* mixed forest in Daxing'an Mountains. *J. For. Res.* **2001**, *12*, 136–138. [CrossRef]
5. Ghareghiye, Z.N.; Etemed, V. Estimating the economic value of stored carbon in growing stocks of mixed and pure forest stands. *Int. J. Agrisci.* **2013**, *3*, 543–549.
6. Möller, A. *Der Dauerwaldgedanke—Sein Sinn und Seine Bedeutung;* Springer: Berlin/Heidelberg, Germany, 1922; 136p.
7. Gayer, K. *Der Gemischte Wald—Seine Begründung und Pflege, Insbesondere Durch Horst-und Gruppenwirtschaft;* Paul Parey Verlag: Berlin/Heidelberg, Germany, 1886.
8. Schütz, J.P. *Der Plenterwald und Weitere Formen strukturierter und Gemischter Wälder;* Parey: Berlin/Heidelberg, Germany, 2001; 207p.
9. Zhang, Y.T.; Ding, K.; Yrjala, K.; Liu, H.Y.; Tong, Z.K.; Zhang, J.H. Introduction of broadleaf species into monospecific *Cunninghamia lanceolata* plantations changed the soil Acidobacteria subgroups composition and nitrogen-cycling gene abundances. *Plant Soil* **2021**, *467*, 29–46. [CrossRef]
10. Ding, K.; Zhang, Y.T.; Yrjala, K.; Tong, Z.K.; Zhang, J.H. The introduction of *Phoebe bournei* into *Cunninghamia lanceolata* monoculture plantations increased mi-crobial network complexity and shifted keystone taxa. *For. Ecol. Manag.* **2022**, *509*, 120072. [CrossRef]
11. Brassard, B.W.; Chen, H.Y.H. Stand Structural Dynamics of North American Boreal Forests. *Crit. Rev. Plant Sci.* **2006**, *25*, 115–137. [CrossRef]
12. Staudhammer, C.L.; Lemay, V.M. Introduction and evaluation of possible indices of stand structural diversity. *Can. J. For. Res.* **2001**, *31*, 1105–1115. [CrossRef]
13. Ishii, H.T.; Tanabe, S.; Hiura, T. Exploring the Relationships Among Canopy Structure, Stand Productivity, and Bio-diversity of Temperate Forest Ecosystems. *For. Sci.* **2004**, *50*, 342–355. [CrossRef]
14. Yin, G.F.; Li, J.; Liu, Q.H.; Fan, W.L.; Xu, B.D.; Zeng, Y.L.; Zhao, J. Regional Leaf Area Index Retrieval Based on Remote Sensing: The Role of Radiative Transfer Model Selection. *Remote Sens.* **2015**, *7*, 4604–4625. [CrossRef]
15. Zhou, H.M.; He, B.T.; Peng, H.; Shen, B.; Wu, K.L.; Lin, F.; Liu, C.H. Research on Spatial Structure of *Cunninghamia lanceolata* Coppice Forest. *For. Res.* **2015**, *28*, 686–690. [CrossRef]
16. Yin, Z.S.; Sun, C.Z.; Zhao, M.Y. Study on canopy interception model and its parameter characteristics of *Pinus tabu-laeformis* artificial plantation in the loess plateau. *For. Res.* **2015**, *28*, 261–264.
17. Shi, B.K.; Gao, W.F.; Cai, H.Y.; Jin, G.Z. Spatial variation of soil respiration is linked to the forest structure and soil parameters in an old-growth mixed broadleaved-Korean pine forest in northeastern China. *Plant Soil* **2016**, *400*, 263–274. [CrossRef]
18. Hu, Y.J.; Li, J.P.; Cao, X.Y.; Chen, J. Dynamic changes and their relationship of spatial structure and soil's water conservation function in *Cunninghamia lanceolata* forest after stand improvement. *J. Cent. South Univ. For. Technol.* **2018**, *38*, 103–109. [CrossRef]
19. Yasuhiro, K. Effects of disturbance and size structure on the regeneration process in a sub-boreal coniferous forest, northern Japan. *Ecol. Res.* **1995**, *10*, 135–142. [CrossRef]
20. Ryan, M.G.; Stape, J.L.; Binkley, D.; Fonseca, S.; Loos, R.A.; Takahashi, E.N.; Silva, C.R.; Silva, S.R.; Hakamada, R.E.; Ferreira, J.M.; et al. Factors controlling Eucalyptus productivity: How water availability and stand structure alter production and carbon allocation. *For. Ecol. Manag.* **2010**, *259*, 1695–1703. [CrossRef]
21. Pretzsch, H.; Biber, P.; Uhl, E.; Dauber, E. Long-term stand dynamics of managed spruce–fir–beech mountain forests in Central Europe: Structure, productivity and regeneration success. *Forestry* **2015**, *88*, 407–428. [CrossRef]
22. Saha, S.; Rajwar, G.S.; Kumar, M. Forest structure, diversity and regeneration potential along altitudinal gradient in Dhanaulti of Garhwal Himalaya. *For. Syst.* **2016**, *25*, e058. [CrossRef]
23. Zhao, Y.; Duan, X.; Shu, S.H. Relationship between space structure characteristics and site environment of Pinus Yunnanensis secondary forests on Mopan Mountain in the middle of Yunnan, Southern China. *Bulg. Chem. Commun.* **2017**, *49*, 83–88.
24. Wang, D.M.; Li, J.P.; Tang, T. Determining the Optimal Density of Phoebe bournei Plantations Based on Dynamic Programming under Close-to-Nature Management Measures. *Sustainability* **2022**, *14*, 1. [CrossRef]

25. Tang, Q.; Li, J.P.; Tang, T.; Liao, P.C.; Wang, D.M. Construction of a Forest Ecological Network Based on the Forest Ecological Suitability Index and the Morphological Spatial Pattern Method: A Case Study of Jindong Forest Farm in Hunan Province. *Sustainability* **2022**, *14*, 3082. [CrossRef]
26. Grime, J.P. Benefits of plant diversity to ecosystems: Immediate, filter and founder effects. *J. Ecol.* **1998**, *86*, 902–910. [CrossRef]
27. Wang, S.Y.; Mi, F.; Pan, W.J.; Zhang, C.H. The Research on the Influential Factors of Forest Biomass Energy Industry Chain. In Proceedings of the International Conference On New Energy and Renewable Resources, Guangzhou, China, 2 June 2015; pp. 181–189.
28. Yamamoto, T.; Ikeda, S. Relationship between changes of the forests and runoff property in soil and water conservation function reinforcement synthesis model basin: Direction of forest management for improving of water conservation func-tion. *Bull. Hiroshima Prefect. For. Res. Cent.* **2005**, *37*, 15–33.
29. Sherk, J.T.; Fu, W.Y.; Neal, J.C. Site Conditions, Maintenance Costs, and Plant Performance of 10 Extensive Green Roofs in the Research Triangle Area of Central North Carolina. *HortTechnology* **2020**, *30*, 761–769. [CrossRef]
30. Elliott, K.J.; Miniat, C.F.; Pederson, N.; Laseter, S.H. Forest tree growth response to hydroclimate variability in the southern Appalachians. *Glob. Chang. Biol.* **2015**, *21*, 4627–4641. [CrossRef]
31. Yang, X.C.; Liu, D.P.; Fu, Q.; Li, T.X.; Hou, R.J.; Li, Q.L.; Li, M.; Meng, F.X. Characteristics of greenhouse gas emissions from farmland soils based on a structural equation model: Regulation mechanism of biochar. *Environ. Res.* **2022**, *206*, 112303. [CrossRef]
32. Singhai, S.; Singh, R.; Sardana, H.K.; Madhukar, A. Analysis of Factors Influencing Technology Transfer: A Structural Equation Modeling Based Approach. *Sustainability* **2021**, *13*, 5600. [CrossRef]
33. Howard, J.L.; Gagne, M.; Morin, A.J.S.; Forest, J. Using Bifactor Exploratory Structural Equation Modeling to Test for a Continuum Structure of Motivation. *J. Manag.* **2016**, *44*, 2638–2664. [CrossRef]
34. Guo, J.S.; Marsh, H.W.; Parker, P.D.; Dicke, T.; Lüdtke, O.; Diallo, T.M.O. A Systematic Evaluation and Comparison Between Exploratory Structural Equation Modeling and Bayesian Structural Equation Modeling. *Struct. Equ. Model. A Multidiscip. J.* **2019**, *26*, 529–556. [CrossRef]
35. Fan, X.; Thompson, B.; Wang, L. Effects of sample size, estimation methods, and model specification on structural equation modeling fit indexes. *Struct. Equ. Model. A Multidiscip. J.* **1999**, *6*, 56–83. [CrossRef]
36. Santangelo, G.M.; Drubin, D.G. Article-level assessment of influence and translation in biomedical research. *Mol. Biol. Cell* **2017**, *28*, 1401–1408. [CrossRef] [PubMed]
37. Kirkman, L.K.; Mitchell, R.J.; Helton, R.C.; Drew, M.B. Productivity and species richness across an environmental gradient in a fire-dependent ecosystem. *Am. J. Bot.* **2001**, *88*, 2119–2128. [CrossRef] [PubMed]
38. Kahmen, A.; Perner, J.; Audorff, V.; Weisser, W.; Buchmann, N. Effects of plant diversity, community composition and environmental parameters on productivity in montane European grasslands. *Ecosyst. Ecol.* **2005**, *142*, 606–615. [CrossRef]
39. Georges, M.; Anne, F.; Cyril, F.; Bodil, B.; Claire, M.; Stine, H.K.; Ola, M.; Florence, P. 11q13 Alterations in two cases of hibernoma: Large heterozygous deletions and rearrangement breakpoints near GARP in 11q13.5. *Aust. J. Bot.* **2003**, *37*, 389–395. [CrossRef]
40. Caspersen, J.P.; Pacala, S.W. Successional diversity and forest ecosystem function. *Ecol. Res.* **2001**, *16*, 895–903. [CrossRef]
41. Palik, B.J.; Montgomery, R.A.; Reich, P.B.; Boyden, S.B. Biomass growth response to spatial pattern of variable-retention harvesting in a northern Minnesota pine ecosystem. *Ecol. Appl.* **2014**, *24*, 2078–2088. [CrossRef]
42. Liu, Y.; Yu, D.; Xun, B.; Sun, Y.; Hao, R. The potential effects of climate change on the distribution and productivity of *Cunninghamia lanceolatain* China. *Environ. Monit. Assess.* **2014**, *186*, 135–149. [CrossRef]
43. Bustamante, R.; Badano, E.; Pickett, S. Impacts of land use change on seed removal patterns of native and exotic species in a forest landscape. *Community Ecol.* **2012**, *13*, 171–177. [CrossRef]
44. Geir, S.V.; Colin, F.; Loubère, M. Influence of Progeny and Initial Stand Density on the Relationship between Diameter at Breast Height and Knot Diameter of Picea abies. *Scand. J. For. Res.* **1999**, *14*, 470–480. [CrossRef]
45. Zhang, J.; Bruelheide, H.; Chen, X.; Eichenberg, D.; Kröber, W.; Xu, X.; Xu, L.; Schuldt, A. Tree diversity promotes generalist herbivore community patterns in a young subtropical forest experiment. *Oecologia* **2007**, *183*, 455–467. [CrossRef] [PubMed]
46. Ma, Z.Q.; Liu, Q.J.; Xu, W.J.; Li, X.R.; Liu, Y.C. Carbon Storage of Artificial Forest in Qianyanzhou, Jiangxi Province. *Sci. Silvae Sin.* **2007**, *43*, 1–7. [CrossRef]
47. Dai, W.; Fu, W.J.; Jiang, P.K.; Zhao, K.L.; Li, Y.H.; Tao, J.X. Spatial pattern of carbon stocks in forest ecosystems of a typical subtropical region of southeastern China. *For. Ecol. Manag.* **2018**, *409*, 288–297. [CrossRef]
48. Wang, Z.C.; Du, H.; Song, T.Q.; Peng, W.X.; Zhang, H. Allometric models of major tree species and forest biomass in Guangxi. *Acta Ecol. Sin.* **2015**, *35*, 4462–4472. [CrossRef]
49. Luo, Y.; Zhang, X.; Wang, X.; Fei, L. Biomass and its allocation of Chinese forest ecosystems. *Ecology* **2016**, *95*, 2026. [CrossRef]
50. Lu, J.; Feng, J.K.; Zhu, Y. Estimation of Forest Biomass and Carbon Storage in China Based on Forest Resources Inventory Data. *Forests* **2019**, *10*, 650. [CrossRef]
51. Laughlin, D.C.; Abella, S.R.; Covington, W.W.; Grace, J.B. Species richness and soil properties in Pinus ponderosa forests: A structural equation modeling analysis. *J. Veg. Sci.* **2007**, *18*, 231–242. [CrossRef]
52. Schmidt, T.S.; Van, M.P.C.; Carlisle, M.M.; Carlisle, D.M. Linking the Agricultural Landscape of the Midwest to Stream Health with Structural Equation Modeling. *Environ. Sci. Technol.* **2019**, *53*, 452–462. [CrossRef]
53. Feng, M.I.; Pan, W.; Chen, K. Study on Forest Biomass Energy Industry Chain External Driving Force Based on Structural Equation Modeling. *Sci. Technol. Manag. Res.* **2015**, *35*, 128–132. [CrossRef]

54. Zhu, Y.W.; Bathelt, H.; Zeng, G. Learning in Context: A Structural Equation Modeling Approach to Analyze Knowledge Acquisition at Trade Fairs. *Z. Für Wirtsch.* **2020**, *64*, 165–179. [CrossRef]
55. Goodboy, A.K.; Bolkan, S.; Brisini, K.; Solomon, D.H. Relational Uncertainty Within Relational Turbulence Theory: The Bifactor Exploratory Structural Equation Model. *J. Commun.* **2021**, *71*, 403–430. [CrossRef]
56. Zhang, J.H.; Tong, H.; Jiao, G.; Zhao, L. Research on soil organic carbon spatial distribution in Qilian Mountain based on geographic infor-mation system and spatial analysis technology. In Proceedings of the SPIE—The International Society for Optical Engineering, San Diego, CA, USA, 12–16 August 2012; Volume 8334, p. 95. [CrossRef]
57. Tian, X.; Zheng, J.; Wang, D.; Huang, X.R.; Zhang, Z.D. Structural equation modeling analysis of the response of herbaceous species richness to landscape factors in a forest-steppe zone. *Acta Ecol. Sin.* **2018**, *38*, 4649–4656. [CrossRef]
58. Wang, S.L.; Han, Y.H.C. Diversity of northern plantations peaks at intermediate management intensity. *For. Ecol. Manag.* **2010**, *259*, 360–366. [CrossRef]
59. Torresan, C.; del Río, M.; Hilmers, T.; Notarangelo, M.; Bielak, K.; Binder, F.; Boncina, A.; Bosela, M.; Forrester, D.I.; Hobi, M.L.; et al. Importance of tree species size dominance and heterogeneity on the productivity of spruce-fir-beech mountain forest stands in Europe. *For. Ecol. Manag.* **2020**, *457*, 117716. [CrossRef]
60. Wolf, E.J.; Harrington, K.M.; Clark, S.L.; Miller, M.W. Sample Size Requirements for Structural Equation Models: An Evaluation of Power, Bias, and Solution Propriety. *Educ. Psychol. Meas.* **2013**, *73*, 913–934. [CrossRef] [PubMed]
61. Shi, F.; Liu, S.; Sun, Y.; An, Y.; Zhao, S.; Liu, Y.; Li, M. Ecological network construction of the heterogeneous agro-pastoral areas in the upper Yellow River basin. *Agric. Ecosyst. Environ.* **2020**, *302*, 107069. [CrossRef]

plants

MDPI

Article

Do Tree Size and Tree Shade Tolerance Affect the Photosynthetic Capacity of Broad-Leaved Tree Species?

Yuhan Song [1] and Guangze Jin [1,2,3,*]

[1] Center for Ecological Research, Northeast Forestry University, Harbin 150040, China
[2] Key Laboratory of Sustainable Forest Ecosystem Management-Ministry of Education, Northeast Forestry University, Harbin 150040, China
[3] Northeast Asia Biodiversity Research Center, Northeast Forestry University, Harbin 150040, China
* Correspondence: taxus@126.com

Abstract: (1) Background: leaf structure traits are closely related to leaf photosynthesis, reflecting the ability of trees to obtain external resources in the process of growth. (2) Methods: We studied the morphological, chemical, anatomical, stomatal traits and maximum net photosynthetic rate of six broad-leaf species in northern temperate mixed broad-leaved Korean pine (*Pinus koraiensis*) forest. (3) Aim: To investigate whether there are differences in leaf structural traits of trees with different shade tolerances and different sizes and the effects of these differences on leaf photosynthetic capacity. (4) Results: the effects of leaf structure traits on leaf photosynthesis were different among trees with different shade tolerances or different sizes. Under the condition of light saturation, the net photosynthetic rate, nitrogen use efficiency, phosphorus use efficiency and stomatal conductance of shade-intolerant trees or small trees were higher than those of shade-tolerant trees or large trees. (5) Conclusions: the shade tolerance of tree species or the size of trees affect the traits of leaf structure and indirectly affect the photosynthetic ability of plants. When constructing the leaf trait–photosynthesis model, the shade tolerance and tree size of tree species should be taken into account.

Keywords: mixed broad-leaved Korean pine forest; photosynthesis; shade tolerance; tree size; woody plants; functional traits

Citation: Song, Y.; Jin, G. Do Tree Size and Tree Shade Tolerance Affect the Photosynthetic Capacity of Broad-Leaved Tree Species? *Plants* **2023**, *12*, 523. https://doi.org/10.3390/plants12030523

Academic Editor: Cecilia Gotor

Received: 29 December 2022
Revised: 15 January 2023
Accepted: 20 January 2023
Published: 23 January 2023

1. Introduction

The essence of the trait differences of different trees in different life history stages is the trade-off in the capacity of trees to obtain light, nutrients, water and other resources under specific environments and physiological conditions to better grow and reproduce [1,2]. Photosynthesis is one of the most important functions of trees. The construction of an appropriate photosynthesis model is helpful for us to predict the photosynthetic capacity of trees under different conditions. As the most important organ of tree photosynthesis, the structural traits of leaves are closely related to tree photosynthetic capacity [3]. In the past few decades, some studies have explored the relationship between leaf photosynthetic capacity and different leaf structure traits [4–6]. At the same time, some studies have shown that tree shade tolerance or tree size affect the distribution of leaf resources among structural traits [7–9]. However, few studies have linked tree shade tolerance or tree size, leaf structure traits and photosynthesis to explain the supporting mechanism of shade tolerance on plant photosynthesis. The relationship between the structure and function of plant leaves is a research hotspot for ecologists in recent years. The establishment of the relationship model between leaf structure and function is helpful for us to understand the forest ecosystem. Previous studies have pointed out the relationship between some structural traits and the photosynthetic capacity of leaves. For example, some studies have shown that the photosynthetic capacity of leaves is related to the content of nitrogen (N) in leaves. Higher N in leaves usually means that trees can allocate more N to photosynthesis-related enzymes, so plants with higher N in leaves tend to have higher photosynthetic rates [10–12]. Some studies have also constructed a model to estimate the carboxylation rate of the Rubisco enzyme using leaf N, so as to predict the photosynthesis rate of trees [13,14]. However, plant functional traits are affected not only by single structural traits, but also by the synergistic regulation of multiple structural traits [15]. N in leaves is not the only factor affecting leaf photosynthetic capacity. Many studies have shown that many structural characters of plant leaves are closely related to the photosynthetic ability of plant leaves [10,16,17]. The most representative theory is the leaf economic spectrum (LES) theory put forward by Wright et al. in 2004, which holds that trees with higher photosynthetic rates are located at the end of resource acquisition and tend to have higher leaf N, leaf phosphorus content (P) and lower leaf mass per unit area (LMA) [17]. Over the next twenty years, this trade-off reflected by LES theory has been verified by a large number of reports and widely recognized [18,19]. However, some structural traits not included in LES are also closely related to photosynthesis. For example, leaf photosynthesis is usually closely related to leaf anatomical structures; thicker palisade tissue promotes a more uniform distribution of light in leaves and usually means more chlorophyll in leaves [20,21], so palisade tissue thickness (PT) tends to be positively correlated with leaf photosynthetic capacity. Plants can better absorb light through the scattering of cells in sponge tissue, so the sponge tissue thickness (ST) of leaves is also closely related to photosynthesis ability [22]. In addition, as the main channel of CO_2 diffusion, stomata also have an important effect on the photosynthetic ability of leaves. Previous studies have shown that the increase in stomatal conductance (G_s) can increase the photosynthetic rate of leaves when other restrictions are not significant [23,24]. Understanding and quantifying the relationship between plant leaf traits and the photosynthesis rate is of great significance for the establishment of a plant photosynthesis model. However, for plants, the trait–trait relationship is not constant either between species or within species [1,25]. Therefore, it is necessary to explore the intraspecific and interspecific factors that cause the variation of leaf characters, and to understand the effects of leaf characters on photosynthesis under different conditions.

Within the same tree species, there is often variation in leaf traits. Tree size is one of the decisive factors affecting intraspecific variation [18,26]. Tree size affects the difficulty of obtaining light, water, nutrients and other resources [27] and then changes the relationship between the cost and benefit of plant traits [28]. Therefore, the variation in traits within trees of different sizes reflects the trade-off in resource allocation made by trees for better growth. Previous studies have shown that the change in N and P contents in leaves is a typical characteristic with increasing tree size [29]. Higher growth rates usually require higher N and P contents to maintain [30]. A large amount of resources are allocated to leaves for photosynthesis to promote the rapid growth of young trees; as trees grow, more resources are allocated to stems and roots rather than photosynthetic tissues to enhance competitiveness [1,31,32], resulting in lower N and P contents in leaves of larger trees. Additionally, a previous study showed that after N and P addition, differently sized trees had different responses, in which small trees showed higher growth rates, while the growth rate of large trees did not change [33]. This heterogeneous response to nutrient addition reflects the different nutrient utilization strategies of trees of different sizes. In addition to the differences in chemical traits, there are also differences in the leaf morphological characters of trees of different sizes. Even in the same forest, smaller trees may develop larger or thinner leaves than large trees of the same species, or develop leaves earlier than large trees [34,35] to resist the shade of canopy trees and obtain more light. In addition, there are great differences in stomatal conductance between large trees and small trees. With the growth of trees, the height of the tree leads to an increase in water transport resistance from roots to leaves, which leads to stronger water restriction in large trees, so large trees usually have lower stomatal conductance than small ones, thus reducing their own transpiration rate to cope with water restriction [36,37]. The variation of these leaf structure traits can affect the photosynthetic capacity of trees. Therefore, to explore the effect of leaf traits on plant photosynthesis, we cannot ignore the intraspecific variation in leaf traits with increasing plant size.

The variation of leaf traits among different tree species is mainly caused by genetic differences. Shade tolerance is one of the factors affecting the interspecific variation of leaf traits. The difference of shade tolerance among species reflects the response of species to different light environments [38], determines whether species can grow and reproduce in new habitats [39], and has an important influence on the formation of stand structure in the process of natural forest succession [40,41]. There is evidence that shade-tolerant species tend to have wider crowns and more fixed branching patterns than shade-intolerant species, reducing their self-occlusion and helping them better capture light in low light conditions [42]. The main reason why shade-tolerant species have this crown structure is that their growth environment is often more restricted by light than shade-intolerant species, which makes shade-tolerant species different from shade-intolerant species in leaf traits. For example, in order to make up for the higher construction cost, shade-tolerant trees tend to have longer leaf life to maintain long-term photosynthesis. The leaves of shade-tolerant trees also showed higher mechanical strength, lower N and photosynthetic ability [8,43]. Some studies have described the effects of plant shade tolerance on partial leaf traits [9,44]. However, there are few studies on whether the difference of shade tolerance will lead to the variation of leaf anatomical traits. The formation of PT in leaves usually depends on the light environment of the previous year or the light environment of current mature leaves. Usually, the better the previous light conditions, the higher the PT of new leaves [22]. Therefore, we predict that the PT of shade-tolerant species is lower than that of shade-intolerant species. Previous research also supports the hypothesis that trees with higher PT tend to have better light conditions and can better absorb light [45].

At present, there are few studies on the relationship between tree shade tolerance or tree size, leaf structure traits and photosynthesis ability. In order to explore the mechanism of the effect of tree shade tolerance of tree species or tree size on leaf photosynthesis, we selected six different shade-tolerant broad-leaved tree species in mixed broad-leaved Korean pine (*Pinus koraiensis*) forest in Northeast China, measured their leaf structural traits and maximum net photosynthetic rate in two life cycle stages, and explored the variation of leaf traits and their effects on leaf photosynthetic capacity. We tested two interrelated hypotheses: (1) there were significant differences in leaf traits between trees of different sizes and shade tolerances; (2) tree size and shade tolerance can affect leaf photosynthetic capacity indirectly by affecting leaf structure traits.

2. Results

2.1. Variations in Leaf Traits of Different Tree Types

Across all tree species, the range of epidermis thickness (ET) was 16.81–40.62 μm, the range of palisade tissue thickness (PT) was 25.58–103.93 μm, and the range of spongy tissue thickness (ST) was 18.94–84.16 μm (Table S1). All anatomical traits except ST were significantly different between shade-tolerant trees and shade-intolerant trees (Table 1). The PT, palisade–spongy tissue ratio (PT/ST) and palisade tissue–leaf thickness ratio (PT/LT) of shade-intolerant trees were significantly higher than those of shade-tolerant trees, while the ET and spongy tissue-leaf thickness ratio (ST/LT) were significantly lower than those of shade-tolerant trees. There was no significant difference in ST between shade-intolerant and shade-tolerant trees (Table 1). The maximum net photosynthetic rate based on mass (P_n), carbon content (C), nitrogen content (N), utilization efficiency of photosynthetic nitrogen (PNUE), phosphorus content (P), utilization efficiency of photosynthetic phosphorus (PPUE) and stomatal conductance based on mass (G_s) of shade-intolerant trees were higher than those of shade-tolerant trees, while the SLA of shade-intolerant trees was lower than that of shade-tolerant trees (Table 1).

Table 1. T-test results of leaf traits between shade-intolerant trees and shade-tolerant tree groups.

Leaf Traits	Shade-Tolerant Groups		Leaf Traits	Shade-Tolerant Groups	
	T	*p*-Value		T	*p*-Value
P_n	7.387	**<0.001**	PT/LT	13.756	**<0.001**
SLA	−3.901	**<0.001**	ST/LT	−5.598	**<0.001**
Gs	5.331	**<0.001**	C	9.410	**<0.001**
ET	−4.379	**<0.001**	N	3.504	**0.001**
PT	11.435	**<0.001**	P	8.375	**<0.001**
ST	−0.036	0.971	PNUE	5.687	**<0.001**
PT/ST	9.572	**<0.001**	PPUE	3.620	**<0.001**

Bold indicates a significant difference ($p < 0.05$). P_n, maximum net photosynthetic rate based on mass; ET, epidermis thickness; PT, palisade tissue thickness; ST, spongy tissue thickness; PT/ST, palisade–spongy tissue ratio; PT/LT, palisade tissue–leaf thickness ratio; ST/LT, spongy tissue–leaf thickness ratio; C, carbon content; N, nitrogen content; P, phosphorus content; PNUE, utilization efficiency of photosynthetic nitrogen; PPUE, utilization efficiency of photosynthetic phosphorus; (PPUE) SLA, specific leaf area; G_s, stomatal conductance based on mass.

Tree species had significant effects on all leaf traits ($p < 0.05$); species with strong shade tolerance tended to have lower P_n, Gs and higher SLA (Figure S1, Table 2). Tree size had significant effects on all leaf traits except PT/ST, PT/LT and ST/LT ($p < 0.05$). The P_n, specific leaf area (SLA), G_s, N, P, PNUE and PPUE of small trees were significantly higher than those of large trees, while the ET, PT, ST and C of small trees were significantly lower than those of large trees (Table 2). The interaction between tree species and tree size had significant effects on P_n, SLA, G_s, PT, ST, N, P and PNUE (Table 2).

Table 2. Effect of species and tree size on leaf traits.

Source	P_n		SLA		G_s	
	F	P	F	P	F	P
Species	34.664	**<0.001**	20.054	**<0.001**	15.391	**<0.001**
Tree size	87.059	**<0.001**	48.959	**<0.001**	49.800	**<0.001**
Species × Tree size	6.868	**<0.001**	7.844	**<0.001**	7.454	**<0.001**
	ET		**PT**		**ST**	
	F	P	F	P	F	P
Species	29.440	**<0.001**	79.129	**<0.001**	70.469	**<0.001**
Tree size	9.296	**0.003**	14.469	**<0.001**	17.348	**<0.001**
Species × Tree size	2.043	0.079	3.077	**0.012**	2.704	**0.024**
	PT/ST		**PT/LT**		**ST/LT**	
	F	P	F	P	F	P
Species	137.328	**<0.001**	126.407	**<0.001**	108.730	**<0.001**
Tree size	0.133	0.716	0.270	0.604	1.476	0.227
Species × Tree size	0.471	0.797	0.426	0.830	1.043	0.396
	C		**N**		**P**	
	F	P	F	P	F	P
Species	155.950	**<0.001**	4.792	**0.001**	26.268	**<0.001**
Tree size	18.293	**<0.001**	4.074	**0.046**	8.894	**0.004**
Species × Tree size	1.058	0.388	3.761	**0.004**	4.505	**0.001**
	PNUE		**PPUE**			
	F	P	F	P		
Species	16.893	**<0.001**	8.050	**<0.001**		
Tree size	46.159	**<0.001**	40.123	**<0.001**		
Species × Tree size	3.647	**0.004**	1.807	0.118		

Bold indicates a significant difference ($p < 0.05$). P_n, maximum net photosynthetic rate based on mass; ET, epidermis thickness; PT, palisade tissue thickness; ST, spongy tissue thickness; PT/ST, palisade–spongy tissue ratio; PT/LT, palisade tissue–leaf thickness ratio; ST/LT, spongy tissue–leaf thickness ratio; C, carbon content; N, nitrogen content; P, phosphorus content; PNUE, utilization efficiency of photosynthetic nitrogen; PPUE, utilization efficiency of photosynthetic phosphorus; (PPUE) SLA, specific leaf area; G_s, stomatal conductance based on mass.

The two principal axes of principal component analysis can explain 63.1% and 16.7% of the total variance, respectively. With the increase in tree size, the leaf traits developed along the direction of SLA decrease, and with the increase in shade tolerance, leaf traits developed along the direction of G_s decrease (Figure 1).

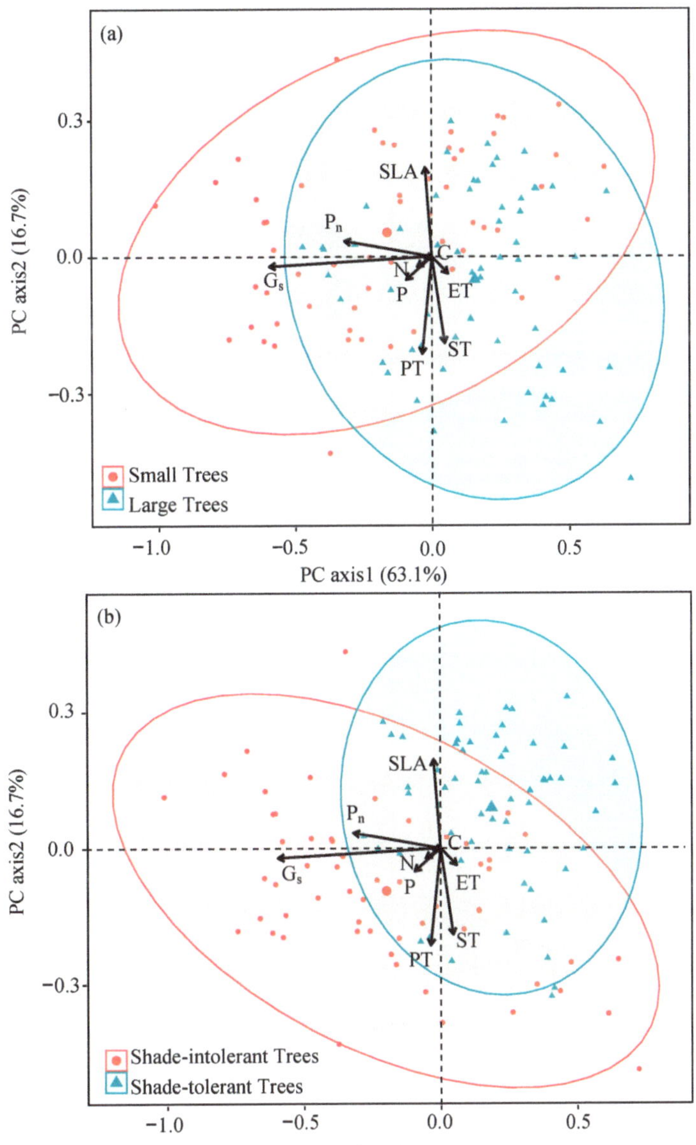

Figure 1. Principal component analysis of 8 leaf traits of different tree sizes or shade-tolerances. The meanings of abbreviations are shown in Table 2. (**a**) Distribution of trees of different sizes with respect to leaf traits; (**b**) Distribution of different shade tolerance trees with leaf traits.

2.2. Relationship between Leaf Structural Traits and Leaf Net Photosynthetic Rate

The results of HP analysis showed that although the morphological, anatomical, chemical and stomatal traits of leaves contributed to the variation of P_n, the contribution rate of G_s to the total variation of P_n was the largest, which was much higher than that of other traits to P_n variation (Figure 2). In different groups, the chemical traits of leaves made a greater contribution to the P_n of small trees and shade-intolerant groups, while the anatomical traits of leaves made a greater contribution to the P_n of large trees and shade-tolerant species (Figure 2).

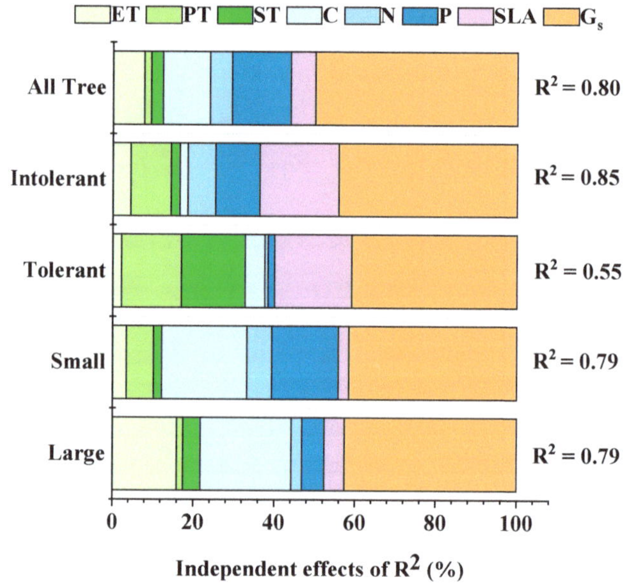

Figure 2. Contribution of leaf traits to P_n variation. The meanings of abbreviations are shown in Table 2.

In the SEM, both tree size and shade tolerance had indirect effects on leaf P_n through leaf structural traits (Figure 3), while leaf shade tolerance had a negative direct effect on P_n (Figure 3b). The increase in tree size had a significant negative direct effect on G_s and SLA (Figure 3a). With the change in tree size, G_s had a direct effect on P_n. ET and N not only have a direct effect on P_n, but also have an indirect effect on P_n by directly affecting Gs (Figure 3a). The increase in tree size had a significant negative direct effect on G_s, N and SLA (Figure 3b). SLA and G_s had a direct effect on P_n, PT and N, not only has a direct effect on P_n, but also has an indirect effect on P_n by directly affecting G_s (Figure 3b).

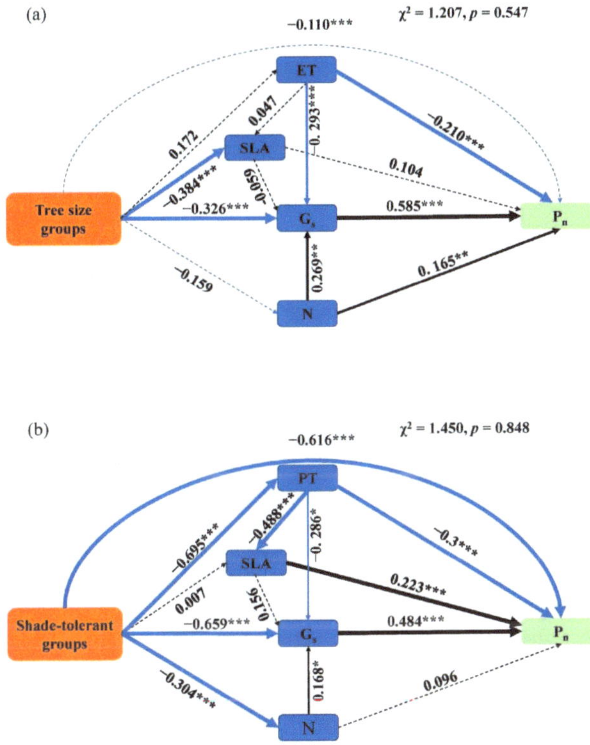

Figure 3. Structural equation model of tree type influencing P_n. The solid lines indicate significant paths, and the thickness of the solid line represents the degree of significant correlation (*, $p < 0.05$; **, $p < 0.01$; ***, $p < 0.001$.), while the dotted lines indicate nonsignificant paths ($p > 0.05$). The meanings of abbreviations are shown in Table 2. (**a**) Structural equation model of tree sizes influencing P_n; (**b**) Structural equation model of shade tolerances influencing P_n.

3. Discussion

3.1. Variations in Leaf Traits between Different Tree Size Groups or Shade-Tolerant Groups

Tree species had significant effects on all leaf traits ($p < 0.05$); species with strong shade tolerance tended to have lower P_n, Gs and higher SLA (Figure S1, Table 2). The results showed that the photosynthetic capacity of different tree species was different, and photosynthetic capacity was related to shade tolerance. Similarly, according to the relationship between leaf traits of shade-tolerant species and shade-intolerant species in Table 1, as well as the relationship between leaf traits of different tree species in Figure S2, we can assume that the interspecific differences represented by different shade tolerance tree species in this study are similar to the interspecific variation based on genetic differences.

Consistent with our first hypothesis, there were significant differences in leaf traits among trees of different shade tolerances in this study (Table 1). Our study shows that shade tolerance is a good predictor of leaf anatomical traits. Most of the anatomical traits are significantly different between different shade-tolerant trees. Shade-intolerant species tend to have thicker PT and larger PT/LT (Table 1), which is in line with our expectations. Shade-intolerant trees live in a better light environment, and thicker PT can guide the leaves to better absorb light [46]. The PT/ST of shade-intolerant trees is higher, which is beneficial to better gas exchange in leaves [47], and can further enhance their photosynthetic ability. The shade-intolerant trees had lower ET than shade-tolerant trees. This represents a trade-off in the allocation of plant leaf resources. ET is often related to plant drought resistance [48,49]. Shade-intolerant trees allocate more resources to palisade tissue, which means that shade-intolerant species tend more towards obtaining resources rather than conservative growth. This growth strategy accordingly sacrifices part of the investment in the epidermis that helps to resist stress [9]. Therefore, when the water is sufficient, the shade-intolerant trees have higher carbon assimilation ability and can achieve faster growth, but when the drought is serious, the shade-tolerant trees have a stronger ability to resist water stress.

Similar to previous studies, the N and P of shade-intolerant trees were higher than those of shade-tolerant species (Table 1) in this study. This is because shade-intolerant trees are at the end of resource acquisition in LES. The higher N in the leaves, the greater the carbon gain without light restriction. Shade-tolerant trees are located at the conservative end of resource use. Compared with shade-intolerant trees, their maximum carbon yield is smaller, but they have better tolerance to light resource stress, so they have a larger niche [50]. The G_s of shade-intolerant trees is higher than that of shade-tolerant trees. There are three possible explanations: (1) The availability of N can affect the stomatal response. G_s in plant leaves usually increases with the increase in N content [51]; (2) G_s is anatomically regulated, and the PT/ST of leaves is inversely proportional to the number of intercellular spaces, so the higher the PT/ST of leaves, the more need for gas exchange, which will indirectly lead to the increase in G_s [47]; (3) The shade tolerance itself will affect the opening and closing of stomata. For example, trees with shade tolerance can open stomata in poor light conditions, and stomata can be opened faster under bright spots than shade-intolerant trees [52]. Therefore, we boldly predict that under the condition of light saturation, stomatal conductance decreases with the increase in shade tolerance, because shade-tolerant trees enhance their ability to absorb carbon dioxide in low light at the expense of maximum gas exchange capacity.

In contrast to LES theory, SLA in this study did not decrease with the increase in shade tolerance (Table 1). We think this may be because this study was conducted in broad-leaved deciduous species; in this case, lower SLA cannot increase the leaf life span of shade-tolerant trees, and the leaf-cost recovery time cannot be prolonged, so shade-tolerant species tend to choose lower quality leaf area in order to reduce the leaf construction cost [53].

The leaf N, P, PNUE and PPUE of trees of different sizes had significant differences ($p < 0.05$), and the four traits of small trees were significantly higher than those of large trees (Table 2). This result not only further verified the LES theory, but also showed that the photosynthetic capacity of trees at different growth stages not only depended on the changes of N and P content in leaves, but also on the utilization efficiency of nitrogen and phosphorus by the photosynthetic structure. Tree size had significant effects on ET, PT and ST ($p < 0.05$), but had no effect on PT/ST, PT/LT and ST/LT (Table 2), indicating that the anatomical trait of leaf thickness changed with the growth of trees, but its proportion in leaves was stable. In this study, the G_s of large trees was significantly lower than that of small trees ($p < 0.05$) (Table 2) Trees respond to changes in water availability by adjusting stomata [54]. According to the hydraulic limitation hypothesis, the hydraulic resistance increases with the increase in tree height, so large trees are more susceptible to drought than small ones [55,56]. To address this hydraulic limitation, large trees will reduce the G_s of leaves to form a compensation mechanism to prevent the plant from losing water too fast [36,37]. The compensation mechanism of large trees for water stress is also reflected in the morphological structure (SLA) (Table 1). To compensate for water stress, the leaves of large trees need to reduce the water potential to obtain enough water, so the leaves of tall trees may increase the investment of solutes such as starch and lipids, resulting in a decrease in SLA [57,58].

We found that the differences in the intraspecific and interspecific traits of leaves were driven by different strategies, and the differences of intraspecific traits among trees of different sizes were driven by SLA (Figure 1). Small trees showed a resource acquisition strategy with high SLA, while large trees tended to grow conservatively. The difference of interspecific traits among different shade-tolerant groups is driven by G_s (Figure 1). The shade-intolerant trees are in the stage of high carbon acquisition, while the shade-tolerant trees are at the carbon-conservative end. The coefficient of variation of each leaf trait also showed that, on the whole, the leaf traits of small trees and shade-intolerant species fluctuated greatly, while those of large trees and shade-tolerant species were more stable (Table S1).

3.2. Effects of Tree Size and Shade Tolerance on P_n

Our study shows that the P_n of shade-tolerant species is lower than that of shade-intolerant species (Table 2). The photosynthetic capacity of plant leaves is mainly determined by two factors: (1) the biochemical carbon sequestration ability of leaves, and (2) the CO_2 concentration in chloroplasts [12], in which the biochemical carbon sequestration capacity of leaves is mainly related to the maximum carboxylation rate, and the concentration of CO_2 in chloroplasts is usually determined by the diffusion rate of CO_2. In this study, the main leaf traits affecting the maximum carboxylation rate of leaves were PT, N and P, because these three traits could affect the content of chlorophyll and Rubisco in leaves, or directly participate in the process of photosynthesis [4,59]. The main leaf trait affecting CO_2 diffusion rate was G_s, because G_s is the main factor affecting gas exchange inside and outside leaves [24]. SLA affects these two factors at the same time; on the one hand, SLA is related to the percentage of cell wall in leaves, and leaves with low SLA usually have a large proportion of cell wall and stratum corneum, which increases the difficulty of carbon dioxide diffusion [5]. On the other hand, SLA reflects the trade-off between leaf mass and area, and higher SLA often indicates that leaves tend to use larger leaf area to obtain carbon resources [60]. For all types of trees, the contribution rate of G_s to P_n variation is the highest (Figure 2), indicating that the photosynthetic ability of plant leaves is mainly limited by the ability to obtain CO_2. However, for different types of trees, the contribution of other leaf traits to the photosynthetic capacity of the tree is also different. The chemical traits of leaves make a greater contribution to the photosynthetic ability of small trees and shade-intolerant species, while the anatomical traits of leaves make a greater contribution to the photosynthetic capacity of large trees and shade-tolerant species (Figure 2). This is because small trees and shade-intolerant trees are at the end of resource acquisition of LES, and more nutrients are put into the leaves to promote the rapid growth of trees. Large trees and shade-tolerant trees are at the conservative end of resource acquisition and pay more attention to the investment in leaf toughness in order to obtain stronger resistance. The results of HP showed that the variation of P_n depended not only on the single leaf structure trait, but also was commonly limited by the morphological, chemical, anatomical and stomatal traits of leaves.

Shade tolerance of trees can directly affect leaf photosynthetic capacity, but also indirectly affect leaf photosynthetic capacity by affecting leaf structural traits (Figure 3b). Shade-intolerant trees can enhance G_s, improve gas exchange capacity [61], increase the investment in leaf N, increase the maximum carboxylation rate of leaves [12], and further improve the efficiency of carbon sequestration. Shade-tolerant trees adopt a more conservative strategy of resource acquisition, devoting more resources to supporting organs such as roots and branches [62], in order to enhance the competitiveness of trees. The specific manifestation of this resource trade-off is that the N and PT of shade-tolerant trees are lower than that of shade-intolerant trees. Similarly, compared with small trees, large trees adopt a more conservative resource acquisition strategy, which reduces the area receiving light per unit mass (SLA) and carbon dioxide absorption capacity (G_s) (Figure 3a), devotes more resources to non-photosynthetic tissues [31,32], reduces the growth rate of large trees, but improves the competitiveness of large trees.

Our study explains how shade tolerance and tree size affect tree photosynthesis indirectly by affecting leaf structure traits on a local scale, which makes up for the data gaps in related fields and provides new evidence for the relationship between tree traits. However, it is undeniable that all our experiments were conducted on a limited number of tree species in broad-leaved Korean pine forests in northern China, which means limitations, because leaf traits of trees usually vary with different plant functional groups or environmental conditions [19,45]. In future research, we hope to make up for the deficiency of the current research through more experiments.

4. Materials and Methods

4.1. Research Site

All research was performed in the Liangshui National Nature Reserve (47°6′~47°16′ N, 128°47′~128°57′ E) in Heilongjiang Province, Northeast China. The site is a hilly region with a temperate continental monsoon climate. The mean annual temperature is −3 °C, and the mean annual precipitation and evaporation are 676 mm and 805 mm, respectively. The zonal soil of the region is dark brown forest soil, and the zonal vegetation of the region is mixed broad-leaved Korean pine (*Pinus koraiensis*) forest, and the dominant tree species are *Pinus koraiensis, Abies nephrolepis, Acer pictum* subsp. *mono, Betula platyphylla* and *Fraxinus mandschurica* [63,64].

4.2. Sampling

Six major broadleaf species were selected at the sampling site, including *Acer pictum* subsp. *mono, Acer tegmentosum, Betula platyphylla, Fraxinus mandschurica, Juglans mandshurica* and *Ulmus laciniata*. Six tree species were divided into two types (shade-tolerant trees and shade-intolerant trees) based on their relative shade tolerance, among which shade-tolerant trees included *Acer pictum* subsp. *mono, Acer tegmentosum, Ulmus laciniata*, and shade-intolerant trees included *Betula platyphylla, Fraxinus mandschurica, Juglans mandshurica* [65,66]. Each tree species was divided into two sizes based on tree height (these trees are of different ages, so they are of different sizes). For each species and each size, ten individuals were randomly sampled in August 2021, including 7 small individuals from *Juglans mandshurica*, for a total of 117 sample trees (the basic traits of the trees are shown in Table 3). The sampling lasted for a total of 12 days and was only taken on sunny days from 8 to 11 o 'clock. All trees were sampled in the upper canopy on the sunny side. The mature leaves were collected from the sunny side of each sample tree for the determination of photosynthetic rate. The light intensity of the Li-6400 (LI-COR, Lincoln, USA) photosynthesis system was set to 1500 ($\mu mol \cdot m^{-2} \cdot s^{-1}$), and the concentration of CO_2 was set to 400 ($\mu mol\ mol^{-1}$). According to past experience, under these conditions, the leaf can reach the maximum net photosynthetic rate [6]. Three leaves of each tree were selected to measure the area-based maximum net photosynthetic rate ($P_{n\text{-}area}$, $\mu mol\ m^{-2}\ s^{-1}$) and stomatal conductance ($G_{s\text{-}area}$, $\mu mol\ m^{-2}\ s^{-1}$). Photosynthetic traits were measured in the field. The leaves with measured photosynthetic traits were kept fresh and sent to the laboratory as soon as possible to measure their morphological traits. Ten leaves were selected from each individual and preserved in formalin-acetic acid-alcohol (FAA) solution for the analyses of anatomical traits [67]. The rest of the leaves were dried to constant weight in a baking oven at 65 °C and were used to measure chemical properties.

Table 3. Means and standard errors (SE) of basic traits of trees in mixed broad-leaved Korean pine forests.

Species	Small Tree		Large Tree	
	DBH (cm)	H (m)	DBH (cm)	H (m)
All species	3.16 ± 0.12	4.11 ± 0.16	26.15 ± 1.19	19.18 ± 0.63
Acer pictum subsp. *mono*	2.74 ± 0.27	4.04 ± 0.32	24.22 ± 1.98	16.76 ± 1.16
Acer tegmentosum	3.1 ± 0.35	3.92 ± 0.15	14.20 ± 1.25	13.13 ± 1.01
Ulmus laciniata	2.58 ± 0.12	3.38 ± 0.21	26.18 ± 3.05	16.75 ± 0.61
Betula platyphylla	4.05 ± 0.16	5.40 ± 0.43	28.47 ± 2.25	21.00 ± 0.95
Fraxinus mandschurica	2.75 ± 0.27	3.77 ± 0.44	33.14 ± 2.60	24.19 ± 1.22
Juglans mandshurica	3.89 ± 0.37	4.36 ± 0.52	31.22 ± 2.20	23.62 ± 0.88

DBH: diameter at breast height of tree; H: height of tree.

4.3. Leaf Trait Measures

4.3.1. Leaf Morphological Traits

The leaves were scanned into pictures by a Canon LiDE 400 scanner (Canon Inc., Tokyo, Japan), and the leaf area calculation program was used to calculate the leaf area (LA, cm^2) through the pixels of the picture. Then the leaf samples were dried in a baking oven at 65 °C to constant weight, and the leaf dry weight (DW, g) was measured with an electronic balance (accuracy 0.0001 g). The formula for calculating LMA (g m^{-2}) and SLA (m^2 g^{-1}) are as follows:

$$\mathrm{LMA} = 10{,}000 \times \mathrm{DM}/\mathrm{LA} \tag{1}$$

$$\mathrm{SLA} = 1/\mathrm{LMA} \tag{2}$$

The formula for converting leaf $P_{n\text{-area}}$ and $G_{s\text{-area}}$ into mass-based maximum net photosynthetic rate (P_n, μmol g^{-1} s^{-1}) and stomatal conductance (Gs, μmol g^{-1} s^{-1}) is as follows:

$$P_n = P_{n\text{-area}}/\mathrm{LMA} \tag{3}$$

$$G_s = G_{s\text{-area}}/\mathrm{LMA} \tag{4}$$

4.3.2. Leaf Chemical Traits

Leaf samples used for measuring chemical traits were oven-dried to constant weight at 65 °C for grinding. The total nitrogen content (N, mg g^{-1}) and total phosphorus content (P, mg g^{-1}) of leaf samples were measured by CleverChem380 (DeChem-Tech. GmbH, Hamburg, Germany) automatic discontinuous chemical analyzer after H_2SO_4-H_2O_2 digestion. The total carbon contents (C, mg g^{-1}) of leaf samples were measured by multiN/C3000 (Analytik Jena AG, Jena, Germany) carbon and nitrogen element analyzer.

The formulas for calculating the utilization efficiency of photosynthetic nitrogen (PNUE) and phosphorus (PPUE) are as follows:

$$\mathrm{PNUE} = P_n/\mathrm{N} \tag{5}$$

$$\mathrm{PPUE} = P_n/\mathrm{P} \tag{6}$$

4.3.3. Leaf Anatomical Traits

Three intact leaves per individual were selected from the FAA solution, and the leaf sections with a thickness of 6 μm were obtained by the paraffin section technique and stained [9]. The leaf sections were observed by a light microscope (Olympus Electronics, Inc., Tsukuba, Japan), and photographed by cellSens Standard 1.11 software (Olympus Electronics Inc., Tsukuba, Japan) and measured by ImageJ 1.53a software (National Institutes of Health, Bethesda, Rockville, MD, USA). Three photographs were selected per leaf section for measuring the adaxial epidermis thickness (AD, μm), abaxial epidermis thickness (AB, μm), PT (μm) and ST (μm). The summed value of AT and AB was taken as epidermis thickness (ET, μm).

4.4. Data Analysis

An independent samples t-test in SPSS 21.0 was applied to examine the differences in leaf traits between shade-tolerant trees and shade-intolerant trees. The influence of tree species and tree sizes on leaf traits were tested by two-factor analysis of variance, and the influence of tree species on leaf traits was tested after the fact. The following data analyses were carried out in R 4.2.1. Principal component analysis was used to analyze the relationship among leaf traits. The 'hier.part' package was used to perform hierarchical partitioning (HP) analysis of leaf traits and quantify the explanation rate of different traits to P_n [68]. The effect of shade tolerances and tree sizes and leaf traits on P_n were studied by structural equation modeling (SEM) constructed by the 'lavaan' package [69]. All traits were log-transformed before calculation.

5. Conclusions

With the increase in shade tolerance of tree species or tree size, the net photosynthetic rate of tree leaves decreased. This variation is not determined by single leaf traits, but caused by the differences of many leaf traits. Generally speaking, the N, P (or their use efficiency) and G_s in the leaves of shade-intolerant species or small trees are higher, which means that shade-intolerant species or small trees have more nutrients to provide for photosynthesis, and at the same time, the resistance of CO_2 diffusion in leaves is lower, which is helpful to improve the photosynthetic capacity of leaves. At the same time, the leaves of shade-tolerant trees usually have higher epidermal tissue thickness, although the photosynthetic ability of plants is further reduced, but the greater leaf toughness makes their leaves better resistant to external water stress and physical damage, and enhance their competitiveness. Therefore, the shade tolerance of tree species and tree sizes plays a key role in the variation of leaf photosynthetic capacity, and should be considered in the study of leaf photosynthetic capacity.

Supplementary Materials: The following supporting information can be downloaded at: https://www.mdpi.com/article/10.3390/plants12030523/s1, Table S1: Basic information table for leaf traits; Figure S1: Variation in leaf traits between different tree species.

Author Contributions: Y.S.: Data Curation, Formal analysis, Investigation, Methodology, Writing–Original Draft; G.J.: Conceptualization, Funding acquisition, Resources, Supervision, Writing–Review and Editing. All authors have read and agreed to the published version of the manuscript.

Funding: This work was supported by the National Natural Science Foundation of China (No. 32071533) and the Fundamental Research Funds for the Central Universities (grant no. 2572022DS13).

Institutional Review Board Statement: Not applicable.

Informed Consent Statement: Not applicable.

Data Availability Statement: The data presented in this study are available upon request from the corresponding author. The data are not publicly available due to privacy reasons.

Acknowledgments: We thank all team members for their support and help in sample collection. We also thank the reviewers and editors for helpful comments that led to an overall improvement of the manuscript.

Conflicts of Interest: The authors declare no conflict of interest.

References

1. Gibert, A.; Gray, E.F.; Westoby, M.; Wright, I.J.; Falster, D.S.; Wilson, S. On the link between functional traits and growth rate: Meta-analysis shows effects change with plant size, as predicted. *J. Ecol.* **2016**, *104*, 1488–1503. [CrossRef]
2. Falster, D.S.; Duursma, R.A.; FitzJohn, R.G. How functional traits influence plant growth and shade tolerance across the life cycle. *Proc. Natl. Acad. Sci. USA* **2018**, *115*, E6789–E6798. [CrossRef] [PubMed]
3. Coble, A.P.; Cavaleri, M.A. Vertical leaf mass per area gradient of mature sugar maple reflects both height-driven increases in vascular tissue and light-driven increases in palisade layer thickness. *Tree Physiol.* **2017**, *37*, 1337–1351. [CrossRef] [PubMed]
4. Ellsworth, D.S.; Crous, K.Y.; Lambers, H.; Cooke, J. Phosphorus recycling in photorespiration maintains high photosynthetic capacity in woody species. *Plant Cell Environ.* **2015**, *38*, 1142–1156. [CrossRef] [PubMed]
5. Onoda, Y.; Wright, I.J.; Evans, J.R.; Hikosaka, K.; Kitajima, K.; Niinemets, U.; Poorter, H.; Tosens, T.; Westoby, M. Physiological and structural tradeoffs underlying the leaf economics spectrum. *New Phytol.* **2017**, *214*, 1447–1463. [CrossRef] [PubMed]
6. Mo, Q.F.; Li, Z.A.; Sayer, E.J.; Lambers, H.; Li, Y.W.; Zou, B.; Tang, J.W.; Heskel, M.; Ding, Y.Z.; Wang, F.; et al. Foliar phosphorus fractions reveal how tropical plants maintain photosynthetic rates despite low soil phosphorus availability. *Funct. Ecol.* **2019**, *33*, 503–513. [CrossRef]
7. Woodruff, D.R.; Meinzer, F.C. Water stress, shoot growth and storage of non-structural carbohydrates along a tree height gradient in a tall conifer. *Plant Cell Environ.* **2011**, *34*, 1920–1930. [CrossRef]
8. Salgado-Luarte, C.; Gianoli, E. Shade tolerance and herbivory are associated with RGR of tree species via different functional traits. *Plant Biol.* **2017**, *19*, 413–419. [CrossRef]
9. Zhang, X.S.; Jin, G.Z.; Liu, Z.L. Contribution of leaf anatomical traits to leaf mass per area among canopy layers for five coexisting broadleaf species across shade tolerances at a regional scale. *For. Ecol. Manag.* **2019**, *452*, 117569. [CrossRef]
10. Evans, J.R. Nitrogen and photosynthesis in the flag leaf of wheat (*Triticum aestivum* L.). *Plant Physiol.* **1983**, *72*, 297–302. [CrossRef]

11. Sun, J.; Yao, F.; Wu, J.; Zhang, P.; Xu, W. Effect of nitrogen levels on photosynthetic parameters, morphological and chemical characters of saplings and trees in a temperate forest. *J. For. Res.* **2017**, *29*, 1481–1488. [CrossRef]
12. Xiong, D.; Flexas, J. Leaf anatomical characteristics are less important than leaf biochemical properties in determining photosynthesis responses to nitrogen top-dressing. *J. Exp. Bot.* **2021**, *72*, 5709–5720. [CrossRef] [PubMed]
13. Rogers, A. The use and misuse of $V_{c,max}$ in Earth System Models. *Photosynth. Res.* **2014**, *119*, 15–29. [CrossRef] [PubMed]
14. Friend, A.D. Terrestrial plant production and climate change. *J. Exp. Bot.* **2010**, *61*, 1293–1309. [CrossRef]
15. Pierce, S.; Maffi, D.; Faoro, F.; Cerabolini, B.E.L.; Spada, A. The leaf anatomical trade-offs associated with plant ecological strategy variation. *Plant Ecol.* **2022**, *223*, 1233–1246. [CrossRef]
16. Flexas, J.; Carriqui, M. Photosynthesis and photosynthetic efficiencies along the terrestrial plant's phylogeny: Lessons for improving crop photosynthesis. *Plant J.* **2020**, *101*, 964–978. [CrossRef]
17. Wright, I.J.; Reich, P.B.; Westoby, M.; Ackerly, D.D.; Baruch, Z.; Bongers, F.; Cavender-Bares, J.; Chapin, T.; Cornelissen, J.H.; Diemer, M.; et al. The worldwide leaf economics spectrum. *Nature* **2004**, *428*, 821–827. [CrossRef]
18. Liu, Z.L.; Hikosaka, K.; Li, F.R.; Jin, G.Z.; Ostertag, R. Variations in leaf economics spectrum traits for an evergreen coniferous species: Tree size dominates over environment factors. *Funct. Ecol.* **2020**, *34*, 458–467. [CrossRef]
19. Visakorpi, K.; Block, S.; Pellissier, L.; Levine, J.M.; Alexander, J. Eco-physiological and morphological traits explain alpine plant species' response to warming. *Funct. Ecol.* **2022**. *early view*. [CrossRef]
20. Vogelmann, T.C.; Martin, G. The functional significance of palisade tissue: Penetration of directional versus diffuse light. *Plant Cell Environ.* **1993**, *16*, 65–72. [CrossRef]
21. Knapp, A.K.; Carter, G.A. Variability in leaf optical properties among 26 species from a broad range of habitats. *Am. J. Bot.* **1998**, *85*, 940–946. [CrossRef]
22. Terashima, I.; Hanba, Y.T.; Tholen, D.; Niinemets, U. Leaf functional anatomy in relation to photosynthesis. *Plant Physiol.* **2011**, *155*, 108–116. [CrossRef]
23. Ryan, M.G.; Bond, B.J.; Law, B.E.; Hubbard, R.M.; Woodruff, D.; Cienciala, E.; Kucera, J. Transpiration and whole-tree conductance in ponderosa pine trees of different heights. *Oecologia* **2000**, *124*, 553–560. [CrossRef]
24. Gago, J.; Carriqui, M.; Nadal, M.; Clemente-Moreno, M.J.; Coopman, R.E.; Fernie, A.R.; Flexas, J. Photosynthesis Optimized across Land Plant Phylogeny. *Trends Plant Sci.* **2019**, *24*, 947–958. [CrossRef]
25. Liu, Z.; Jiang, F.; Li, F.; Jin, G. Coordination of intra and inter-species leaf traits according to leaf phenology and plant age for three temperate broadleaf species with different shade tolerances. *For. Ecol. Manag.* **2019**, *434*, 63–75. [CrossRef]
26. Tredennick, A.T.; Teller, B.J.; Adler, P.B.; Hooker, G.; Ellner, S.P. Size-by-environment interactions: A neglected dimension of species' responses to environmental variation. *Ecol. Lett.* **2018**, *21*, 1757–1770. [CrossRef]
27. Mencuccini, M.; Martinez-Vilalta, J.; Vanderklein, D.; Hamid, H.A.; Korakaki, E.; Lee, S.; Michiels, B. Size-mediated ageing reduces vigour in trees. *Ecol. Lett.* **2005**, *8*, 1183–1190. [CrossRef]
28. Falster, D.S.; Brännström, Å.; Dieckmann, U.; Westoby, M. Influence of four major plant traits on average height, leaf-area cover, net primary productivity, and biomass density in single-species forests: A theoretical investigation. *J. Ecol.* **2011**, *99*, 148–164. [CrossRef]
29. Niinemets, U. Stomatal conductance alone does not explain the decline in foliar photosynthetic rates with increasing tree age and size in Picea abies and Pinus sylvestris. *Tree Physiol.* **2002**, *22*, 515–535. [CrossRef] [PubMed]
30. Sterner, R.W.; Elser, J.J. *Ecological Stoichiometry: The Biology of Elements from Molecules to the Biosphere*; Princeton University Press: Princeton, NJ, USA, 2002.
31. Mensah, S.; Glèlè Kakaï, R.; Seifert, T. Patterns of biomass allocation between foliage and woody structure: The effects of tree size and specific functional traits. *Ann. For. Res.* **2016**, *59*, 1–12. [CrossRef]
32. Carranca, C.; Brunetto, G.; Tagliavini, M. Nitrogen nutrition of fruit trees to reconcile productivity and environmental concerns. *Plants* **2018**, *7*, 4. [CrossRef] [PubMed]
33. Alvarez-Clare, S.; Mack, M.C.; Brooks, M. A direct test of nitrogen and phosphorus limitation to net primary productivity in a lowland tropical wet forest. *Ecology* **2013**, *94*, 1540–1551. [CrossRef]
34. Augspurger, C.K.; Bartlett, E.A. Differences in leaf phenology between juvenile and adult trees in a temperate deciduous forest. *Tree Physiol.* **2003**, *23*, 517–525. [CrossRef]
35. Marchand, L.J.; Dox, I.; Gricar, J.; Prislan, P.; Leys, S.; Van den Bulcke, J.; Fonti, P.; Lange, H.; Matthysen, E.; Penuelas, J.; et al. Inter-individual variability in spring phenology of temperate deciduous trees depends on species, tree size and previous year autumn phenology. *Agric. For. Meteorol.* **2020**, *290*, 108031. [CrossRef]
36. Ryan, M.G.; Binkley, D.; Fownes, J.H. Age-related decline in forest productivity: Pattern and process. *Adv. Ecol. Res.* **1997**, *27*, 213–262. [CrossRef]
37. Ryan, M.G.; Phillips, N.; Bond, B.J. The hydraulic limitation hypothesis revisited. *Plant Cell Environ.* **2006**, *29*, 367–381. [CrossRef]
38. Portsmuth, A.; Niinemets, Ü. Structural and physiological plasticity in response to light and nutrients in five temperate deciduous woody species of contrasting shade tolerance. *Funct. Ecol.* **2007**, *21*, 61–77. [CrossRef]
39. Koleszar, G.; Lukacs, B.A.; Nagy, P.T.; Szabo, S. Shade tolerance as a key trait in invasion success of submerged macrophyte *Cabomba caroliniana* over *Myriophyllum spicatum*. *Ecol. Evol.* **2022**, *12*, e9306. [CrossRef]
40. Kaber, Y.; Meyer, P.; Stillhard, J.; De Lombaerde, E.; Zell, J.; Stadelmann, G.; Bugmann, H.; Bigler, C. Tree recruitment is determined by stand structure and shade tolerance with uncertain role of climate and water relations. *Ecol. Evol.* **2021**, *11*, 12182–12203. [CrossRef]

41. Yu, Z.C.; Lin, W.; Zheng, X.T.; Cai, M.L.; Zhang, T.J.; Luo, Y.N.; Peng, C.L. Interpretation of the difference in shade tolerance of two subtropical forest tree species of different successional stages at the transcriptome and physiological levels. *Tree Physiol.* **2021**, *41*, 1669–1684. [CrossRef]
42. Seiwa, K.; Kikuzawa, K.; Kadowaki, T.; Akasaka, S.; Ueno, N. Shoot life span in relation to successional status in deciduous broad-leaved tree species in a temperate forest. *New Phytol.* **2006**, *169*, 537–548. [CrossRef] [PubMed]
43. He, P.; Wright, I.J.; Zhu, S.; Onoda, Y.; Liu, H.; Li, R.; Liu, X.; Hua, L.; Oyanoghafo, O.O.; Ye, Q. Leaf mechanical strength and photosynthetic capacity vary independently across 57 subtropical forest species with contrasting light requirements. *New Phytol.* **2019**, *223*, 607–618. [CrossRef] [PubMed]
44. Chmura, D.J.; Modrzynski, J.; Chmielarz, P.; Tjoelker, M.G. Plasticity in seedling morphology, biomass allocation and physiology among ten temperate tree species in response to shade is related to shade tolerance and not leaf habit. *Plant Biol.* **2017**, *19*, 172–182. [CrossRef] [PubMed]
45. He, N.; Liu, C.; Tian, M.; Li, M.; Yang, H.; Yu, G.; Guo, D.; Smith, M.D.; Yu, Q.; Hou, J.; et al. Variation in leaf anatomical traits from tropical to cold-temperate forests and linkage to ecosystem functions. *Funct. Ecol.* **2017**, *32*, 10–19. [CrossRef]
46. Ollinger, S.V. Sources of variability in canopy reflectance and the convergent properties of plants. *New Phytol.* **2011**, *189*, 375–394. [CrossRef] [PubMed]
47. Krober, W.; Heklau, H.; Bruelheide, H. Leaf morphology of 40 evergreen and deciduous broadleaved subtropical tree species and relationships to functional ecophysiological traits. *Plant Biol.* **2015**, *17*, 373–383. [CrossRef]
48. Binks, O.; Meir, P.; Rowland, L.; da Costa, A.C.; Vasconcelos, S.S.; de Oliveira, A.A.; Ferreira, L.; Mencuccini, M. Limited acclimation in leaf anatomy to experimental drought in tropical rainforest trees. *Tree Physiol.* **2016**, *36*, 1550–1561. [CrossRef]
49. Fu, P.L.; Zhu, S.D.; Zhang, J.L.; Finnegan, P.M.; Jiang, Y.J.; Lin, H.; Fan, Z.X.; Cao, K.F. The contrasting leaf functional traits between a karst forest and a nearby non-karst forest in south-west China. *Funct. Plant Biol.* **2019**, *46*, 907–915. [CrossRef]
50. Sterck, F.; Markesteijn, L.; Schieving, F.; Poorter, L. Functional traits determine trade-offs and niches in a tropical forest community. *Proc. Natl. Acad. Sci. USA* **2011**, *108*, 20627–20632. [CrossRef]
51. Chen, B.; Wang, C.; Tian, Y.; Chu, Q.; Hu, C. Anatomical characteristics of young stems and mature leaves of dwarf pear. *Sci. Hortic.* **2015**, *186*, 172–179. [CrossRef]
52. Woods, D.B.; Turner, N.C. Stomatal Response to Changing Light by Four Tree Species of Varying Shade Tolerance. *New Phytol.* **1971**, *70*, 77–84. [CrossRef]
53. Janse-Ten Klooster, S.H.; Thomas, E.J.P.; Sterck, F.J. Explaining interspecific differences in sapling growth and shade tolerance in temperate forests. *J. Ecol.* **2007**, *95*, 1250–1260. [CrossRef]
54. Farquhar, G.D.; Wong, S.C. An empirical model of stomatal conductance. *Aust. J. Plant Physiol.* **1984**, *11*, 191–210. [CrossRef]
55. Rowland, L.; da Costa, A.C.; Galbraith, D.R.; Oliveira, R.S.; Binks, O.J.; Oliveira, A.A.; Pullen, A.M.; Doughty, C.E.; Metcalfe, D.B.; Vasconcelos, S.S.; et al. Death from drought in tropical forests is triggered by hydraulics not carbon starvation. *Nature* **2015**, *528*, 119–122. [CrossRef]
56. Stovall, A.E.L.; Shugart, H.; Yang, X. Tree height explains mortality risk during an intense drought. *Nat. Commun.* **2019**, *10*, 4385. [CrossRef]
57. Rijkers, T.; Pons, T.L.; Bongers, F. The effect of tree height and light availability on photosynthetic leaf traits of four neotropical species differing in shade tolerance. *Funct. Ecol.* **2000**, *14*, 77–86. [CrossRef]
58. Sala, A.; Hoch, G. Height-related growth declines in ponderosa pine are not due to carbon limitation. *Plant Cell Environ.* **2009**, *32*, 22–30. [CrossRef]
59. Oguchi, R.; Hikosaka, K.; Hirose, T. Leaf anatomy as a constraint for photosynthetic acclimation: Differential responses in leaf anatomy to increasing growth irradiance among three deciduous trees. *Plant Cell Environ.* **2005**, *28*, 916–927. [CrossRef]
60. Xu, H.; Wang, H.; Prentice, I.C.; Harrison, S.P.; Wright, I.J. Coordination of plant hydraulic and photosynthetic traits: Confronting optimality theory with field measurements. *New Phytol.* **2021**, *232*, 1286–1296. [CrossRef]
61. Aasamaa, K.; Sõber, A. Stomatal sensitivities to changes in leaf water potential, air humidity, CO2 concentration and light intensity, and the effect of abscisic acid on the sensitivities in six temperate deciduous tree species. *Environ. Exp. Bot.* **2011**, *71*, 72–78. [CrossRef]
62. Poorter, L. Leaf traits show different relationships with shade tolerance in moist versus dry tropical forests. *New Phytol.* **2009**, *181*, 890–900. [CrossRef]
63. Liu, Z.; Li, B.; Jin, G. Scale-dependent changes in the contributions of biotic and abiotic factors to leaf area index in a natural forest in northeast China. *For. Ecol. Manag.* **2021**, *479*, 118540. [CrossRef]
64. Pu, X.; Weemstra, M.; Jin, G.; Umana, M.N. Tree mycorrhizal type mediates conspecific negative density dependence effects on seedling herbivory, growth, and survival. *Oecologia* **2022**, *199*, 907–918. [CrossRef] [PubMed]
65. Zhou, Y.L. *Heilongjiang Tree Records*; Heilongjiang Science & Technology Press: Heilongjiang, China, 1986.
66. Niinemets, Ü.; Valladares, F. Tolerance To shade, drought, and waterlogging of femperate northern hemisphere trees and shrubs. *Ecol. Monogr.* **2006**, *76*, 521–547. [CrossRef]
67. Fletcher, L.R.; Cui, H.; Callahan, H.; Scoffoni, C.; John, G.P.; Bartlett, M.K.; Burge, D.O.; Sack, L. Evolution of leaf structure and drought tolerance in species of Californian Ceanothus. *Am. J. Bot.* **2018**, *105*, 1672–1687. [CrossRef]

68. Nally, R.M.; Walsh, C.J. Hierarchical partitioning public-domain software. *Biodivers. Conserv.* **2004**, *13*, 659–660. [CrossRef]
69. Rosseel, Y. lavaan: An R Package for Structural Equation Modeling. *J. Stat. Softw.* **2012**, *48*, 1–36. [CrossRef]

plants

MDPI

Article

Top Canopy Height and Stem Size Variation Enhance Aboveground Biomass across Spatial Scales in Seasonal Tropical Forests

Zhenhua Sun [1,2,†], Arunkamon Sonsuthi [1,3,†], Tommaso Jucker [4], Arshad Ali [5], Min Cao [1], Feng Liu [6], Guanghong Cao [7], Tianyu Hu [8], Qin Ma [3,8], Qinghua Guo [9] and Luxiang Lin [1,2,*]

1 CAS Key Laboratory of Tropical Forest Ecology, Xishuangbanna Tropical Botanical Garden, Chinese Academy of Sciences, Mengla 666303, China
2 Center of Plant Ecology, Core Botanical Gardens, Chinese Academy of Sciences, Mengla 666303, China
3 University of Chinese Academy of Sciences, Beijing 100049, China
4 School of Biological Sciences, University of Bristol, Bristol BS8 1QU, UK
5 Forest Ecology Research Group, College of Life Sciences, Hebei University, Baoding 071002, China
6 Yunnan Academy of Forestry and Grassland, Kunming 650201, China
7 Administration Bureau of Naban River Watershed National Nature Reserve, Jinghong 666100, China
8 State Key Laboratory of Vegetation and Environmental Change, Institute of Botany, Chinese Academy of Sciences, Beijing 100093, China
9 Institute of Ecology, College of Urban and Environmental Science, Peking University, Beijing 100871, China
* Correspondence: linluxa@xtbg.ac.cn; Tel.: +86-0871-68125324
† These authors contributed equally to this work.

Abstract: Tropical forests are biologically diverse and structurally complex ecosystems that can store a large quantity of carbon and support a great variety of plant and animal species. However, tropical forest structure can vary dramatically within seemingly homogeneous landscapes due to subtle changes in topography, soil fertility, species composition and past disturbances. Although numerous studies have reported the effects of field-based stand structure attributes on aboveground biomass (AGB) in tropical forests, the relative effects and contributions of UAV LiDAR-based canopy structure and ground-based stand structural attributes in shaping AGB remain unclear. Here, we hypothesize that mean top-of-canopy height (TCH) enhances AGB directly and indirectly via species richness and horizontal stand structural attributes, but these positive relationships are stronger at a larger spatial scale. We used a combined approach of field inventory and LiDAR-based remote sensing to explore how stand structural attributes (stem abundance, size variation and TCH) and tree species richness affect AGB along an elevational gradient in tropical forests at two spatial scales, i.e., 20 m × 20 m (small scale), and 50 m × 50 m (large scale) in southwest China. Specifically, we used structural equation models to test the proposed hypothesis. We found that TCH, stem size variation and abundance were strongly positively associated with AGB at both spatial scales, in addition to which increasing TCH led to greater AGB indirectly through increased stem size variation. Species richness had negative to negligible influences on AGB, but species richness increased with increasing stem abundance at both spatial scales. Our results suggest that light capture and use, modulated by stand structure, are key to promoting high AGB stocks in tropical forests. Thus, we argue that both horizontal and vertical stand structures are important for shaping AGB, but the relative contributions vary across spatial scales in tropical forests. Importantly, our results highlight the importance of including vertical forest stand attributes for predicting AGB and carbon sequestration that underpins human wellbeing.

Keywords: aboveground biomass (AGB); stand structural attributes; UAV LiDAR-based canopy structure; mean top of canopy height (TCH); tropical forest

Citation: Sun, Z.; Sonsuthi, A.; Jucker, T.; Ali, A.; Cao, M.; Liu, F.; Cao, G.; Hu, T.; Ma, Q.; Guo, Q.; et al. Top Canopy Height and Stem Size Variation Enhance Aboveground Biomass across Spatial Scales in Seasonal Tropical Forests. *Plants* **2023**, *12*, 1343. https://doi.org/10.3390/plants12061343

Academic Editor: Fernando Henrique Reboredo

Received: 6 December 2022
Revised: 25 January 2023
Accepted: 11 February 2023
Published: 16 March 2023

1. Introduction

Forests play a critical role in global carbon cycling while conserving terrestrial biodiversity at the same time [1,2]. Among forest biomes, tropical forests are not only sequestering a large amount of carbon in standing aboveground biomass (AGB) but also form dense complex stand structures through higher species richness and tree stem variation [3,4]. Forest stand structural complexity defines how species capture and use available resources through variation in both vertical and horizontal tree sizes within a community [5,6], which in turn may greatly influence AGB [7]. Thus, understanding the ecological mechanisms underlying the relationships between forest stand structural complexity and AGB is critical to predicting how forests will respond to anthropogenic impact as well as to managing forests in the context of global climate changes [8,9].

During the last few decades, the relationships between forest stand structural attributes and forest functioning (i.e., AGB or productivity) have been widely reported but remain highly debated [7]. More specifically, recent studies found a positive relationship between stand structural complexity (i.e., tree stem size variation) and AGB in subtropical and tropical forests [3,10]. The observed relationships in previous studies have been linked to the niche complementary mechanisms, which assumes that higher variation in tree stem size variation could lead to an increased resource-use complementarity by allowing the formation of multiple leaf layers and/or a highly packed canopy [5,11]. Nevertheless, while resource-use complementarity may explain the positive relationship between forest diversity, structure and function [12], forest structural complexity will not necessarily always lead to greater AGB due to asymmetric competition for light [7]. Thus, the interplay between stand structural attributes (e.g., tree stem size variation, canopy height, species richness and stem abundance) matters in explaining variation in AGB across tropical landscapes. For example, species richness could indirectly affect AGB through the mediation of tree stem size variation and/or stem abundance, or vice versa [13]. Thus, it has been reported that species richness increases AGB indirectly via promoting stem abundance and/or tree stem size variation in tropical forests [3,14]. Alternatively, the selection hypothesis suggests that the presence of a few productive or highly functioning species may contribute to AGB better than other species within a community [15]. For example, it has been reported that the presence of few large trees may overrule the effects of forest stand structural complexity on AGB in forest communities [16,17].

Among forest stand structural attributes, mean top canopy height (TCH) has been recognized as one of the robust predictors of AGB in tropical forests due to the spatial crown variability in coexisting tree species within a community [18]. However, different forest communities at a local scale could have similar TCH but they may differ in tree stem abundance, stem size variation, and species richness [18,19]. This implies that including information from the canopy should improve our understanding of the relationships between stand structural complexity and AGB in forests. Yet, while horizontal forest stand structure attributes (e.g., tree stem size variation in diameters) have extensively been studied to explain AGB, we lack a complete picture of how vertical stand structural attributes (e.g., TCH) covary with horizontal attributes, and how they together shape AGB in tropical forests. A key challenge is that traditional forest inventory is a ground-based approach, lacking a full assessment of TCH in forests. In this regard, airborne laser scanning (i.e., LiDAR) allows us to measure the forest biophysical parameters at high spatial resolution [9]. Previous studies have shown that LiDAR-based canopy structural attributes are the key determinants of AGB in forests [20,21].

Forest stand structural attributes can vary considerably across tropical landscapes due to subtle changes in topography, soil nutrients and past disturbance, which in turn drives local variation in AGB [22–24]. At fine scales, topography affects microclimatic and soil nutrient availability which could have both direct and indirect effects on AGB [25–28]. For example, thermal and hydrological variations could control tree species abundance and spatial distribution [24,29,30]), which could further shape tree size distribution, leaf trait variation and leaf spatial arrangement [31]. Moreover, nutrient-rich soils could lead to higher plant growth but may also lead to higher plant mortality rates due to species' competition for resources, which in turn could shape the stem abundance and stem size variation within a community [16,32], thereby shaping AGB directly and indirectly via forest stand structural complexity [33].

In this study, we used tropical forest inventory data at two spatial scales, i.e., 20 m × 20 m (small scale), and 50 m × 50 m (large scale) in southwest China for the purpose of determining the effects of both horizontal and vertical stand structural attributes on AGB across spatial scales while considering the direct and indirect effects of topography. By using a conceptual model (Figure 1), we ask the following research questions. (1) How does TCH affect tree species richness, stem abundance, and stem size variation directly, and how do they together influence AGB directly and indirectly? (2) How does topography affect AGB directly and indirectly via stand structural complexity attributes? (3) What is the relative contribution of stand structural complexity attributes and topography to AGB, and what is the main direct driver of AGB? (4) Do the relationships of AGB with stand structural complexity attributes and topography vary across spatial scales? We hypothesize that TCH enhances AGB directly and indirectly via species richness and horizontal stand structural attributes, but these positive relationships are stronger at a larger spatial scale.

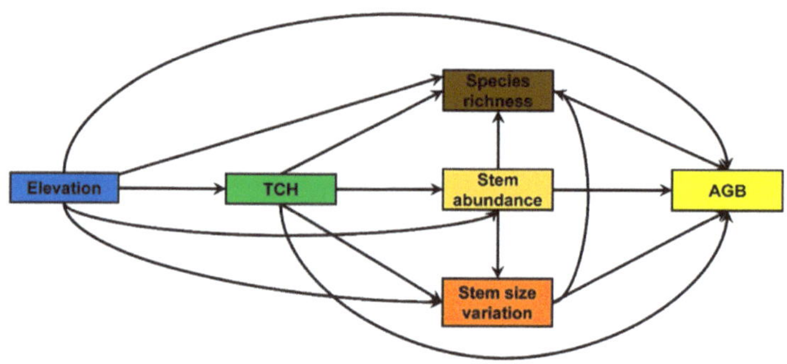

Figure 1. A conceptual model for linking elevation, mean top canopy height (TCH), stem size variation, stem abundance, species richness and aboveground biomass (AGB) across spatial scales in tropical forests.

2. Materials and Methods

2.1. The Study Sites and Forest Plots

This study was conducted in the tropical seasonal rain forests of Yunnan Province located in southwestern China. We collected data from two forest dynamic plots (each plot size = 20-ha, Figure 2), namely, the Nabanhe plot (NBH; 100.601° N, 22.249° E) and Xishuangbanna plots (XSBN; 101.574° N, 21.611° E), which were established according to the standard guidelines issued by the ForestGEO network (http://www.forestgeo.si.edu/, accessed data 10 August 2021). Each forest plot was subdivided into non-overlapping quadrats at two spatial scales: 20 m × 20 m (500 quadrats) and 50 m × 50 m (80 quadrats) which allow us to account for the possible scale-dependence of forest structural patterns and processes and to test whether scale matters in the relationships amongst species diversity, stand structure and AGB. Both forest plots are formed under similar climatic and geographic conditions [34,35].

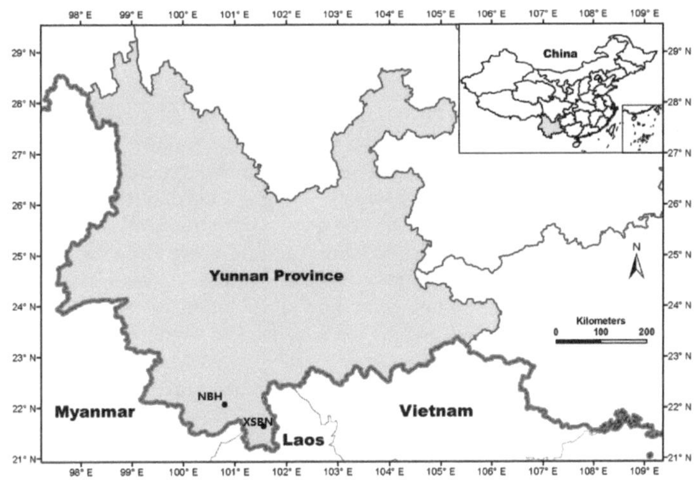

Figure 2. Location of the study area and the sampling points. NBH and XSBN stand for Nabanhe plot and Xishuangbanna plot, respectively.

2.2. Forest Inventory and Quantification of Variables

In each plot, all freestanding woody stems with a diameter at breast height (DBH) $\geq$ 1 cm were identified to species, tagged, measured, and mapped. We used the latest forest inventory data (censused in 2017) and measured species' woody density values to calculate AGB for each tree, using the pantropical biomass allometric equation (Equation (1)) [36] in the BIOMASS package [37]. The AGB values across individual trees within each quadrat were summed and scaled up to Mg/ha. Species-level wood density was measured by collecting wood core samples from 3–5 individuals per species, following the standard measurement protocols in both field and laboratory [38].

$$AGB = \exp(-2.024 - 0.896 \times E + 0.920 \times \ln(WD) + 2.795 \times \ln(DBH) - 0.0461 \times (\ln(DBH)^2)) \quad (1)$$

where DBH is the diameter at breast height (cm), WD is wood density (g cm^{-3}), and E is the environmental stress factor (i.e., 0.336 for our study area).

Within each quadrat, we quantified tree species richness as the observed number of tree species using the vegan package [39]. Stem size variation, as a proxy of horizontal stand structure, was quantified by the coefficient of variation in stem DBHs within a quadrat [40]. For the quantification of the vertical stand structure, we used the UAV LiDAR data which were collected in September 2017 using a Greenvalley International LiHawk system (GreenValley International, Beijing, China). The system is equipped with a RIEGL VUX-1 UAV laser scanner, which has a maximum ranging capability of 1000 m and provides high-speed data acquisition (550 kHz) using a narrow near-infrared laser beam. The collected UAV LiDAR data of each study site were then pre-processed following the same protocol, including denoising, filtering, and normalization. The filtering steps classified ground points and generated a digital terrain model (DTM) from the ground points. An improved progressive triangulated irregular network densification filtering algorithm integrated into LiDAR360 was used to extract ground points [41], and a DTM in 5 m resolution was interpolated using the ordinary kriging method for each study site. From the DTM, we also extracted the mean elevation of each quadrat. Finally, the normalization step was used to remove the influence of terrain elevation on LiDAR point clouds by subtracting the DTM value from the original point height at the corresponding location. Based on the normalized LiDAR point clouds, a canopy height model (CHM) was produced, and we then calculated the vertical stand structural attribute, i.e., mean top-of-canopy height (TCH) as the mean height of pixels making up the surface of the CHM.

2.3. Conceptual Model Development and Statistical Analyses

For the development of a conceptual model to test the proposed hypothesis in this study (Figure 1), we assumed that elevation, TCH, stem size variation, stem abundance and species richness shape AGB directly and indirectly via each other at both spatial scales in tropical forests. For the interplay (i.e., the indirect effects of stand structural attributes on AGB) between stand structural attributes, we assumed that higher stem abundance is expected to lead to greater variation in stem size as well as higher species richness [40]. Moreover, stem size variation is expected to shape species coexistence [42]. As such, higher TCH may allow more tree species to coexist through differential light capture and use, which may lead to higher stem size variation with a large number of stems, and/or by forming a densely packed canopy structure [43,44], thereby shaping AGB simultaneously. In addition, to tease apart how topography influences AGB directly and indirectly via species richness and stand structural attributes [9], elevation was included as the exogenous variable in the model (Figure 1). Thus, we tested the conceptual model using the structural equation models (SEMs) across small and large scales, as it allows us to test the direct and indirect pathways in one integrative model [45].

The SEM fit evaluation was determined by using the following statistical parameters [46]: the chi-square test ($p > 0.05$ shows an accepted SEM), the comparative fit index (CFI), the goodness-of-fit index (GFI) (> 0.90 shows a satisfactory SEM fit), and the standardized root mean square residual (SRMR < 0.08 shows SEM fit with less error). To get the best model fit, we excluded the path between TCH and stem abundance, as this relationship was not significant at a small scale whereas it was weakly negative at a large scale. The direct effect was quantified by considering the standardized regression coefficient of the predictor on the response variable, whereas the indirect effect was quantified by multiplying the direct effects of the predictor on the mediator and then on the response variable in one route. The total effect was quantified by summing the direct and indirect effects of predictors on the response variable. In addition, we calculated the relative contributions (in percentage) of predictors to AGB through the ratio of the standardized coefficient of a given predictor to the sum of all coefficients in SEM. SEMs were evaluated using the *lavaan* package [47].

To meet the assumptions of data normality and homoscedasticity [48], all continuous variables including AGB, stand abundance, stem size variation, species richness and TCH were log-transformed and then standardized (by subtracting the variable's mean and dividing by the standard deviation) prior to statistical analyses. Elevation was transformed between 0 to 1 using the function of (elevation—mean (elevation))/(max(elevation)-min(elevation)). To complement the results from SEMs, we tested bivariate relationships and Pearson correlations amongst tested predictors across spatial scales. All statistical analyses were conducted in R.3.6.0 [49]. Note that, during statistical analyses, we used the combined data from two forest dynamic plots at two different spatial scales, i.e., 1000 quadrats at a small scale and 160 quadrats at a large scale. A summary of variables used in the analyses is provided in Table S1.

3. Results

The tested SEMs had the best fit to the data and explained variation of 76% and 82% of the variance in AGB at 20 m × 20 m and 50 m × 50 m scales, respectively (Figures 3 and 4). At both spatial scales, TCH (β = 0.21 to 0.55), stem size variation (β = 0.71 to 0.44) and stem abundance (β = 0.17 to 0.30) increased AGB directly (Figures 3a and 4a; Tables S2 and S3). Species richness possessed a negligible positive effect on AGB at a 20 m × 20 m scale (β = 0.04, Figure 3a, Table S3) but a negative effect at a 50 m × 50 m scale (β = −0.11, Figure 4a, Table S3). However, the strength of the positive effects of TCH and stem abundance on AGB increased, whereas the effect of stem size variation decreased on AGB with increasing spatial scale. As such, the negative direct effect of species richness on

AGB seemed to be important at a large spatial scale. In addition, elevation increased AGB directly across both scales (β = 0.10 to 0.12, Figures 3a and 4a, Tables S2 and S3); however, this effect was a little higher at a larger spatial scale.

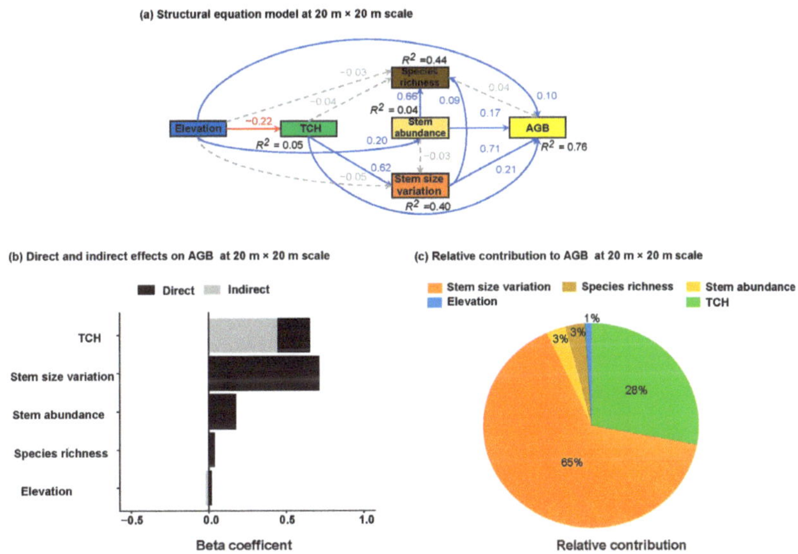

Figure 3. Structural equation model (**a**) for linking elevation, TCH, stem size variation, stem abundance, species richness and AGB at a small scale (i.e., 20 m × 20 m). Blue and red arrows represent significant positive and negative paths, respectively ($p < 0.05$) whereas dashed arrows show non-significant paths ($p > 0.05$). For each path, a standardized regression coefficient is shown. R^2 indicates the total variation in a dependent variable, which is explained by the combined independent variables. See Table S4 for statistics. CFI = 1.000, GFI = 1.000, SRMR = 0.004, Chi-square = 0.190, *p*-value = 0.663. (**b**) Comparison of direct (dark bars) and indirect (grey bars) effects, derived from structural equation model, of elevation, TCH, stem size variation, stem abundance, and species richness on AGB. (**c**) Relative contributions of elevation, TCH, stem size variation, stem abundance, and species richness on AGB.

Most of the indirect effects on AGB were relatively negligible at both spatial scales. However, TCH possessed a strong positive indirect effect on AGB via stem size variation at a scale of 20 m × 20 m (β = 0.44, Figure 3b, Table S2) and 50 m × 50 m (β = 0.33, Figure 4b, Table S3). The indirect effect of stem abundance on AGB via stem size variation was negligible at a scale of 20 m × 20 m (Figure 3b, Table S2) but negative at a scale of 50 m × 50 m (Figure 4b, Table S3) due to the divergent direct effects on species richness (β = 0.60, Figure 4b) and stem size variation (β = −0.11, Figure 4b). In addition, stem size variation and abundance promoted species richness at a scale of 20 m × 20 m (β = 0.09 to 0.66, Figure 3b, Table S2), and the indirect effects on AGB mediated by species richness were negligible. Regarding the indirect effects of elevation, we found negligible effects on AGB (Figures 3b and 4b, Tables S2 and S3). However, elevation decreased TCH but increased stem abundance directly at both spatial scales, yet it did not strongly influence species richness and stem size variation (Figures 3a and 4a).

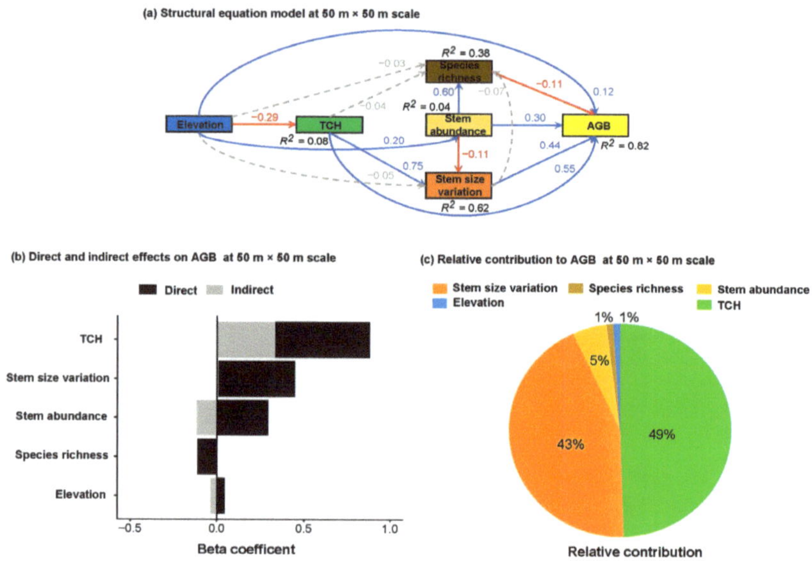

Figure 4. Structural equation model (**a**) for linking elevation, TCH, stem size variation, stem abundance, species richness and AGB at a large scale (i.e., 50 m × 50 m). Blue and red arrows represent significant positive and negative paths, respectively ($p < 0.05$) whereas dashed arrows show non-significant paths ($p > 0.05$). For each path, a standardized regression coefficient is shown. R^2 indicates the total variation in a dependent variable, which is explained by the combined independent variables. See Table S5 for statistics. CFI = 0.997, GFI = 0.998, SRMR = 0.04, Chi-square = 2.292, *p*-value = 0.130. (**b**) Comparison of direct (dark bars) and indirect (grey bars) effects, derived from structural equation model, of elevation, TCH, stem size variation, stem abundance, and species richness on AGB. (**c**) Relative contributions of elevation, TCH, stem size variation, stem abundance, and species richness on AGB.

The relative contributions result showed that stem size variation was the most important predictor, followed by TCH, thereby contributing 65% and 28% of the explained variance in AGB at a small scale (Figure 3c). In contrast, TCH was the most important predictor, followed by stem size variation by contributing 49 % and 43% of the explained variance in AGB at a large scale (Figure 4c). These comparative results indicated that horizontal stand structure is relatively important at a small scale whereas both horizontal and vertical stand structures are almost equally important at a large scale. Although TCH promoted stem size variation at both scales, the reverse relationship (i.e., the effect of stem size variation on TCH) might be also true and consistent.

The bivariate relationships provided support to the tested SEMs where most of the relationships were consistent with the direct effects, as shown in the SEMs (Figure S1). However, we noted a slight positive relationship of species richness with elevation and TCH at 20 m × 20 m, as well as the negative relationship between species richness and TCH that changed to non-significant at a scale of 50 m × 50 m. These small mismatches indicated the necessity of using multiple multivariate analyses for better understanding the complex relationships; for example, species richness was also controlled by many other factors in SEMs.

4. Discussion

In this study, we tease apart the direct and indirect effects of TCH, stem size variation, stem abundance and species richness on AGB along elevational gradients across two spatial scales in tropical forests. We found partial support for our proposed hypothesis that TCH, stem size variation and abundance increased AGB as compared to species richness at both small and large spatial scales. These positive relationships of horizontal and vertical stand structural attributes with AGB are indeed due to light capture and use by component species and interacting individual trees within a forest community, and hence supporting the niche complementarity effect [5,50]. However, we did not find a positive effect of species richness on AGB, but rather a negative relationship between species richness and AGB at a larger scale, indicating the selection or competitive exclusion effect [51].

Our results confirm that the positive relationship between stand structural attributes, especially stem size variation and stem abundance, and AGB can be extended to other forest types, as previously shown in boreal [42], temperate [51], and tropical forests [3]. The contribution of our study emphasizes the importance of TCH in ecological models for predicting AGB in tropical forests [52]. Thus, to the best of our knowledge, our study explores the effects of both UAV-LiDAR-derived canopy structural attributes and census-based horizontal stand structural attributes on AGB in tropical forests, which could further enhance our understanding of carbon sequestration that underpins human wellbeing.

Previously when fine-scale canopy height information was unavailable, the observed strong positive effect of stem size variation on AGB was indirectly ascribed to the higher vertical occupation of available canopy space by various sizes of trees and higher species richness [3,42,51]. Here, we further show that TCH contributed more comprehensively to stem size variation than stem abundance and species richness in tropical forests at both small and large spatial scales. Our findings were not just in accord with early findings in other forests, but also suggest that higher TCH may strengthen the stem size variation by providing more canopy space to fill, which allows more leaf area to intercept light and in turn increases forest productivity [33,44]. The underlying ecological mechanisms appear to be largely dependent on individual plant responses to light availability and crown complementarity among individual stems [5,50,53]. For instance, the growth rate difference between light-intolerant and shade-tolerant species may first define the vertical portioning of canopy height by increasing overall canopy space and then occupy these spaces efficiently through crown complementarity. Besides, we found the effect of TCH on AGB increased with increasing spatial scales, which is consistent with the general notion that large-scale climatic factors related to water and energy balance could shape canopy height and thus influence AGB and carbon sequestration [54].

Counter to the expectation that AGB will weakly increase with species richness [3,55], our study shows that species richness was negatively or negligibly related to AGB across spatial scales. This lack of a significant positive effect of species diversity on AGB might be attributable to the dominance of certain productive species in the studied forests, which might dilute the effect of species diversity [51]. For example, *Parashorea chinensis* was not just an emergent but also the monodominant species of the forest in the XSBN plot, contributing around 22% of the total AGB separately compared to the rest of AGB shared by around 390 other tree species. *P. chinensis* is wind- and gravity-dispersed, and as a result, most of the seeds fall within 10 m of conspecific adults, thereby causing the strongly aggregated distribution pattern in the valley and lower slope [56]. As the selection hypothesis posits that species diversity effects on AGB are more likely driven by the presence of highly productive species or emergent species in diverse communities [15], both the niche complementarity effect and selection effect may together exist as the main mechanisms for shaping AGB in our studied tropical forests across spatial scales. However, we did not find the consistent positive effects of TCH and stem size variation on tree species richness as previously shown for tree [43], liana [57], and different animal groups [58,59]. This result could be due to the reason that higher species richness may occur when TCH is lower, but stem abundance is higher. However, we found that higher stem abundance and lower TCH occurred on higher elevations (ridges) in our studied forests, where seasonal drought may only allow drought-tolerant species to coexist.

Our tested SEMs show that elevation had both direct and indirect effects via stand structural attributes on AGB, but the relative contribution of elevation to AGB was relatively small compared to stand structural attributes. However, we found that TCH and stem abundance rather than species richness and stem size variation mediated the divergent pathways of elevation to AGB, indicating the importance of hydrological controls on forest structural attributes [60]. Moreover, TCH was higher in the gulley of lower elevations than on the ridges, whereas stem abundance increased with increasing elevation, suggesting strong underlying ecological gradients shaped by topography [9]. Thus, despite the fact that the importance of TCH has been linked with AGB, we do acknowledge that including only TCH as a vertical structure attribute versus three horizontal structure attributes is imperfect and further studies are needed to incorporate more proxies of vertical structure [61], e.g., maximum canopy height [62], gap fraction [63], and canopy rugosity [64].

One more caveat of this study is that the local habitat is not represented solely based on the elevation. There are other abiotic factors, such as topography-related hydrological feathers [65], soil properties [22,66], and legacies of human impact [67] that impact species diversity and AGB. These contexts are not explicitly included in the present analyses, but they could have potentially influenced the observed variation in forest stand structure attributes and forest biomass.

5. Conclusions

This study shows that TCH (as a proxy of vertical stand structure) and stem size variation (a proxy of horizontal stand structure) boost AGB across spatial scales in tropical forests. The negative to negligible effects of species richness on AGB suggest the competitive exclusion effect, and hence, it is important to test the influence of large trees in future studies. As such, the positive effects of stem abundance on species richness and AGB, the weak effect on stem size variation and the lack of any relationship with TCH could indicate the role of few productive tree species in the studied forests. In addition, the divergent stem abundance, size variation and TCH pathways mediate the influences of topography on AGB, indicating the differential roles of microclimatic conditions on biotic factors. Although we did not test the actual ecological mechanisms through experimental data, the observed results show that light capture and use, modulated by stand structure, seemed to be important for higher AGB, and these effects were stronger at a large scale. Thus, we argue that both horizontal and vertical stand structures are important for shaping AGB, but the relative contributions vary across spatial scales in tropical forests.

Supplementary Materials: The following supporting information can be downloaded at: https://www.mdpi.com/article/10.3390/plants12061343/s1, Table S1. Summary of the variables used in the study; Table S2. Direct, indirect and total standardized effects of predictors on AGB (aboveground biomass) at scale of 20 m × 20 m, based on the structural equation model; Table S3. Direct, indirect and total standardized effects of predictors on AGB (aboveground biomass) at scale of 50 m × 50 m, based on the structural equation model; Figure. S1. Pearson's correlation coefficient between all pairs of variables at scale used in this meta-analysis.

Author Contributions: Z.S., A.S. and L.L. conceived and designed this study; Z.S. analyzed the data with help of T.J. and A.A.; and Z.S. wrote and revised the manuscript and all authors contributed critically to the draft and gave final approval for publication. All authors have read and agreed to the published version of the manuscript.

Funding: This study is supported by the Strategic Priority Research Program of Chinese Academy of Sciences (Grant No. XDB31000000), the Joint Fund of the National Natural Science Foundation of China-Yunnan Province (Grant No. U1902203), Southeast Asia Biodiversity Research Institute, Chinese Academy of Sciences (Grant No. 151C53KYSB20200019), and the CAS 135 program (Grant No. 2017XTBG-T01). Arshad Ali is supported by Hebei University (Grant No. 521100221033).

Data Availability Statement: The tree census data used in this study can be accessed by contacting the plot PI through ForestGEO Data Portal website (http://ctfs.si.edu/datarequest/).

Acknowledgments: The authors thank two anonymous reviewers for their constructive comments, which improved the quality of earlier versions of the manuscript. The authors also thank members of the field team in XSBN and NBH plot for their field support.

Conflicts of Interest: The authors declare no conflict of interest.

References

1. Pan, Y.; Birdsey, R.A.; Fang, J.; Houghton, R.; Kauppi, P.E.; Kurz, W.A.; Phillips, O.L.; Shvidenko, A.; Lewis, S.L.; Canadell, J.G.; et al. A Large and Persistent Carbon Sink in the World's Forests. *Science* **2011**, *333*, 988–993. [CrossRef]
2. De Frenne, P.; Lenoir, J.; Luoto, M.; Scheffers, B.R.; Zellweger, F.; Aalto, J.; Ashcroft, M.B.; Christiansen, D.M.; Decocq, G.; De Pauw, K.; et al. Forest microclimates and climate change: Importance, drivers and future research agenda. *Glob. Change Biol.* **2021**, *27*, 2279–2297. [CrossRef] [PubMed]
3. Poorter, L.; van der Sande, M.T.; Thompson, J.; Arets, E.J.M.M.; Alarcón, A.; Álvarez-Sánchez, J.; Ascarrunz, N.; Balvanera, P.; Barajas-Guzmán, G.; Boit, A.; et al. Diversity enhances carbon storage in tropical forests. *Glob. Ecol. Biogeogr.* **2015**, *24*, 1314–1328. [CrossRef]
4. Meakem, V.; Tepley, A.J.; Gonzalez-Akre, E.B.; Herrmann, V.; Muller-Landau, H.C.; Wright, S.J.; Hubbell, S.P.; Condit, R.; Anderson-Teixeira, K.J. Role of tree size in moist tropical forest carbon cycling and water deficit responses. *New Phytol.* **2018**, *219*, 947–958. [CrossRef]
5. Yachi, S.; Loreau, M. Does complementary resource use enhance ecosystem functioning? A model of light competition in plant communities. *Ecol. Lett.* **2007**, *10*, 54–62. [CrossRef] [PubMed]
6. Van Pelt, R.; Sillett, S.C.; Kruse, W.A.; Freund, J.A.; Kramer, R.D. Emergent crowns and light-use complementarity lead to global maximum biomass and leaf area in Sequoia sempervirens forests. *For. Ecol. Manag.* **2016**, *375*, 279–308. [CrossRef]
7. Ali, A. Forest stand structure and functioning: Current knowledge and future challenges. *Ecol. Indic.* **2019**, *98*, 665–677. [CrossRef]
8. Piponiot, C.; Anderson-Teixeira, K.J.; Davies, S.J.; Allen, D.; Bourg, N.A.; Burslem, D.F.R.P.; Cárdenas, D.; Chang-Yang, C.-H.; Chuyong, G.; Cordell, S.; et al. Distribution of biomass dynamics in relation to tree size in forests across the world. *New Phytol.* **2022**, *234*, 1664–1677. [CrossRef]
9. Jucker, T.; Bongalov, B.; Burslem, D.F.R.P.; Nilus, R.; Dalponte, M.; Lewis, S.L.; Phillips, O.L.; Qie, L.; Coomes, D.A. Topography shapes the structure, composition and function of tropical forest landscapes. *Ecol. Lett.* **2018**, *21*, 989–1000. [CrossRef]
10. Duran, S.M.; Sanchez-Azofeifa, G.A.; Rios, R.S.; Gianoli, E. The relative importance of climate, stand variables and liana abundance for carbon storage in tropical forests. *Glob. Ecol. Biogeogr.* **2015**, *24*, 939–949. [CrossRef]
11. Taubert, F.; Jahn, M.W.; Dobner, H.J.; Wiegand, T.; Huth, A. The structure of tropical forests and sphere packings. *Proc. Natl. Acad. Sci. USA* **2015**, *112*, 15125–15129. [CrossRef] [PubMed]
12. Cardinale, B.J.; Wright, J.P.; Cadotte, M.W.; Carroll, I.T.; Hector, A.; Srivastava, D.S.; Loreau, M.; Weis, J.J. Impacts of plant diversity on biomass production increase through time because of species complementarity. *Proc. Natl. Acad. Sci. USA* **2007**, *104*, 18123–18128. [CrossRef] [PubMed]
13. Chiang, J.-M.; Spasojevic, M.J.; Muller-Landau, H.C.; Sun, I.-F.; Lin, Y.; Su, S.-H.; Chen, Z.-S.; Chen, C.-T.; Swenson, N.G.; McEwan, R.W. Functional composition drives ecosystem function through multiple mechanisms in a broadleaved subtropical forest. *Oecologia* **2016**, *182*, 829–840. [CrossRef] [PubMed]
14. Ouyang, S.; Xiang, W.H.; Wang, X.P.; Xiao, W.F.; Chen, L.; Li, S.G.; Sun, H.; Deng, X.W.; Forrester, D.I.; Zeng, L.X.; et al. Effects of stand age, richness and density on productivity in subtropical forests in China. *J. Ecol.* **2019**, *107*, 2266–2277. [CrossRef]
15. Loreau, M.; Hector, A. Partitioning selection and complementarity in biodiversity experiments. *Nature* **2001**, *412*, 72–76. [CrossRef]
16. Paoli, G.D.; Curran, L.M.; Slik, J.W.F. Soil nutrients affect spatial patterns of aboveground biomass and emergent tree density in southwestern Borneo. *Oecologia* **2008**, *155*, 287–299. [CrossRef]
17. Lutz, J.A.; Furniss, T.J.; Johnson, D.J.; Davies, S.J.; Allen, D.; Alonso, A.; Anderson-Teixeira, K.J.; Andrade, A.; Baltzer, J.; Becker, K.M.L.; et al. Global importance of large-diameter trees. *Glob. Ecol. Biogeogr.* **2018**, *27*, 849–864. [CrossRef]
18. Asner, G.P.; Mascaro, J. Mapping tropical forest carbon: Calibrating plot estimates to a simple LiDAR metric. *Remote Sens. Environ.* **2014**, *140*, 614–624. [CrossRef]
19. Mascaro, J.; Detto, M.; Asner, G.P.; Muller-Landau, H.C. Evaluating uncertainty in mapping forest carbon with airborne LiDAR. *Remote Sens. Environ.* **2011**, *115*, 3770–3774. [CrossRef]
20. Patenaude, G.; Hill, R.A.; Milne, R.; Gaveau, D.L.A.; Briggs, B.B.J.; Dawson, T.P. Quantifying forest above ground carbon content using LiDAR remote sensing. *Remote Sens. Environ.* **2004**, *93*, 368–380. [CrossRef]
21. Cao, L.; Coops, N.C.; Innes, J.L.; Sheppard, S.R.J.; Fu, L.; Ruan, H.; She, G. Estimation of forest biomass dynamics in subtropical forests using multi-temporal airborne LiDAR data. *Remote Sens. Environ.* **2016**, *178*, 158–171. [CrossRef]
22. Baldeck, C.A.; Harms, K.E.; Yavitt, J.B.; John, R.; Turner, B.L.; Valencia, R.; Navarrete, H.; Davies, S.J.; Chuyong, G.B.; Kenfack, D.; et al. Soil resources and topography shape local tree community structure in tropical forests. *Proc. R. Soc. B-Biol. Sci.* **2013**, *280*, 20122532. [CrossRef] [PubMed]
23. Feret, J.-B.; Asner, G.P. Microtopographic controls on lowland Amazonian canopy diversity from imaging spectroscopy. *Ecol. Appl.* **2014**, *24*, 1297–1310. [CrossRef] [PubMed]
24. Fortunel, C.; Lasky, J.R.; Uriarte, M.; Valencia, R.; Joseph Wright, S.; Garwood, N.C.; Kraft, N.J.B. Topography and neighborhood crowding can interact to shape species growth and distribution in a diverse Amazonian forest. *Ecology* **2018**, *99*, 2272–2283. [CrossRef]

25. Daws, M.I.; Mullins, C.E.; Burslem, D.; Paton, S.R.; Dalling, J.W. Topographic position affects the water regime in a semideciduous tropical forest in Panama. *Plant Soil* **2002**, *238*, 79–90. [CrossRef]
26. Dobrowski, S.Z.; Abatzoglou, J.T.; Greenberg, J.A.; Schladow, S.G. How much influence does landscape-scale physiography have on air temperature in a mountain environment? *Agric. For. Meteorol.* **2009**, *149*, 1751–1758. [CrossRef]
27. Werner, F.A.; Homeier, J. Is tropical montane forest heterogeneity promoted by a resource-driven feedback cycle? Evidence from nutrient relations, herbivory and litter decomposition along a topographical gradient. *Funct. Ecol.* **2015**, *29*, 430–440. [CrossRef]
28. Chadwick, K.D.; Asner, G.P. Tropical soil nutrient distributions determined by biotic and hillslope processes. *Biogeochemistry* **2016**, *127*, 273–289. [CrossRef]
29. Long, W.; Xiong, M.; Zang, R.; Schamp, B.S.; Yang, X.; Ding, Y.; Huang, Y.; Xiang, Y. Changes in Patterns of Species Co-occurrence across Two Tropical Cloud Forests Differing in Soil Nutrients and Air Temperature. *Biotropica* **2015**, *47*, 416–423. [CrossRef]
30. Murphy, S.J.; Audino, L.D.; Whitacre, J.; Eck, J.L.; Wenzel, J.W.; Queenborough, S.A.; Comita, L.S. Species associations structured by environment and land-use history promote beta-diversity in a temperate forest. *Ecology* **2015**, *96*, 705–715. [CrossRef]
31. Long, W.; Zang, R.; Ding, Y. Air temperature and soil phosphorus availability correlate with trait differences between two types of tropical cloud forests. *Flora* **2011**, *206*, 896–903. [CrossRef]
32. Quesada, C.; Phillips, O.; Schwarz, M.; Czimczik, C.; Baker, T.; PatiÃ±o, S.; Fyllas, N.; Hodnett, M.; Herrera, R.; Almeida, S. Basin-wide variations in Amazon forest structure and function are mediated by both soils and climate. *Biogeosciences* **2012**, *9*, 2203–2246. [CrossRef]
33. Coomes, D.A.; Kunstler, G.; Canham, C.D.; Wright, E. A greater range of shade-tolerance niches in nutrient-rich forests: An explanation for positive richness–productivity relationships? *J. Ecol.* **2009**, *97*, 705–717. [CrossRef]
34. Lan, G.; Zhu, H.; Cao, M. Tree species diversity of a 20-ha plot in a tropical seasonal rainforest in Xishuangbanna, southwest China. *J. For. Res.* **2012**, *17*, 432–439. [CrossRef]
35. Shi, G.; Feng, L.; Dian, C.; Yun, D.; Lu-Xiang, L. Species composition and community classification of the 20-ha tropical seasonal rainforest dynamics monitoring plot in the Naban River, Yunnan. *Biodivers. Sci.* **2021**, *29*, 10–20. [CrossRef]
36. Chave, J.; Rejou-Mechain, M.; Burquez, A.; Chidumayo, E.; Colgan, M.S.; Delitti, W.B.C.; Duque, A.; Eid, T.; Fearnside, P.M.; Goodman, R.C.; et al. Improved allometric models to estimate the aboveground biomass of tropical trees. *Glob. Change Biol.* **2014**, *20*, 3177–3190. [CrossRef]
37. Rejou-Mechain, M.; Tanguy, A.; Piponiot, C.; Chave, J.; Herault, B. BIOMASS: An R package for estimating above-ground biomass and its uncertainty in tropical forests. *Methods Ecol. Evol.* **2017**, *8*, 1163–1167. [CrossRef]
38. Perez-Harguindeguy, N.; Diaz, S.; Garnier, E.; Lavorel, S.; Poorter, H.; Jaureguiberry, P.; Bret-Harte, M.S.; Cornwell, W.K.; Craine, J.M.; Gurvich, D.E.; et al. New handbook for standardised measurement of plant functional traits worldwide. *Aust. J. Bot.* **2013**, *61*, 167–234. [CrossRef]
39. Oksanen, J.; Blanchet, F.G.; Kindt, R.; Legendre, P.; Minchin, P.R.; O'Hara, R.B.; Simpson, G.L.; Solymos, P.; Stevens, M.H.H.; Wagner, H. *Vegan: Community Ecology Package*, R package version 2.3-1; R package: Vienna, Austria, 2015.
40. Chu, C.J.; Lutz, J.A.; Kral, K.; Vrska, T.; Yin, X.; Myers, J.A.; Abiem, I.; Alonso, A.; Bourg, N.; Burslem, D.; et al. Direct and indirect effects of climate on richness drive the latitudinal diversity gradient in forest trees. *Ecol. Lett.* **2019**, *22*, 245–255. [CrossRef]
41. Zhao, X.; Guo, Q.; Su, Y.; Xue, B. Improved progressive TIN densification filtering algorithm for airborne LiDAR data in forested areas. *J. Photogramm. Remote Sens.* **2016**, *117*, 79–91. [CrossRef]
42. Zhang, Y.; Chen, H.Y.H. Individual size inequality links forest diversity and above-ground biomass. *J. Ecol.* **2015**, *103*, 1245–1252. [CrossRef]
43. Wolf, J.A.; Fricker, G.A.; Meyer, V.; Hubbell, S.P.; Gillespie, T.W.; Saatchi, S.S. Plant Species Richness is Associated with Canopy Height and Topography in a Neotropical Forest. *Remote Sens.* **2012**, *4*, 4010–4021. [CrossRef]
44. Morin, X. Species richness promotes canopy packing: A promising step towards a better understanding of the mechanisms driving the diversity effects on forest functioning. *Funct. Ecol.* **2015**, *29*, 993–994. [CrossRef]
45. Grace, J.B.; Anderson, T.M.; Seabloom, E.W.; Borer, E.T.; Adler, P.B.; Harpole, W.S.; Hautier, Y.; Hillebrand, H.; Lind, E.M.; Partel, M.; et al. Integrative modelling reveals mechanisms linking productivity and plant species richness. *Nature* **2016**, *529*, 390–393. [CrossRef]
46. Hoyle, R.H. *Handbook of Structural Equation Modeling*; Guilford Press: New York, NY, USA, 2012.
47. Rosseel, Y. lavaan: An R Package for Structural Equation Modeling. *J. Stat. Softw.* **2012**, *48*, 1–36. [CrossRef]
48. Zuur, A.; Ieno, E.N.; Walker, N.; Saveliev, A.A.; Smith, G.M. *Mixed Effects Models and Extensions in Ecology With R.*; Springer: New York, NY, USA, 2009.
49. R Core Team. *R: A Language and Environment for Statistical Computing*; R Foundation for Statistical Computing: Vienna, Austria, 2019.
50. Williams, L.J.; Paquette, A.; Cavender-Bares, J.; Messier, C.; Reich, P.B. Spatial complementarity in tree crowns explains overyielding in species mixtures. *Nat. Ecol. Evol.* **2017**, *1*, 0063. [CrossRef]
51. Fotis, A.T.; Murphy, S.J.; Ricart, R.D.; Krishnadas, M.; Whitacre, J.; Wenzel, J.W.; Queenborough, S.A.; Comita, L.S. Above-ground biomass is driven by mass-ratio effects and stand structural attributes in a temperate deciduous forest. *J. Ecol.* **2018**, *106*, 561–570. [CrossRef]
52. Jubanski, J.; Ballhorn, U.; Kronseder, K.; Franke, J.; Siegert, F. Detection of large above-ground biomass variability in lowland forest ecosystems by airborne LiDAR. *Biogeosciences* **2012**, *10*, 3917–3930. [CrossRef]
53. Pretzsch, H. Canopy space filling and tree crown morphology in mixed-species stands compared with monocultures. *For. Ecol. Manag.* **2014**, *327*, 251–264. [CrossRef]
54. Keith, H.; Mackey, B.G.; Lindenmayer, D.B. Re-evaluation of forest biomass carbon stocks and lessons from the world's most carbon-dense forests. *Proc. Natl. Acad. Sci. USA* **2009**, *106*, 11635–11640. [CrossRef]

55. Chisholm, R.A.; Muller-Landau, H.C.; Abdul Rahman, K.; Bebber, D.P.; Bin, Y.; Bohlman, S.A.; Bourg, N.A.; Brinks, J.; Bunyavejchewin, S.; Butt, N.; et al. Scale-dependent relationships between tree species richness and ecosystem function in forests. *J. Ecol.* **2013**, *101*, 1214–1224. [CrossRef]
56. Lan, G.; Getzin, S.; Wiegand, T.; Hu, Y.; Xie, G.; Zhu, H.; Cao, M. Spatial Distribution and Interspecific Associations of Tree Species in a Tropical Seasonal Rain Forest of China. *PLoS ONE* **2012**, *7*, e46074. [CrossRef]
57. Meyer, L.; Diniz-Filho, J.A.F.; Lohmann, L.G.; Hortal, J.; Barreto, E.; Rangel, T.; Kissling, W.D. Canopy height explains species richness in the largest clade of Neotropical lianas. *Glob. Ecol. Biogeogr.* **2020**, *29*, 26–37. [CrossRef]
58. Roll, U.; Geffen, E.; Yom-Tov, Y. Linking vertebrate species richness to tree canopy height on a global scale. *Glob. Ecol. Biogeogr.* **2015**, *24*, 814–825. [CrossRef]
59. Feng, G.; Zhang, J.; Girardello, M.; Pellissier, V.; Svenning, J.-C. Forest canopy height co-determines taxonomic and functional richness, but not functional dispersion of mammals and birds globally. *Glob. Ecol. Biogeogr.* **2020**, *29*, 1350–1359. [CrossRef]
60. Fricker, G.A.; Wolf, J.A.; Saatchi, S.S.; Gillespie, T.W. Predicting spatial variations of tree species richness in tropical forests from high-resolution remote sensing. *Ecol. Appl.* **2015**, *25*, 1776–1789. [CrossRef]
61. Atkins, J.W.; Bohrer, G.; Fahey, R.T.; Hardiman, B.S.; Morin, T.H.; Stovall, A.E.L.; Zimmerman, N.; Gough, C.M. Quantifying vegetation and canopy structural complexity from terrestrial LiDAR data using the forestr r package. *Methods Ecol. Evol.* **2018**, *9*, 2057–2066. [CrossRef]
62. Hardiman, B.S.; Bohrer, G.; Gough, C.M.; Vogel, C.S.; Curtis, P.S. The role of canopy structural complexity in wood net primary production of a maturing northern deciduous forest. *Ecology* **2011**, *92*, 1818–1827. [CrossRef]
63. Zhao, F.; Strahler, A.H.; Schaaf, C.L.; Yao, T.; Yang, X.; Wang, Z.; Schull, M.A.; Román, M.O.; Woodcock, C.E.; Olofsson, P.; et al. Measuring gap fraction, element clumping index and LAI in Sierra Forest stands using a full-waveform ground-based lidar. *Remote Sens. Environ.* **2012**, *125*, 73–79. [CrossRef]
64. Hardiman, B.S.; Bohrer, G.; Gough, C.M.; Curtis, P.S. Canopy Structural Changes Following Widespread Mortality of Canopy Dominant Trees. *Forests* **2013**, *4*, 537–552. [CrossRef]
65. Detto, M.; Muller-Landau, H.C.; Mascaro, J.; Asner, G.P. Hydrological Networks and Associated Topographic Variation as Templates for the Spatial Organization of Tropical Forest Vegetation. *PLoS ONE* **2013**, *8*, e76296. [CrossRef] [PubMed]
66. Unger, M.; Homeier, J.; Leuschner, C. Effects of soil chemistry on tropical forest biomass and productivity at different elevations in the equatorial Andes. *Oecologia* **2012**, *170*, 263–274. [CrossRef] [PubMed]
67. Martínez-Ramos, M.; Ortiz-Rodríguez, I.A.; Piñero, D.; Dirzo, R.; Sarukhán, J. Anthropogenic disturbances jeopardize biodiversity conservation within tropical rainforest reserves. *Proc. Natl. Acad. Sci. USA* **2016**, *113*, 5323–5328. [CrossRef] [PubMed]

plants

MDPI

Article

Seasonal Influence of Biodiversity on Soil Respiration in a Temperate Forest

Mengxu Zhang [1,2,3], Emma J. Sayer [4,5], Weidong Zhang [1], Ji Ye [1,3], Zuoqiang Yuan [6], Fei Lin [1,3], Zhanqing Hao [6], Shuai Fang [1,3], Zikun Mao [1,3], Jing Ren [1,3] and Xugao Wang [1,3,*]

[1] CAS Key Laboratory of Forest Ecology and Management, Institute of Applied Ecology, Chinese Academy of Sciences, Shenyang 110016, China
[2] University of Chinese Academy of Sciences, Beijing 100049, China
[3] Key Laboratory of Terrestrial Ecosystem Carbon Neutrality, Shenyang 110016, China
[4] Lancaster Environment Centre, Lancaster University, Lancaster LA1 4YQ, UK
[5] Smithsonian Tropical Research Institute, Panama City 32402, Panama
[6] School of Ecology and Environment, Northwestern Polytechnical University, Xi'an 710072, China
[*] Correspondence: wangxg@iae.ac.cn; Tel.: +86-24-83-970-207 (ext. 10016)

Abstract: Soil respiration in forests contributes to significant carbon dioxide emissions from terrestrial ecosystems but it varies both spatially and seasonally. Both abiotic and biotic factors influence soil respiration but their relative contribution to spatial and seasonal variability remains poorly understood, which leads to uncertainty in models of global C cycling and predictions of future climate change. Here, we hypothesize that tree diversity, soil diversity, and soil properties contribute to local-scale variability of soil respiration but their relative importance changes in different seasons. To test our hypothesis, we conducted seasonal soil respiration measurements along a local-scale environmental gradient in a temperate forest in Northeast China, analyzed spatial variability of soil respiration and tested the relationships between soil respiration and a variety of abiotic and biotic factors including topography, soil chemical properties, and plant and soil diversity. We found that soil respiration varied substantially across the study site, with spatial coefficients of variation (CV) of 29.1%, 27.3% and 30.8% in spring, summer, and autumn, respectively. Soil respiration was consistently lower at high soil water content, but the influence of other factors was seasonal. In spring, soil respiration increased with tree diversity and biomass but decreased with soil fungal diversity. In summer, soil respiration increased with soil temperature, whereas in autumn, soil respiration increased with tree diversity but decreased with increasing soil nutrient content. However, soil nutrient content indirectly enhanced soil respiration via its effect on tree diversity across seasons, and forest stand structure indirectly enhanced soil respiration via tree diversity in spring. Our results highlight that substantial differences in soil respiration at local scales was jointly explained by soil properties (soil water content and soil nutrients), tree diversity, and soil fungal diversity but the relative importance of these drivers varied seasonally in our temperate forest.

Keywords: soil fungal diversity; soil water content; soil CO_2 efflux; spatial heterogeneity; tree diversity

Citation: Zhang, M.; Sayer, E.J.; Zhang, W.; Ye, J.; Yuan, Z.; Lin, F.; Hao, Z.; Fang, S.; Mao, Z.; Ren, J.; et al. Seasonal Influence of Biodiversity on Soil Respiration in a Temperate Forest. *Plants* **2022**, *11*, 3391. https://doi.org/10.3390/plants11233391

Academic Editor: Frank M. Thomas

Received: 2 November 2022
Accepted: 2 December 2022
Published: 5 December 2022

1. Introduction

Forests store ~45% of terrestrial biomass carbon and play a crucially important role in the global carbon cycle [1]. Forests not only sequester and store carbon in biomasses and soils, but they release large amounts of carbon dioxide (CO_2) back into the atmosphere through respiration from plants and soil. Quantifying soil respiration is particularly important for accurate predictions of global C cycling because it represents 60–80% of photosynthetic production (80–98 Pg C yr^{-1}) and accounts for 40–90% of the global CO_2 emissions from terrestrial ecosystems [2]. Indeed, forest soil respiration releases ten times more CO_2 into the atmosphere than current human fossil fuel consumption [3]. However, there is considerable uncertainty around estimates of soil respiration from forest ecosystems due to high spatial and temporal variation. Most studies of soil respiration in forests have focused on temporal variation [4], which is often quantified using continuous automated measurements at a few sampling points within a given forest. However, soil respiration can vary two- to eight-fold within tens of meters [5,6], making it difficult to fully capture the spatial variability of soil respiration without continuous large-scale observations. Current knowledge of the underlying mechanisms of spatial variability of soil respiration in forests is relatively limited, which creates great uncertainty when estimates of soil respiration are upscaled to the ecosystem, regional, or global scale [7].

The high temporal and spatial variability of soil respiration results from numerous individual organisms and processes that contribute to total CO_2 efflux from the soil. Soil respiration is often partitioned into heterotrophic respiration from microorganisms in the bulk soil and rhizosphere respiration, comprising the CO_2 efflux from plant roots, their microbial symbionts, and other rhizosphere microorganisms [8]. These two major components of soil respiration each involve various biological processes from numerous organisms, which respond differently to environmental conditions, such as topography, microclimates, and soil elemental concentrations. Soil elemental concentrations influence soil respiration by determining the availability of nutrients and carbon for plant and microbial metabolism [9]. Soil respiration is also strongly regulated by microclimates, including soil temperature and water content. Heterotrophic respiration generally increases exponentially with temperature within typical soil temperature ranges due to greater microbial enzyme activity, greater substrate affinity, and enhanced substrate diffusion rates [10]. By contrast, soil respiration is often highest at intermediate soil water content because CO_2 efflux is limited by substrate transport as well as microbial physiology and plant activity at low soil moisture levels, but limited by oxygen availability at high soil moisture levels [11]. Many forests have high soil water content, which can result in anaerobic conditions, thereby reducing soil CO_2 emissions from roots and soil microorganisms [12]. However, topography can modify the hydrological conditions and other biophysical variables, which complicates assessments of the spatial and temporal variation of soil respiration [13]. Importantly, as plant growth and microbial activities are influenced by seasonal temperature and precipitation, the drivers of soil respiration can also vary across seasons [14,15]. Indeed, compared with abiotic factors, the relative contribution of biotic factors, such as plants and soil microbes, to soil variation could be more variable in both space and time.

Plant and microbial communities play decisive roles in regulating soil respiration since they are the principal circular pathways through which carbon enters the soil and is released back into the atmosphere. Plant diversity increases soil autotrophic respiration by enhancing metabolic rates and fine root biomass [16], but it also stimulates soil heterotrophic respiration due to a greater amount and variety of carbon and nutrient resources available for soil microorganisms [17]. In addition, stand structural complexity (e.g., individual tree size variation) could indirectly affect soil respiration by altering understory light environments, understory plant diversity [18], soil temperature variability [19], and soil microbial activity [20]. However, the influence of plant diversity on soil respiration is likely to vary substantially in time, as root growth and plant litter inputs are inextricably linked to plant growth and therefore often show seasonal patterns [21]. Besides differences in root respiration, plant diversity primarily influences soil respiration via the quality and quantity of plant inputs available to microbes [22]. It is highly likely that soil microbial diversity plays critical roles in soil C cycling in terrestrial ecosystems. However, empirical evidence demonstrating the role of microbial community composition in driving C fluxes such as soil respiration is limited, especially in natural forests. Some experimental evidence has shown that loss of microbial diversity led to higher rates of soil microbial respiration [23]. High soil microbial diversity not only limits community activity by increasing interspecific competition [24], but diverse communities can also contain species that contribute less to heterotrophic respiration [25]. Importantly, soil microbial diversity and community composition are closely associated with plant diversity but also display seasonal patterns that follow changes in soil temperature and soil water content [26]. However, to date, we do not know how above- and belowground diversity collectively shape seasonal differences in soil respiration at local scales.

Large forest dynamics plots provide a great opportunity to understand how plant attributes (e.g., plant biomass, species diversity, and stand structure) and soil diversity regulate soil respiration in forests while accounting for environmental conditions (microclimate and soil nutrients). Here, we conducted field measurements of soil respiration at 150 sampling points uniformly covering a 25 ha permanent temperate mixed forest plot in Northeast China in different seasons (spring, summer, and autumn). We hypothesize that: (1) The relative importance of biotic and abiotic factors in determining local variation in soil respiration varies among seasons; and (2) Above- and belowground species diversity and soil properties jointly regulate local variation in soil respiration, whereby soil respiration would increase with plant diversity but decline with increasing soil diversity.

2. Results

2.1. Seasonal and Spatial Variability of Soil Respiration

Soil respiration rates differed among seasons, with the highest rates in summer (4.82 ± 1.32 μmol CO_2 m^{-2} s^{-1}), followed by spring (3.21 ± 0.93 μmol CO_2 m^{-2} s^{-1}), and the lowest in autumn (2.25 ± 0.69 μmol CO_2 m^{-2} s^{-1}). In addition, the range of soil respiration rates was also largest in summer (1.65~8.34 μmol CO_2 m^{-2} s^{-1}), somewhat smaller in spring (1.34~7.16 μmol CO_2 m^{-2} s^{-1}), and smallest in autumn, with much lower minimum values (0.69~4.95 μmol CO_2 m^{-2} s^{-1}; Table 1). Nonetheless, spatial variability of soil respiration was similarly high in all seasons, with a spatial variation coefficient (CV) of 29.1%, 27.3%, and 30.8% in spring, summer, and autumn, respectively.

Table 1. Descriptive statistics for the soil respiration, above- and below-ground community structure and soil properties in a 25 ha temperate forest dynamics plot in Northeast China, showing maximum (Max.), minimum (Min.), and median (Med.) values as well as standard deviations (SD) and the coefficient of variation (CV) for n = 120 subplots. Rs is the mean soil respiration rate, ST is mean soil temperature, and SWC is mean soil water content, given for spring, summer, and autumn. Soil PCA1 and Soil PCA2 are ordination axes representing eight indexes of total soil elements or extractable nutrients and pH, respectively (Figure S1); $Tree_{BA}$ is the total basal area of aboveground plants; $Tree_{SV}$ is tree size variation; $Tree_{SR}$ is the species richness of the plant community; $Bacteria_{SWI}$ is the Shannon–Wiener index for soil bacteria; $Fungi_{SWI}$ is the Shannon–Wiener index for soil fungi; $Nematode_{SR}$ is the species richness of nematodes.

Variable	Mean	Max.	Median	Min.	SD	CV (%)
Rs_{Spring} (μmol CO_2 m^{-2} s^{-1})	3.21	7.16	3.08	1.34	0.93	29.1
Rs_{Summer} (μmol CO_2 m^{-2} s^{-1})	4.82	8.34	4.81	1.65	1.32	27.3
Rs_{Autumn} (μmol CO_2 m^{-2} s^{-1})	2.25	4.95	2.26	0.67	0.69	30.8
ST_{Spring} (°C)	12.3	14.7	12.3	11.0	0.70	5.7
ST_{Summer} (°C)	16.8	18.2	16.8	15.4	0.75	4.5
ST_{Autumn} (°C)	12.3	14.8	12.3	8.6	1.35	11.0
SWC_{Spring} (%)	39.79	53.40	40.33	13.15	7.99	20.1
SWC_{Summer} (%)	36.31	52.63	35.07	11.53	8.93	24.6
SWC_{Autumn} (%)	41.88	52.95	44.11	20.85	7.25	17.3
Soil PCA1	0	5.43	−0.24	−4.49	1.77	-
Soil PCA2	0	3.41	−0.08	−4.61	1.26	-
$Tree_{BA}$ (m^2)	1.73	2.88	1.72	0.33	0.52	29.8
$Tree_{SV}$	1.66	2.26	1.65	1.11	0.26	15.7
$Tree_{SR}$	11	20	11	5	2.41	21.9
$Bacteria_{SWI}$	6.54	6.82	6.55	6.27	0.12	1.8
$Fungi_{SWI}$	3.84	5.26	3.91	2.40	0.53	13.7
$Nematode_{SR}$	5.24	7.42	5.22	3.23	0.82	15.6
Elevation (m)	803.5	809.0	804.2	796.0	3.23	0.40
Slope (°)	2.91	16.03	2.32	0.28	2.38	81.81
Convexity (m)	0.02	2.69	0.02	−3.59	0.68	-

2.2. Abiotic and Biotic Factors Influencing Spatial Variation of Soil Respiration

Of all measured potential predictors of soil respiration, seasonal variation was highest for soil temperature, with values of 12.3 ± 0.7 °C in spring, 16.8 ± 0.8 °C in summer, and 12.3 ± 1.4 °C in autumn. However, the spatial CV for soil temperature was low (5.7%, 4.5%, and 11.0%, in spring, summer, and autumn, respectively). By contrast, soil water content varied less across seasons, with values of 39.8 ± 8.0% in spring, 36.3 ± 8.9% in summer, and 41.9 ± 7.3% in autumn, but the spatial CV of soil water content was high in each season (20.1%, 24.6%, and 17.3%, in spring, summer, and autumn, respectively). Of the tree community parameters, spatial variation in thetree basal area was highest, with a spatial CV of 29.8%, compared to 21.9% for tree species richness and 15.7% for tree size variation. Finally, of the soil diversity parameters, bacteria had higher diversity (6.54 ± 0.12) but lower spatial variation (1.8%) than fungi (3.84 ± 0.53; CV 13.7%) or nematodes (5.24 ± 0.82; CV 15.6%).

Multiple linear regression showed that soil respiration was related to distinct abiotic and biotic factors depending on the season (Figure 1a–c). Soil respiration was not related to topography in any season but declined strongly with increasing water content in all seasons, especially in summer (β = −0.22, $p < 0.01$, Table S5). Soil respiration increased with temperature only in summer (β = 0.05, $p < 0.05$; Figure 1), and declined with increasing total soil elements in autumn (β = −0.09, $p = 0.01$).

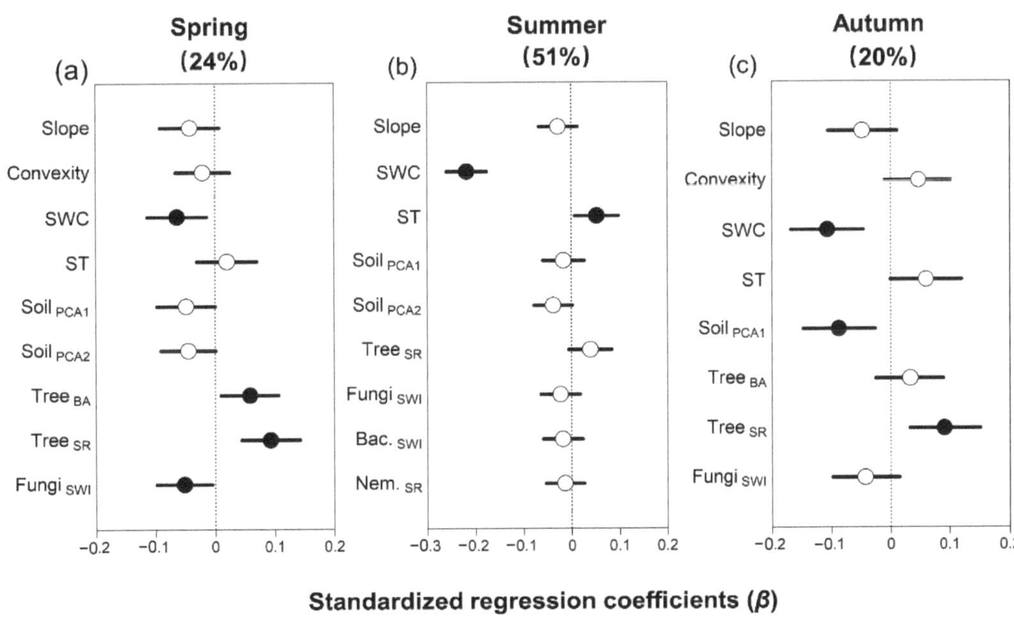

Figure 1. Standardized regression coefficients (β) of the biotic and abiotic predictors of soil respiration across a 25 ha forest dynamics plot in Northeast China, derived from multiple linear regression models for each season. The total R^2 is given in parentheses for each model (See also Table S5). Closed circles indicate significant relationships with soil respiration at $p < 0.05$, and lines indicate standard errors. All abbreviations follow Table 1.

Overall, soil respiration increased with tree biomass and diversity, but the strength of the relationship varied among seasons. In spring, soil respiration was highest in subplots with high tree basal area (β = 0.06, $p < 0.05$) and tree diversity (β = 0.09, $p < 0.01$) but there was no relationship between soil respiration and basal area in summer or autumn. Soil respiration generally declined with increasing soil diversity, but the relationship was only significant for fungal diversity in spring (β = −0.05, $p < 0.05$). In summary, low rates of soil respiration were associated with increasing soil moisture, total soil elements, and fungal diversity, whereas high respiration rates were associated with increasing soil temperature, and high tree biomass and diversity, but the strength of the relationships varied markedly among seasons.

2.3. Interactive Effects of Main Factors on Soil Respiration

The direct paths in the SEMs conformed to the results of the multiple linear regressions. Among the measured abiotic factors, soil water content had the greatest direct effect, as lower rates of soil respiration were associated with high soil water content in all seasons (Figure 2a–c). In autumn, soil respiration was also lower in subplots with high concentrations of total soil elements (Figure 2c) but none of the other abiotic factors were associated with local-scale differences in soil respiration rates. Of the measured biotic factors, soil respiration was most strongly associated with tree diversity, with higher rates of soil respiration in subplots with high tree diversity in all seasons (Figure 3a,b). In spring, subplots with greater tree biomass had higher rates of soil respiration (Figure 3a), but subplots with high soil fungal diversity had lower rates of soil respiration.

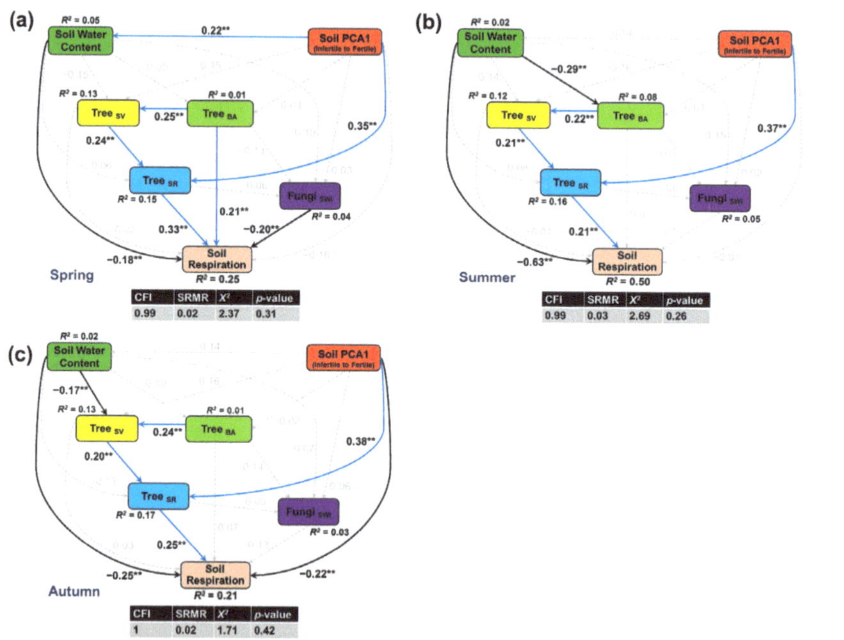

Figure 2. Structural equation models showing differences in soil respiration in spring (**a**), summer (**b**), and autumn (**c**) associated with soil moisture (soil water content), total soil elements (soil PCA1), plant biomass ($Tree_{BA}$), stand structural complexity ($Tree_{SV}$), plant diversity ($Tree_{SR}$), and soil fungal diversity ($Fungi_{SWI}$). Blue and black solid arrows represent significant positive or negative paths, respectively, at $p < 0.05$ (** $p < 0.05$); grey dashed arrows indicate non-significant paths. Values adjacent to arrows represent standardized coefficients (see Table S6). All abbreviations follow Table 1.

Indirect paths in the SEMs revealed that higher rates of respiration were associated with total soil elements in all seasons via tree diversity (Table S6). In summer and autumn, high soil water content was also associated with reduced tree biomass and stand structural complexity (Figure 2b,c). Although stand structure had no direct effect on soil respiration, it was associated with higher respiration rates via tree diversity in spring (Figure 2a). Overall, soil water content and tree diversity were identified as the two strongest predictors of spatial variation in soil respiration (Figure 3). The relative contribution of soil water content to variability in soil respiration in spring, summer, and autumn was 17.2%, 60.5%, and 28.4%, respectively, (Figure 3a), whereas tree diversity contributed 27.4.1%, 18.7%, and 23.9%.

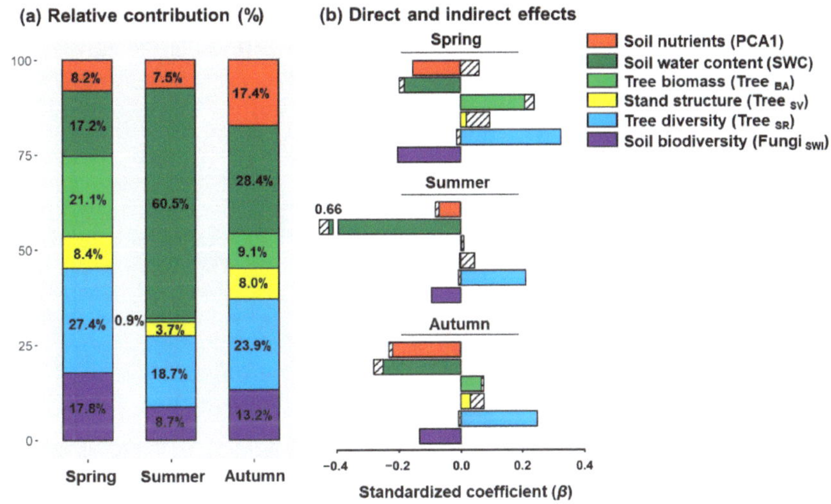

Figure 3. Relative contributions (**a**), and direct and indirect effects (**b**) of significant predictors to soil respiration rates in different seasons. Solid colours represent direct effects derived from structural equation models, whereas striped bars represent indirect effects (see also Table S6).

3. Discussion

Our study evaluated the relative importance of plant and soil diversity in explaining local-scale variability in soil respiration in different seasons in a temperate forest. We demonstrate that high rates of soil respiration were associated with tree community characteristics (i.e., tree biomass, stand structural complexity, and tree species richness), whereas lower respiration rates were associated with belowground abiotic (soil water content and total soil elements) and biotic factors (soil fungal diversity). Importantly, the strength of these above- and below-ground factors as predictors of soil respiration varied greatly among seasons. How soil respiration might be influenced by complex relationships among abiotic conditions and multiple trophic groups throughout the year has rarely been investigated [27]. Thus, our work adds considerably to our understanding of local-scale variability in soil respiration.

3.1. Seasonal and Spatial Variability of Soil Respiration

The clear seasonal dynamics of soil respiration, with a peak during the summer, is characteristic of temperate forest systems, where high temperatures in summer stimulate the growth and metabolic rates of trees and microbes [28]. Compared with soil water content, the seasonal dynamics in soil respiration was predominantly determined by soil temperature (Figure S3), probably because temperature substantially influences plant phenology and soil microbial seasonal activity in temperate forests [29].

Soil respiration was highly variable across the local spatial scales measured in our study, with spatial CVs ranging from 27% to 31%. The spatial variability in our forest fell within the wide ranges previously recorded in Chinese forests (17–62.6%; [30]). Other studies in similar forests have observed both higher and lower spatial variability [31,32], demonstrating the importance of sampling design to gain accurate estimates of soil respiration in forests [7]. High spatial variability in soil respiration can result from excessively dense sampling. For example, Shi et al. [32] found that 87–91% of the spatial variance was explained by an autocorrelation over a range of 15 to 23 m. By contrast, the minimum sampling distance in our study was >28 m, so we only detected a weak spatial autocorrelation of soil respiration in spring.

3.2. Relationships between Abiotic Factors and Soil Respiration

In support of our first hypothesis, soil water content played the most important role in inhibiting soil respiration during the growing season, which was probably due to high rainfall and the high water holding capacity of the forest floor. Previous studies have revealed a threshold value of soil water content (approximately 20%), at which the relationship between soil respiration and soil moisture changes [33]. At soil water content above the threshold value, soil respiration is limited due to low CO_2 transport, oxygen (O_2) availability, and thus reduced biological activity. By contrast, soil water contents below the threshold promote root and soil microbial respiration through the diffusion of soluble substrates in an aerobic environment [33]. In our study, 98% of the measurements showed a soil water content greater than 20%, which explains why soil respiration rates declined with increasing soil water content. The limiting effect of high soil water content was strongest in summer, due to the high autotrophic and heterotrophic respiration and thus greater requirement of oxygen availability and soil porosity [34]. Indeed, soil water content in summer explained a much higher proportion of the variability in soil respiration (60.5%) than any other abiotic or biotic factors.

Surprisingly, soil properties, including organic matter and total and extractable nitrogen, phosphorus, and potassium, were not strong predictors of soil respiration in our study. In general, high-nutrient substrates should increase soil heterotrophic respiration by enhancing microbial biomass and fungal abundance [9]. However, late-successional forests such as ours often generally have high soil nutrient concentrations, and nutrient availability is therefore less likely to limit soil respiration [35]. Instead, the indirect association between total soil elements and soil respiration in our SEMs (Figure 2) suggests that nutrient availability increased soil respiration by promoting plant diversity [36].

3.3. Contrasting Relationships between Soil Respiration and Above- or Belowground Diversity

Higher rates of soil respiration at sites with greater tree species richness have also been observed in other ecosystems and on a global scale [37]. Plant species richness is usually associated with greater chemical diversity of litter and root exudates [16], which stimulates soil heterotrophic respiration by providing a greater range of carbon and nutrient resources to soil microorganisms [38]. Although we found no relationship between tree species richness and soil diversity (Table S4), previous work at the same study site demonstrated that soil microbial diversity was linked to the functional diversity of the tree community, rather than species diversity [39]. Thus, the strong role of tree species diversity in explaining spatial variability in soil respiration is likely a combination of species differences in the autotrophic component of soil respiration, as well as differences in tree litter quality and quantity influencing decomposer organisms and heterotrophic respiration.

The stronger relationship between soil respiration and fungal diversity compared to soil bacteria and nematodes, is consistent with previous studies [40], and is often explained by the key role of fungi in decomposing recalcitrant plant materials [41]. However, in our study, soil respiration in spring declined with increasing fungal diversity (Figure 1), supporting our second hypothesis. The negative relationship between soil respiration and soil fungal diversity could be the result of higher carbon use efficiency by a diverse fungal community benefitting from distinct resource niches [42], which does not necessarily increase overall community activity and heterotrophic respiration [43]. Alternatively, differences in the abundances and diversity of distinct fungal functional guilds could influence the relationship between soil respiration and overall fungal diversity. In our study, ectomycorrhizal (EcM) fungi accounted for 40.8% of the total fungal abundance, but only contributed 14.8% to fungal species richness (Table S7). Furthermore, ectomycorrhizal fungi account for a large proportion of total soil respiration (15–26%; [44,45]), so the dominance of ectomycorrhizal fungi could explain why soil relationship rates were highest in subplots with low fungal diversity.

Although greater tree biomass is thought to stimulate soil microbial growth and activity by increasing plant inputs, such as litter and root exudates to the soil [46], we found only a weak relationship between basal area, as a proxy of tree biomass, and soil respiration in spring (Figure 1). It is possible that the relationship between soil respiration and tree biomass in summer and autumn was obscured by the overriding influence of soil water content [47], when soil microbial activity or heterotrophic respiration was probably limited by anoxic soil conditions. Although soil respiration was also not directly related to stand structure (represented by tree size variation), our SEMs revealed that a more complex stand structure was associated with higher respiration rates via tree diversity (Figure 2). Structurally complex forests have a greater range of niches and more diverse light conditions, which increase understory plant abundance and richness, thereby enhancing autotrophic and heterotrophic respiration [48].

4. Materials and Methods

4.1. Study Site Description and Experimental Design

The study site was located in the Changbai Nature Reserve in Northeast China, extending from 41°42′ to 42°26′ N and 127°42′ to 128°17′ E. As one of the largest biosphere reserves in China, the Changbai Nature Reserve was established in 1960 and joined the World Biosphere Reserve Network in 1980. Mean annual precipitation is approximately 700 mm and most rainfall occurs from June to August (450–500 mm). Mean annual temperature is 2.8 °C, with monthly means of −13.7 and 19.6 °C in January and July, respectively [49]. Previous work in the study forest indicates a total annual soil respiration of 1017 g m^{-2}, accounting for 76% of ecosystem respiration [50].

Our study was conducted in the 25 ha Changbaishan (CBS) Forest Dynamics Plot (FDP), which is one of the sites in the worldwide CTFS-ForestGEO forest monitoring network (http://www.forestgeo.si.edu, accessed on 5 January 2021). All free-standing individual woody plants with a diameter at breast height (DBH) ≥1 cm were mapped, measured, and identified to species in 2004, 2009, 2014, and 2019. Based on the first census data in 2004, there are 38,902 individuals belonging to 52 species, 32 genera, and 18 families [51]. The 25 ha plot is dominated by the late-successional deciduous broadleaved Korean pine (*Pinus koraiensis*) mixed forest with common tree species including *P. koraiensis*, *Tilia amurensis*, *Quercus mongolica*, *Fraxinus mandshurica*, *Ulmus japonica*, and *Acer mono*. Following a standard field protocol [52], the 25 ha plot was divided into 625 subplots (20 m × 20 m). We measured three topographical variables (elevation, slope, and convexity) for each subplot following Harms et al. [53]. Elevation of each subplot was estimated from the mean elevation at each corner. The slope was defined as the mean angular deviation from horizontal of each of the four triangular planes formed by connecting three subplot corners. Subplot convexity was calculated as the elevation of the subplot minus the mean elevation of the eight surrounding subplots; for edge subplots, convexity was calculated as the elevation of the central point minus the mean of the four corners.

To capture the spatial variability of soil respiration, 150 sampling points were established as evenly as possibly across the 25 ha plot (Figure 4). In April 2020, permanent soil collars made of polyvinyl chloride (20 cm diameter and 10 cm height) were inserted 4 cm into the soil at each sampling point and left in situ throughout the soil respiration measurements. Soil collars set at this depth were stable and caused minimal disturbance to shallow fine roots. To avoid the confounding effects of above-ground plants on soil respiration, we removed small living plants growing inside the collars one day before each measurement.

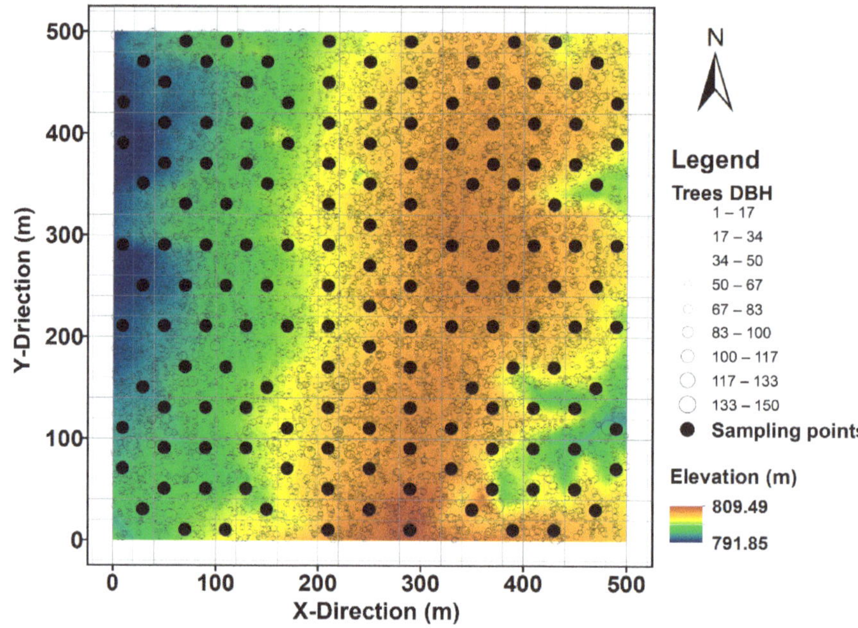

Figure 4. Contour map of the 25 ha Changbaishan Forest Dynamics Plot in Northeast China, showing the sampling points of soil respiration as solid circles and the positions of all living trees as open circles, whereby symbol sizes indicate tree diameter at breast height (DBH). Colours indicate differences in elevation.

4.2. Measurements of Soil Respiration and Soil Microclimates

Soil respiration, temperature, and water content were measured seven times during the growing season of 2020 (May to October): twice in spring (May–June), three times in summer (July–August), and twice in autumn (September–October). Soil respiration was measured using an automated soil CO_2 flux system, which consisted of a dynamic soil chamber (3140 cm^3 volume) attached to an infrared gas analyzer (Li-8100A, Li-COR Inc., Lincoln, NE, USA). Each measurement lasted 90 s, recording CO_2 and water concentrations and air temperature inside the chamber every second. Soil temperature and volumetric water content at 0–5 cm soil depth were determined adjacent to each collar using a type E thermocouple (8100-201, Li-COR Inc., Lincoln, NE, USA) and a ML2x soil moisture probe (Li-COR Inc., Lincoln, NE, USA), respectively. During each sampling campaign, measurements took two days to complete. As diel variation in soil CO_2 efflux is low in heavily shaded forested areas [54], all measurements were performed in a random order between 10 a.m. and 4 p.m. on each measurement day. We did not measure soil respiration in winter, as low temperatures strongly constrain microbial activity.

4.3. Sampling and Analysis of Soil Physicochemical Properties

To represent the average soil properties associated with plant communities, we randomly selected two soil sampling sites at the midpoints between the central point and the four corners in the above-mentioned 150 subplots (20 m × 20 m). At each site, we took five soil cores (3.8 cm in diameter and 10 cm in depth) at random locations near the 150 sampling points using a soil auger after removing the litter layer from the ground surface. Subplots located at the edge of 25 ha plot were not included to eliminate potential edge effects resulting in a total of 120 quadrats. We mixed the cores to create one composite sample per measurement point. Each soil sample was then divided into two parts after sieving the sample through a 2 mm mesh to remove the roots and stones: one for soil nutrient analysis and the other for soil diversity measurement (i.e., bacteria, fungi, and nematodes). All processing was completed within 12 h of collection, and the subsamples for soil diversity analysis were stored at −80 °C.

We measured eight soil nutrient variables known to influence soil respiration: soil pH, organic matter (SOM), extractable nitrogen (N_{EX}), extractable phosphorus (P_{EX}), extractable potassium (K_{EX}), total nitrogen (TN), total phosphorus (TP), and total potassium (TK). Following Lu [55], soil pH was determined in water (1:1 soil:solution ratio) using a glass electrode, SOM was determined colorimetrically following dichromate oxidation, N_{EX} was determined by extraction in 1 mol NaOH L^{-1} and subsequent titration with 0.01 $mL \cdot L^{-1}$ sulfuric acid, TN was estimated colorimetrically after KCl extraction, using the Kjeldahl method, and P_{EX} and TP were determined by molybdate colorimetry, after Mehlich 3 extraction or digestion in H_2SO_4–$HClO_4$, respectively; K_{EX} and TK were determined by atomic absorption spectrometry after extraction with 1 mol L^{-1} NH_4Ac or digestion in hydrofluoric acid, respectively.

4.4. Soil Diversity

Soil bacterial and fungal diversity and community composition were determined by sequencing on an Illumina Miseq platform (Illumina, San Diego, CA, USA). Soil genomic DNA was isolated from 0.25 g of fresh soil using the MoBio PowerSoil® DNA Isolation extraction kit according to the manufacturer's instructions (MoBio Laboratories Inc., Carlsbad, CA, USA). The quality of the DNA was assessed based on 260/280nm and 260/230 nm absorbance ratios obtained using a Spectrophotometer (NanoDrop Technologies Inc., Wilmington, DE, USA). Extracted DNA samples were stored at −20 °C until further use.

The universal bacterial V4~V5 region of 16S rRNA gene was amplified by using primers 515 F (5′-GTGCCAGCMGCCGCGG-3′) and 907 R (5′-CCGTCAATTCMTTTRAGTTT-3′) [56]. The fungal ITS sequence of 18S rRNA genes was amplified using primers ITS_1737F (5′-GGAAGTAA AAGTCGTAACAAGG-3′) and ITS_2043R (5′-ATGCAGGCTGCGTTCTTCA TCGATGC-3′) [57]. Amplification by polymerase chain reaction (PCR) was conducted in triplicate using a 20 µL mixture containing 4 µL of 5× FastPfu Buffer, 2 µL of 2.5 mM dNTPs, 0.8 µL of each primer (5 µM), 0.4 µL of FastPfu Polymerase, and 10 ng of template DNA was performed. The PCR analyses were carried out on a Gene Amp PCR-System 9700 (Applied Biosystems, Foster City, CA, USA) using thermal cycling conditions, as follows: initial denaturation step at 95 °C for 3 min, followed by 27 (16S rRNA) or 35 (ITS) cycles at 95 °C for 30 s, annealing at 55 °C for 30 s and extension72 °C for 45 s, and a final extension at 72 °C for 10 min. The PCR products were sequenced using 300PE MiSeq (Illumina, San Diego, CA, USA), at the Shanghai Majorbio Bio-pharm Biotechnology Co., Ltd. (Shanghai, China).

Obtained DNA sequences were processed using the QIIME 2 software [58], discarding sequences shorter than 200 bp with a mean quality score <25 and ambiguous bases. All quality-filtered sequences were clustered into Operational Taxonomic Units (OTUs) with a 97% identity threshold using UPARSE version 7.1 (http://drive5.com/uparse/, accessed on 14 May 2021). Chimeras were filtered using the Ribosomal Database Project (RDP) and UCHIME [59]. Saprotrophic vs. ectomycorrhizal fungal functional guilds were identified according to Yao et al. [60].

Nematodes were extracted from 200 g of fresh soil by an updated cotton-wool filter method [61]. For each sample, the first nematodes encountered on the slides were identified at genus level at 100× magnification under an inverted microscope. The nematodes were assigned to four trophic groups (bacterivores, fungivores, omnivores-predators, and plant parasites) according to Yeates et al. [62]. The abundance of nematodes was calculated as the number of individuals per 100 g dry soil. The detailed procedure for soil nematode extraction and identification has been described by Guan et al. [63]. Soil diversity in each subplot was represented by species richness calculated using the Shannon–Wiener index for soil fungi ($Fungi_{SWI}$), bacteria ($Bacteria_{SWI}$), and species richness of nematodes ($Nematode_{SR}$).

4.5. Plant Community Characteristics

We described plant community characteristics using the latest census data of the plot in 2019, which measured and identified all woody plants (henceforth 'trees') with a diameter at breast height (DBH) >1 cm. We used species richness to represent tree diversity ($Tree_{SR}$), basal area to represent biomass ($Tree_{BA}$), and the coefficient of variation for DBH as a measure of tree size variation ($Tree_{SV}$), which is a proxy of stand structure [64]. To determine the spatial influence of the tree community on soil respiration, we calculated tree community characteristics for a 5 m, 10 m, and 15 m radius around each sampling point. As the tree community characteristics within a 10 m radius showed the strongest correlation with soil respiration (Table S2), we used the tree community characteristics within a 10 m radius for all subsequent analyses.

4.6. Statistical Analyses

All analyses were conducted in R 4.1.3 (R Development Core Team, Vienna, Austria, 2022), using the vegan package [65] to calculate tree species diversity and soil diversity indices, the MuMIn [66] and lavaan [67] packages for model averaging and structural equation models, respectively, and the spdep package [68] for Moran's *I* test.

To assess spatial autocorrelation of soil respiration among sampling plots, we conducted Moran's *I* test [69] in each season (spring, summer and autumn). We did not find any significant spatial autocorrelation in summer and autumn, and found a slight positive spatial autocorrelation in spring (Table S3).

The eight measured soil properties (soil pH, organic matter content, N_{EX}, P_{EX}, K_{EX}, TN, TP, and TK) were reduced to a set of orthogonal variables using principal component analysis (PCA). The first axis of the PCA (soil PCA1) explained 39% of the total variation and was positively correlated with total soil elements, whereas the second axis (soil PCA2) explained 20% and was positively correlated with extractable soil elements (Figure S1). Thus, both axes represented a soil fertility gradient from infertile to fertile soils (Table S1) and were used in subsequent analyses to represent total soil elements (soil PCA1) and extractable soil nutrients (soil PCA2)

To examine the effects of topography, soil water content, soil temperature, soil properties, and tree or soil biodiversity on soil respiration, we performed multiple linear regressions models for soil respiration in each season. To avoid multicollinearity problems, we first assessed correlations between pairs of predictors within each group of variables (i.e., tree diversity, soil diversity, soil properties, topography) and excluded one predictor per pair if $r > 0.60$ (Table S4); in each case, we retained the predictor that had greater explanatory power for variation in soil respiration. Using this approach, we included two parameters for tree communities (basal area and diversity), one for stand structure (tree size vatiation), three for soil diversity (Shannon's diversity for fungi and bacteria, and species richness for nematodes), three parameters for topography (elevation, slope, and convexity), and four for soil properties (soil water content, soil temperature, total soil elements, and extractable soil nutrients), as predictors of soil respiration in the multiple linear regression models. Then, for each predictor, we performed all subsets regression analysis and selected the optimal model based on the lowest Akaike Information Criterion adjusted for small sample sizes (AICc). However, if the difference in AICc between models was <2 units, we obtained the standardized regression coefficient (β) of each soil respiration predictor using model averaging (Table S5).

To identify the potential direct and indirect pathways and relative contributions through which soil properties and tree and soil communities regulate soil respiration, we built structural equation models (SEMs) following an initial conceptual model (Figure S4). In the SEMs, we used the significant predictors obtained from the optimal regression models: tree biomass, tree diversity, stand structure, and fungal diversity, as well as soil water content and total soil elements. The model fit to the data was evaluated using a Chi-square test ($p > 0.05$ indicates that the model is accepted), Bentler's comparative fit index (CFI close to 1 indicates perfect model), and the standardized root mean square residual (SRMR < 0.08 indicates the most appropriate model) [67].

The indirect effect of each predictor in the final SEMs was calculated through the interaction of the standardized direct effect of a given predictor on a mediator with the direct effect of the mediator on the response variable. More specifically, the total indirect effect was calculated by multiplying the standardized direct effects of a given predictor on soil respiration via mediators for each relevant path, and then we added all indirect effects of the predictor to quantify the total indirect effect. To quantify the relative contribution of different predictors to soil respiration, we calculated the relative importance for each predictor of soil respiration using the ratio between the total effect of a given predictor and the sum of the absolute value of total effects of all predictors (Table S6) [70].

5. Conclusions

Our work demonstrates high spatial variability of soil respiration on a local scale in a temperate forest, but the influence of biotic and abiotic factors on soil respiration changed during the growing season. Our findings enhance our understanding of spatial and temporal variability of soil respiration and can thus improve predictions of soil carbon flux in temperate forests. However, although local-scale differences in soil respiration could be partially attributed to differences in soil water content and tree or soil fungal diversity, a large proportion of variation remains unexplained (75%, 50%, and 79% in spring, summer, and autumn, respectively). Thus, to further reduce uncertainty in estimating soil respiration, more future studies should consider: (a) distinguishing between soil autotrophic and heterotrophic respiration; (b) assessing how plant functional diversity influences autotrophic and heterotrophic respiration; and (c) differences in soil microbial functional groups and their contribution to heterotrophic soil respiration.

Supplementary Materials: The following are available online at https://www.mdpi.com/article/10.3390/plants11233391/s1, Table S1: Principal component analysis (loadings) of the soil nutrients, Table S2: Correlations between plant community structural parameters and soil respiration, Table S3: Spatial autocorrelation analysis of soil respiration, Table S4: Pearson's correlation coefficients of all abiotic and biotic variables used in this study, Table S5: Summary of the multiple linear models for soil respiration, Table S6: The direct, indirect, and total standardized effects of abiotic and biotic factors in SEM, Table S7: The relative contribution of soil fungal components, Figure S1: Principal component analysis (loadings) of the soil nutrients, Figure S2: The pairwise correlations of soil respiration and soil microclimates, Figure S3: Relationships between temporal variation in mean soil respiration, Figure S4: A conceptual model (a) and schematic illustration of ecological hypotheses/theories/mechanisms (b) to test the effects of above- and below-ground community attributes on soil respiration.

Author Contributions: Conceptualization, X.W. and M.Z.; methodology, M.Z.; formal analysis, M.Z.; writing—original draft preparation, M.Z. and X.W.; writing—review and editing, M.Z., E.J.S., W.Z., J.Y., Z.Y., F.L., Z.H., S.F., Z.M., J.R. and X.W.; funding acquisition, X.W. All authors have read and agreed to the published version of the manuscript.

Funding: This research was funded by the National Key Research and Development Program of China (2022YFF1300501), the National Natural Science Foundation of China (Grants 31961133027 and 32001121), the Key Research Program of Frontier Sciences, Chinese Academy of Sciences (Grant ZDBS-LY-DQC019), Major Program of Institute of Applied Ecology, Chinese Academy of Sciences (IAEMP202201), the General Program of China Postdoctoral Science Foundation (2021M703397), and the Special Research Assistant Project of Chinese Academy of Sciences (2022000056).

Data Availability Statement: The data are available upon request from the corresponding author.

Acknowledgments: We sincerely appreciate the incredible support of all teachers and students in the research team throughout the research period.

Conflicts of Interest: The authors declare no conflict of interest.

References

1. Pan, Y.; Birdsey, R.A.; Fang, J.; Houghton, R.; Kauppi, P.E.; Kurz, W.A.; Phillips, O.L.; Shvidenko, A.; Lewis, S.L.; Canadell, J.G.; et al. A Large and Persistent Carbon Sink in the World's Forests. *Science* **2011**, *333*, 988–993. [CrossRef] [PubMed]
2. IPCC. Summary for Policymakers. In *Climate Change 2021: The Physical Science Basis. Contribution of Working Group I to the Sixth Assessment Report of the Intergovernmental Panel on Climate Change*; Masson-Delmotte, V., Zhai, P., Pirani, A., Connors, S.L., Péan, C., Berger, S., Caud, N., Chen, Y., Goldfarb, L., Gomis, M.I., et al., Eds.; Cambridge University Press: Cambridge, UK, 2021; pp. 3–32.
3. Bond-Lamberty, B.; Thomson, A. Temperature-associated increases in the global soil respiration record. *Nature* **2010**, *464*, 579–582. [CrossRef] [PubMed]
4. Lei, J.; Guo, X.; Zeng, Y.; Zhou, J.; Gao, Q.; Yang, Y. Temporal changes in global soil respiration since 1987. *Nat. Commun.* **2021**, *12*, 403. [CrossRef] [PubMed]
5. Ohashi, M.; Gyokusen, K. Temporal change in spatial variability of soil respiration on a slope of Japanese cedar (*Cryptomeria japonica* D. Don) forest. *Soil Biol. Biochem.* **2007**, *39*, 1130–1138. [CrossRef]
6. Wu, X.; Yuan, J.; Ma, S.; Feng, S.; Zhang, X.; Hu, D. Seasonal spatial pattern of soil respiration in a temperate urban forest in Beijing. *Urban For. Urban Green.* **2015**, *14*, 1122–1130. [CrossRef]
7. Rodeghiero, M.; Cescatti, A. Spatial variability and optimal sampling strategy of soil respiration. *For. Ecol. Manag.* **2008**, *255*, 106–112. [CrossRef]
8. Bahn, M.; Janssens, I.A.; Reichstein, M.; Smith, P.; Trumbore, S.E. Soil respiration across scales: Towards an integration of patterns and processes. *New Phytol.* **2010**, *186*, 292–296. [CrossRef]
9. Whitaker, J.; Ostle, N.; Nottingham, A.T.; Ccahuana, A.; Salinas, N.; Bardgett, R.D.; Meir, P.; McNamara, N.P. Microbial community composition explains soil respiration responses to changing carbon inputs along an Andes-to-Amazon elevation gradient. *J. Ecol.* **2014**, *102*, 1058–1071. [CrossRef]
10. Wang, X.; Liu, L.; Piao, S.; Janssens, I.A.; Tang, J.; Liu, W.; Chi, Y.; Wang, J.; Xu, S. Soil respiration under climate warming: Differential response of heterotrophic and autotrophic respiration. *Glob. Change Biol.* **2014**, *20*, 3229–3237. [CrossRef]
11. Moyano, F.E.; Manzoni, S.; Chenu, C. Responses of soil heterotrophic respiration to moisture availability: An exploration of processes and models. *Soil Biol. Biochem.* **2013**, *59*, 72–85. [CrossRef]
12. Keiluweit, M.; Nico, P.S.; Kleber, M.; Fendorf, S. Are oxygen limitations under recognized regulators of organic carbon turnover in upland soils? *Biogeochemistry* **2016**, *127*, 157–171. [CrossRef]
13. Tian, Q.; Wang, D.; Tang, Y.; Li, Y.; Wang, M.; Liao, C.; Liu, F. Topographic controls on the variability of soil respiration in a humid subtropical forest. *Biogeochemistry* **2019**, *145*, 177–192. [CrossRef]
14. Richardson, A.D.; Carbone, M.S.; Keenan, T.F.; Czimczik, C.I.; Hollinger, D.Y.; Murakami, P.; Schaberg, P.G.; Xu, X. Seasonal dynamics and age of stemwood nonstructural carbohydrates in temperate forest trees. *New Phytol.* **2013**, *197*, 850–861. [CrossRef]
15. Voriskova, J.; Brabcova, V.; Cajthaml, T.; Baldrian, P. Seasonal dynamics of fungal communities in a temperate oak forest soil. *New Phytol.* **2014**, *201*, 269–278. [CrossRef]
16. Ma, Z.; Chen, H.Y.H. Positive species mixture effects on fine root turnover and mortality in natural boreal forests. *Soil Biol. Biochem.* **2018**, *121*, 130–137. [CrossRef]
17. Khlifa, R.; Paquette, A.; Messier, C.; Reich, P.B.; Munson, A.D. Do temperate tree species diversity and identity influence soil microbial community function and composition? *Ecol. Evol.* **2017**, *7*, 7965–7974. [CrossRef]
18. Mestre, L.; Toro-Manríquez, M.; Soler, R.; Huertas-Herrera, A.; Martínez-Pastur, G.; Lencinas, M.V. The influence of canopy-layer composition on understory plant diversity in southern temperate forests. *For. Ecosyst.* **2017**, *4*, 6. [CrossRef]
19. Stell, E.; Warner, D.; Jian, J.; Bond-Lamberty, B.; Vargas, R. Spatial biases of information influence global estimates of soil respiration: How can we improve global predictions? *Glob. Change Biol.* **2021**, *27*, 3923–3938. [CrossRef] [PubMed]
20. Kara, Ö.; Bolat, İ.; Çakıroğlu, K.; Öztürk, M. Plant canopy effects on litter accumulation and soil microbial biomass in two temperate forests. *Biol. Fertil. Soils* **2008**, *45*, 193–198. [CrossRef]
21. Vargas, R.; Allen, M.F. Environmental controls and the influence of vegetation type, fine roots and rhizomorphs on diel and seasonal variation in soil respiration. *New Phytol.* **2008**, *179*, 460–471. [CrossRef]
22. Murphy, M.; Balser, T.; Buchmann, N.; Hahn, V.; Potvin, C. Linking tree biodiversity to belowground process in a young tropical plantation: Impacts on soil CO_2 flux. *For. Ecol. Manag.* **2008**, *255*, 2577–2588. [CrossRef]
23. Zhang, F.-G.; Zhang, Q.-G. Microbial diversity limits soil heterotrophic respiration and mitigates the respiration response to moisture increase. *Soil Biol. Biochem.* **2016**, *98*, 180–185. [CrossRef]
24. Allen, B.; Willner, D.; Oechel, W.C.; Lipson, D. Top-down control of microbial activity and biomass in an Arctic soil ecosystem. *Environ. Microbiol.* **2010**, *12*, 642–648. [CrossRef] [PubMed]
25. Jiang, L.; Pu, Z.; Nemergut, D.R. On the importance of the negative selection effect for the relationship between biodiversity and ecosystem functioning. *Oikos* **2008**, *117*, 488–493. [CrossRef]
26. Waldrop, M.P.; Firestone, M.K. Seasonal dynamics of microbial community composition and function in oak canopy and open grassland soils. *Microb. Ecol.* **2006**, *52*, 470–479. [CrossRef]
27. Jiang, Y.; Zhang, B.; Wang, W.; Li, B.; Wu, Z.; Chu, C. Topography and plant community structure contribute to spatial heterogeneity of soil respiration in a subtropical forest. *Sci. Total Environ.* **2020**, *733*, 139287. [CrossRef]

28. Peng, S.; Piao, S.; Wang, T.; Sun, J.; Shen, Z. Temperature sensitivity of soil respiration in different ecosystems in China. *Soil Biol. Biochem.* **2009**, *41*, 1008–1014. [CrossRef]
29. Montgomery, R.A.; Rice, K.E.; Stefanski, A.; Rich, R.L.; Reich, P.B. Phenological responses of temperate and boreal trees to warming depend on ambient spring temperatures, leaf habit, and geographic range. *Proc. Natl. Acad. Sci. USA* **2020**, *117*, 10397–10405. [CrossRef] [PubMed]
30. Zheng, Z.-M.; Yu, G.-R.; Sun, X.-M.; Li, S.-G.; Wang, Y.-S.; Wang, Y.-H.; Fu, Y.-L.; Wang, Q.-F. Spatio-Temporal Variability of Soil Respiration of Forest Ecosystems in China: Influencing Factors and Evaluation Model. *Environ. Manag.* **2010**, *46*, 633–642. [CrossRef] [PubMed]
31. Li, W.; Bai, Z.; Jin, C.; Zhang, X.; Guan, D.; Wang, A.; Yuan, F.; Wu, J. The influence of tree species on small scale spatial heterogeneity of soil respiration in a temperate mixed forest. *Sci. Total Environ.* **2017**, *590–591*, 242–248. [CrossRef]
32. Shi, B.; Gao, W.; Cai, H.; Jin, G. Spatial variation of soil respiration is linked to the forest structure and soil parameters in an old-growth mixed broadleaved-Korean pine forest in northeastern China. *Plant Soil* **2016**, *400*, 263–274. [CrossRef]
33. Balogh, J.; Pintér, K.; Fóti, S.; Cserhalmi, D.; Papp, M.; Nagy, Z. Dependence of soil respiration on soil moisture, clay content, soil organic matter, and CO2 uptake in dry grasslands. *Soil Biol. Biochem.* **2011**, *43*, 1006–1013. [CrossRef]
34. Zhu, Y.; Dyck, M.; Cai, H.-j.; Song, L.-b.; Chen, H. The effects of aerated irrigation on soil respiration, oxygen, and porosity. *J. Integr. Agric.* **2019**, *18*, 2854–2868. [CrossRef]
35. Tan, Z.-H.; Zhang, Y.-P.; Liang, N.; Song, Q.-H.; Liu, Y.-H.; You, G.-Y.; Li, L.-H.; Yu, L.; Wu, C.-S.; Lu, Z.-Y.; et al. Soil respiration in an old-growth subtropical forest: Patterns, components, and controls. *J. Geophys. Res. Atmos.* **2013**, *118*, 2981–2990. [CrossRef]
36. Augusto, L.; Boča, A. Tree functional traits, forest biomass, and tree species diversity interact with site properties to drive forest soil carbon. *Nat. Commun.* **2022**, *13*, 1097. [CrossRef] [PubMed]
37. Chen, X.; Chen, H.Y.H. Plant diversity loss reduces soil respiration across terrestrial ecosystems. *Glob. Change Biol.* **2019**, *25*, 1482–1492. [CrossRef] [PubMed]
38. Lange, M.; Eisenhauer, N.; Sierra, C.A.; Bessler, H.; Engels, C.; Griffiths, R.I.; Mellado-Vázquez, P.G.; Malik, A.A.; Roy, J.; Scheu, S.; et al. Plant diversity increases soil microbial activity and soil carbon storage. *Nat. Commun.* **2015**, *6*, 6707. [CrossRef] [PubMed]
39. Sanaei, A.; Sayer, E.J.; Yuan, Z.; Lin, F.; Fang, S.; Ye, J.; Liu, S.; Hao, Z.; Wang, X. Soil Stoichiometry Mediates Links Between Tree Functional Diversity and Soil Microbial Diversity in a Temperate Forest. *Ecosystems* **2022**, *25*, 291–307. [CrossRef]
40. Zhou, G.; Zhou, X.; Liu, R.; Du, Z.; Zhou, L.; Li, S.; Liu, H.; Shao, J.; Wang, J.; Nie, Y.; et al. Soil fungi and fine root biomass mediate drought-induced reductions in soil respiration. *Funct. Ecol.* **2020**, *34*, 2634–2643. [CrossRef]
41. Delgado-Baquerizo, M.; Fry, E.L.; Eldridge, D.J.; de Vries, F.T.; Manning, P.; Hamonts, K.; Kattge, J.; Boenisch, G.; Singh, B.K.; Bardgett, R.D. Plant attributes explain the distribution of soil microbial communities in two contrasting regions of the globe. *New Phytol.* **2018**, *219*, 574–587. [CrossRef]
42. Finke, D.L.; Snyder, W.E. Niche Partitioning Increases Resource Exploitation by Diverse Communities. *Science* **2008**, *321*, 1488–1490. [CrossRef]
43. Liu, Y.-R.; Delgado-Baquerizo, M.; Wang, J.-T.; Hu, H.-W.; Yang, Z.; He, J.-Z. New insights into the role of microbial community composition in driving soil respiration rates. *Soil Biol. Biochem.* **2018**, *118*, 35–41. [CrossRef]
44. Hagenbo, A.; Hadden, D.; Clemmensen, K.E.; Grelle, A.; Manzoni, S.; Mölder, M.; Ekblad, A.; Fransson, P. Carbon use efficiency of mycorrhizal fungal mycelium increases during the growing season but decreases with forest age across a Pinus sylvestris chronosequence. *J. Ecol.* **2019**, *107*, 2808–2822. [CrossRef]
45. Heinemeyer, A.; Hartley, I.P.; Evans, S.P.; Carreira DE LA Fuente, J.A.; Ineson, P. Forest soil CO_2 flux: Uncovering the contribution and environmental responses of ectomycorrhizas. *Glob. Change Biol.* **2007**, *13*, 1786–1797. [CrossRef]
46. Li, R.; Yang, Q.; Guan, X.; Chen, L.; Wang, Q.; Wang, S.; Zhang, W. High quality litters with faster initial decomposition produce more stable residue remaining in a subtropical forest ecosystem. *CATENA* **2022**, *213*, 106134. [CrossRef]
47. Yoon, T.K.; Noh, N.J.; Han, S.; Lee, J.; Son, Y. Soil moisture effects on leaf litter decomposition and soil carbon dioxide efflux in wetland and upland forests. *Soil Sci. Soc. Am. J.* **2014**, *78*, 1804–1816. [CrossRef]
48. Yuan, Z.; Ali, A.; Ruiz-Benito, P.; Jucker, T.; Mori, A.S.; Wang, S.; Zhang, X.; Li, H.; Hao, Z.; Wang, X. Above-and below-ground biodiversity jointly regulate temperate forest multifunctionality along a local-scale environmental gradient. *J. Ecol.* **2020**, *108*, 2012–2024. [CrossRef]
49. Wang, X.G.; Comita, L.S.; Hao, Z.Q.; Davies, S.J.; Ye, J.; Lin, F.; Yuan, Z.Q. Local-Scale Drivers of Tree Survival in a Temperate Forest. *PLoS ONE* **2012**, *7*, e29469. [CrossRef] [PubMed]
50. Wu, J.; Guan, D.; Wang, M.; Pei, T.; Han, S.; Jin, C. Year-round soil and ecosystem respiration in a temperate broad-leaved Korean Pine forest. *For. Ecol. Manag.* **2006**, *223*, 35–44. [CrossRef]
51. Wang, X.; Wiegand, T.; Hao, Z.; Li, B.; Ye, J.; Lin, F. Species associations in an old-growth temperate forest in north-eastern China. *J. Ecol.* **2010**, *98*, 674–686. [CrossRef]
52. Condit, R.; Foster, R.B.; Hubbell, S.P.; Sukumar, R.; Leigh, E.G.; Manokaran, N.; de Lao, S.L.; LaFrankie, J.V.; Ashton, P.S. Assessing forest diversity on small plots: Calibration using species individual curves from 50-ha plots. In Proceedings of the International Symposium on Measuring and Monitoring Forest Biological Diversity, the International Network of Biodiversity Plots, Washington, DC, USA, 23–25 May 1995; pp. 247–268.
53. Harms, K.E.; Condit, R.; Hubbell, S.P.; Foster, R.B. Habitat associations of trees and shrubs in a 50-ha neotropical forest plot. *J. Ecol.* **2001**, *89*, 947–959. [CrossRef]

54. ArchMiller, A.A.; Samuelson, L.J.; Li, Y. Spatial variability of soil respiration in a 64-year-old longleaf pine forest. *Plant Soil* **2016**, *403*, 419–435. [CrossRef]
55. Lu, R.K. *Analysis Method of Soil Agricultural Chemistry*; China Agricultural Science and Technology Press: Beijing, China, 2000.
56. Mohd Yusoff, M.Z.; Hu, A.; Feng, C.; Maeda, T.; Shirai, Y.; Hassan, M.A.; Yu, C.-P. Influence of pretreated activated sludge for electricity generation in microbial fuel cell application. *Bioresour. Technol.* **2013**, *145*, 90–96. [CrossRef] [PubMed]
57. Zhang, L.; Zhang, H.; Wang, Z.; Chen, G.; Wang, L. Dynamic changes of the dominant functioning microbial community in the compost of a 90-m3 aerobic solid state fermentor revealed by integrated meta-omics. *Bioresour. Technol.* **2016**, *203*, 1–10. [CrossRef] [PubMed]
58. Caporaso, J.G.; Lauber, C.L.; Walters, W.A.; Berg-Lyons, D.; Huntley, J.; Fierer, N.; Owens, S.M.; Betley, J.; Fraser, L.; Bauer, M.; et al. Ultra-high-throughput microbial community analysis on the Illumina HiSeq and MiSeq platforms. *ISME J.* **2012**, *6*, 1621–1624. [CrossRef]
59. Edgar, R.C. UPARSE: Highly accurate OTU sequences from microbial amplicon reads. *Nat. Methods* **2013**, *10*, 996–998. [CrossRef]
60. Yao, F.; Yang, S.; Wang, Z.; Wang, X.; Ye, J.; Wang, X.; DeBruyn, J.M.; Feng, X.; Jiang, Y.; Li, H. Microbial Taxa Distribution Is Associated with Ecological Trophic Cascades along an Elevation Gradient. *Front. Microbiol.* **2017**, *8*, 2071. [CrossRef]
61. Townshend, J.L. A Modification and Evaluation of the Apparatus for the Oostenbrink Direct Cottonwool Filter Extraction Method1). *Nematologica* **1963**, *9*, 106–110. [CrossRef]
62. Yeates, G.W.; Bongers, T.; De Goede, R.G.; Freckman, D.W.; Georgieva, S. Feeding habits in soil nematode families and genera—An outline for soil ecologists. *J. Nematol.* **1993**, *25*, 315.
63. Guan, P.; Zhang, X.; Yu, J.; Cheng, Y.; Li, Q.; Andriuzzi, W.S.; Liang, W. Soil microbial food web channels associated with biological soil crusts in desertification restoration: The carbon flow from microbes to nematodes. *Soil Biol. Biochem.* **2018**, *116*, 82–90. [CrossRef]
64. Ali, A. Forest stand structure and functioning: Current knowledge and future challenges. *Ecol. Indic.* **2019**, *98*, 665–677. [CrossRef]
65. Oksanen, J.; Blanchet, F.G.; Friendly, M.; Kindt, R.; Legendre, P.; McGlinn, D.; Minchin, P.; O'Hara, R.B.; Simpson, G.; Solymos, P.; et al. *Vegan Community Ecology Package Version 2. 5–7 November 2020*; R Project for Statistical Computing: Vienna, Austria, 2020.
66. Bartoń, K. *MuMIn: Multi-Model Inference. R package version 1.43.17–14 April 2020*; R Project for Statistical Computing: Vienna, Austria, 2020.
67. Rosseel, Y. lavaan: An R Package for Structural Equation Modeling. *J. Stat. Softw.* **2012**, *48*, 1–36. [CrossRef]
68. Bivand, R.S.; Wong, D.W.S. Comparing implementations of global and local indicators of spatial association. *TEST* **2018**, *27*, 716–748. [CrossRef]
69. Li, H.; Calder, C.A.; Cressie, N. Beyond Moran's I: Testing for Spatial Dependence Based on the Spatial Autoregressive Model. *Geogr. Anal.* **2007**, *39*, 357–375. [CrossRef]
70. Yuan, Z.; Ali, A.; Jucker, T.; Ruiz-Benito, P.; Wang, S.; Jiang, L.; Wang, X.; Lin, F.; Ye, J.; Hao, Z.; et al. Multiple abiotic and biotic pathways shape biomass demographic processes in temperate forests. *Ecology* **2019**, *100*, e02650. [CrossRef] [PubMed]

plants

MDPI

Article

Differences in Density Dependence among Tree Mycorrhizal Types Affect Tree Species Diversity and Relative Growth Rates

Boliang Wei [1], Lei Zhong [1,2], Jinliang Liu [3], Fangdong Zheng [2], Yi Jin [4], Yuchu Xie [1], Zupei Lei [2], Guochun Shen [5] and Mingjian Yu [1,*]

1 College of Life Sciences, Zhejiang University, Hangzhou 310058, China
2 Zhejiang Wuyanling National Nature Reserve Management Bureau, Taishun 325500, China
3 College of Life and Environmental Science, Wenzhou University, Wenzhou 325035, China
4 State Key Laboratory of Plant Physiology and Development in Guizhou Province, School of Life Sciences, Guizhou Normal University, Guiyang 550025, China
5 Zhejiang Tiantong Forest Ecosystem National Observation and Research Station, Center for Global Change and Ecological Forecasting, School of Ecological and Environmental Sciences, East China Normal University, Shanghai 200241, China
* Correspondence: fishmj@zju.edu.cn; Tel.: +86-0571-88206469

Abstract: Conspecific negative density dependence (CNDD) may vary by tree mycorrhizal type. However, whether arbuscular mycorrhizal (AM)-associated tree species suffer from stronger CNDD than ectomycorrhizal (EcM) and ericoid mycorrhizal (ErM)-associated tree species at different tree life stages, and whether EcM tree species can promote AM and ErM saplings and adults growth, remain to be studied. Based on the subtropical evergreen broad-leaved forest data in eastern China, the generalized linear mixed-effects model was used to analyze the effects of the conspecific density and heterospecific density grouped by symbiont mycorrhizal type on different tree life stages of different tree mycorrhizal types. The results showed that compared to other tree mycorrhizal types at the same growth stage, EcM saplings and AM adults experienced stronger CNDD. Heterospecific EcM density had a stronger positive effect on AM and ErM individuals. Species diversity and average relative growth rate (RGR) first increased and then decreased with increasing basal area (BA) ratios of EcM to AM tree species. These results suggested that the stronger CNDD of EcM saplings and AM adults favored local species diversity over other tree mycorrhizal types. The EcM tree species better facilitated the growth of AM and ErM tree species in the neighborhood, increasing the forest carbon sink rate. Interestingly, species diversity and average RGR decreased when EcM or AM tree species predominated. Therefore, our study highlights that manipulating the BA ratio of EcM to AM tree species will play a nonnegligible role in maintaining biodiversity and increasing forest carbon sink rates.

Keywords: conspecific negative density dependence; tree mycorrhizal types; heterospecific mycorrhizal tree neighbors; species diversity; relative growth rate; forest carbon sink rates

check for updates

Citation: Wei, B.; Zhong, L.; Liu, J.; Zheng, F.; Jin, Y.; Xie, Y.; Lei, Z.; Shen, G.; Yu, M. Differences in Density Dependence among Tree Mycorrhizal Types Affect Tree Species Diversity and Relative Growth Rates. *Plants* **2022**, *11*, 2340. https://doi.org/10.3390/plants11182340

Academic Editor: Xugao Wang

Received: 9 August 2022
Accepted: 5 September 2022
Published: 7 September 2022

1. Introduction

Biodiversity is rapidly decreasing due to human impacts [1–4]. The release of large amounts of CO_2 exacerbates the greenhouse effect and further accelerates the extinction of certain species [5–8]. As important ecosystems on earth, forests play an important role in maintaining species diversity and carbon sequestration and storage. However, research finds that forest productivity declines as species diversity declines [9–11]. Therefore, in the context of global change, protecting and improving tree species diversity is crucial in increasing the forest productivity and slowing down the rate of species extinction.

Ecologists have proposed a plethora of theories and hypotheses to explain the mechanisms by which species diversity is maintained [12–16]. Among them, conspecific negative density dependence (CNDD) is an ecological process in which individuals have increased mortality and slowed growth rates surrounded by high-density conspecific neighbors due to the accumulation of host-specific natural enemies, thus promoting the coexistence of diverse species [17–20]. Research shows that symbiotic fungi that form mycorrhizae with plant roots and soil-borne pathogens that cause plant disease can play important but distinct roles in CNDD [21–24].

Mycorrhizal fungi play an important role in maintaining plant diversity and improving ecosystem function [24,25]. These symbiotic fungi provide up to 80% of the plant's needs for nitrogen and phosphorus to the host plant in exchange for carbohydrates [26,27]. Almost all woody plants form symbiotic relationships with arbuscular mycorrhizal (AM), ectomycorrhizal (EcM), or ericoid mycorrhizal (ErM) fungi [28], which we refer to as AM, EcM, or ErM tree species. Previous studies have found that symbiotic fungi attenuate the strength of CNDD and improve individual survival by promoting plant nutrient acquisition and resistance to soil-borne pathogens [29,30]. Furthermore, trees associated with different mycorrhizal fungi have different capacities for plant nutrient uptake and pathogen defense [27,31–33]. Therefore, different tree mycorrhizal types can play different roles in regulating species diversity and regulating forest carbon sink rates.

Since AM tree species usually experience stronger CNDD than EcM tree species, species diversity around AM tree species is usually higher [34,35], while EcM tree species generally maintain low-diversity, monodominant forests [36,37]. However, Qin et al. [38] found that AM saplings had stronger CNDD than EcM saplings, while AM juveniles had the same CNDD as EcM juveniles. This means that the CNDD of different tree mycorrhizal types varies with tree life stage. However, it is unclear whether this variation affects species diversity.

Unlike the negative effects of conspecific individuals, the existence of heterospecific neighbors alleviates the strength of CNDD and promotes the survival of focal individuals due to herd immunity effects [39–41]. Since the strength and direction of impacts vary by tree-mycorrhizal-type neighbors, mixing all heterospecific individuals together would overlook the different effects of these heterospecific mycorrhizal tree neighbors [34,42,43]. Therefore, it is necessary to group heterospecific species according to mycorrhizal symbionts. While neighborhood EcM tree species can improve seedling survival compared to AM tree species [29], it is unclear how different heterospecific mycorrhizal tree neighbors affect the growth of saplings and adults, and whether this influence will affect the forest carbon sink rate.

Exploring the differences in CNDD among AM, EcM and ErM tree species at sapling and adult stages will help to reveal the role of mycorrhizal associations in maintaining species diversity and regulating forest carbon sink rates. Here, we obtained two census data from a 9 ha (300 m × 300 m) subtropical forest dynamic plot in Wuyanling National Nature Reserve, Zhejiang Province, eastern China, to study the effects of neighborhood factors (conspecific density and heterospecific density grouped by symbiont mycorrhizal types) on the survival and growth of different tree mycorrhizal types along tree life stages. We aimed to explore the following questions: (1) Do AM tree species suffer from stronger CNDD than EcM or ErM tree species along tree life stages? (2) Do neighborhood EcM tree species promote the growth and survival of different tree mycorrhizal types at different tree life stages? (3) How does the variance in the strength of CNDD in different tree mycorrhizal types affect species diversity and forest carbon sink rates?

2. Methods

2.1. Study Site

This study was conducted in Wuyanling National Natural Reserve (119°37′08″–119°50′00″ E, 27°20′52″–27°48′39″ N), Taishun County, Zhejiang Province, eastern China. The reserve is approximately 18,861.5 ha. The mean annual temperature is 15.2 °C. The mean annual precipitation is 2195.8 mm, mostly between May and June. According to the records of the Shangfengxiang Meteorological Station (1040 m above sea level) near the study site, the mean annual temperature in the nearby area is 14.0 °C, with the lowest mean monthly temperature in January (4.0 °C) and the highest in July (23.0 °C), and the extreme lowest temperature is −8.9 °C. The frost period begins in early October and ends in early April of the following year. Frost-free days are about 210 days, and the sunshine rate is 38% [44].

In 2013, we established a 9 ha (300 m × 300 m) forest dynamic plot (119°40′13.73″ E, 27°42′20.27″ N) in an evergreen broad-leaved forest in the reserve (Figure 1). All trees with diameter at breast height (DBH, 1.3 m) ≥ 1 cm were tagged, identified to the species level, mapped and measured according to standard CTFS-ForestGEO protocols [45]. The second census was completed in 2018. The plot is 869 m to 1144 m above sea level. According to the second census, there were 63,158 free-standing woody plant individuals with DBH ≥ 1 cm in the plot, belonging to 52 families, 94 genera and 192 species. The dominant canopy species are *Castanopsis eyrei* (Fagaceae), *Cyclobalanopsis stewardiana* (Fagaceae) and *Schima superba* (Theaceae).

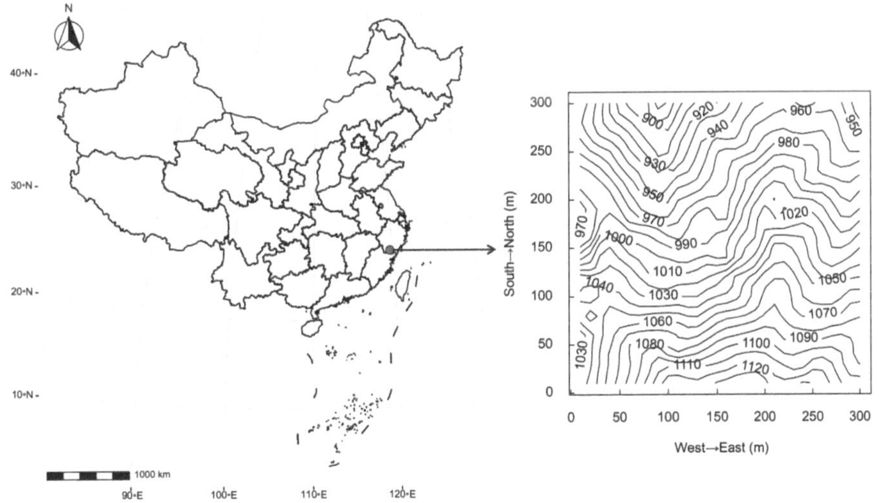

Figure 1. Location and contour map of the 9 ha forest dynamic plot in Wuyanling National Natural Reserve, eastern China.

2.2. Focal Species and Mycorrhizal Associations

We assigned each individual to one of two life stages (saplings or adults) according to LaManna et al. [46], Liu et al. [47] and Pu and Jin [48] in subtropical forests. Saplings were defined as individuals with 1 cm ≤ DBH < 2 cm for shrubs, 1 cm ≤ DBH < 5 cm for understory tree species, and 1 cm ≤ DBH < 10 cm for canopy tree species. Individuals with DBH larger than a sapling were defined as adults. The life forms of these species were classified according to the Flora of China [49] and the Flora of Zhejiang [50] (Table S1). For this study, we focused on the census between 2013 and 2018. Survival information was recorded as 1 if the individual was alive and 0 if the individual was dead. We calculated the relative growth rate (RGR) for each individual in the 5-year census interval from 2013 to 2018. RGR was calculated as $(\log(BA_{t+\Delta t}) - \log(BA_t))/\Delta t$, where BA indicates the sum of the basal area (BA) of an individual at successive time steps t.

Mycorrhizal types of plant species were determined according to published literature and the FungalRoot data set [51]. In the absence of information on the mycorrhizal type of a given species, we referred to the mycorrhizal type of its congeners (Table S1) [52]. In total, we obtained 146 arbuscular mycorrhizal (AM) species, 24 ectomycorrhizal (EcM) species and 9 ericoid mycorrhizal (ErM) species (Table 1).

Table 1. Summary information on species mycorrhizal types.

Mycorrhizal Types	Richness	No. Saplings	No. Adults	Survival Rate	Relative Abundance	Relative Basal Area
AM	146	21,623	9932	85.26%	68.78%	49.06%
EcM	24	2721	3104	84.22%	12.70%	44.34%
ErM	9	6667	1836	92.07%	18.53%	6.60%

2.3. Neighborhood Factors

Four neighborhood factors (NF) were calculated for each focal individual: density of conspecific neighbors (Con), density of heterospecific AM neighbors (HetAM), density of heterospecific EcM neighbors (HetEcM), density of heterospecific ErM neighbors (HetErM). NF was defined as the distance-weighted (Dist) sum of the BAs of conspecific or heterospecific neighbors found within a certain radius (r) of each focal individual, divided by the circular area (πr^2). To account for the potentially nonlinear nature of local biotic interactions, we introduced the exponent c as Equation (20) in Detto et al. [53] to calculate NF, where we set 10 c from 0.1 to 1, and selected the c value with the maximum likelihood value (Figures S1 and S2 and Table S2). NF was calculated as:

$$\mathrm{NF}_i = \left(\frac{1}{\pi r^2} \sum_{j=1}^{n} \frac{BA_j}{Dist_{ij}} \right)^c \tag{1}$$

where n is the number of neighbors within radius r, BA_j is the basal area of neighbor j, $Dist_{ij}$ is the distance between focal individual i and its neighbor j.

2.4. Statistical Analyses

We used generalized linear mixed-effects models (GLMMs) [54] with a binomial error distribution to quantify the effect of local neighbors on individual survival probability. We used linear mixed-effects models (LMMs) [55] to assess the influence of neighborhood density on individual RGR. The fixed effects of models included log-transformed individual size (i.e., DBH measured during the first census) and four scale-dependent neighborhood factors (Con, HetAM, HetEcM, and HetErM). To account for spatial autocorrelation and interspecific differences, quadrat (20 m × 20 m subplots) and species identity of focal individuals were considered as random effects in the model [56]. Since different species have different growth rates and may exhibit different relationships between size and survival, we allowed the effect of initial size to vary by species (i.e., random slopes) [41]. The model was summarized as follows:

$$\mathrm{Survival}_{ij} \sim \mathrm{Binomial}(p_{ij}) \tag{2}$$

$$\mathrm{RGR}_{ij} \sim \mathrm{N}(\lambda_{ij}, \sigma^2{}_{\lambda}) \tag{3}$$

$$\begin{aligned} \mathrm{Logit}\ (p_{ij}) \text{ or } \lambda_{ji} = \beta_{0j} + \beta_{1j} \times \mathrm{DBH}_{ij} + \beta_2 \times \mathrm{Con}_{ij} + \beta_3 \times \mathrm{HetAM}_{ij} + \\ \beta_4 \times \mathrm{HetEcM}_{ij} + \beta_5 \times \mathrm{HetErM}_{ijj} + \Phi_k \end{aligned} \tag{4}$$

where p_{ij} is the predicted survival probability for each individual i from species j, and λ_{ij} is the RGR for each individual i from species j. The parameter β_{0j} represents the intercept, β_{1j} represents the effect of the plant initial size (DBH); β_2, β_3, β_4 and β_5 represent the effect of four scale-dependent neighborhood factors; Φ_k represents the random effect of the quadrat.

We chose 5 m as the minimum neighborhood radius and 30 m as the maximum neighborhood radius based on previous studies [57–59]. We ran the model with 26 different neighborhood radii with a spatial resolution of 1 m (i.e., 5, 6, 7, ..., 30 m from the focal individual). The Akaike's Information Criterion (AIC) value was used to select the best-fit model across a neighborhood radius of 5 to 30 m [60]. The models with the lowest AIC values were given in the main text, and models with a neighborhood radius of 5 to 30 m are shown in Figures S3 and S4. To account for boundary effects, we excluded trees within 30 m of the plot boundary.

From the data of 255 quadrats (20 m × 20 m subplot) in this 9 ha plot, we calculated the ratio of BA of EcM species to BA of AM species (R_{EA}), Shannon–Weiner index (*H*), average RGR, total BA and increment of total BA. Generalized least squares (GLS) models with certain spatial correlation structures were used to eliminate possible influences of spatial autocorrelation [61]. Due to the relatively small proportion of ErM species in the community total BA, we mainly analyzed the effects of AM and EcM tree species on species diversity and average RGR. Due to the nonlinear effect of R_{EA} on species diversity and average RGR, a nonlinear fitting method was used in the GLS model (Figure S5 and Table S3).

All analyses were conducted in R 4.1.3 [62] using the lme4 [63], lmerTest [64] and nlme packages [65].

3. Results

3.1. Neighbor Effects on All Individuals across Tree Life Stages

We found that CNDD had significant effects on both sapling survival and RGR. The negative effects of conspecific neighbors decreased with increasing tree life stage, turning into positive effects on the survival of adults. Different mycorrhizal neighbors showed quite similar positive effects on individual survival at both the sapling and adult stages. Specifically, compared to other heterospecific neighbors, HetAM showed a greater positive effect on adults and HetEcM showed a greater positive effect on saplings (Figure 2a). However, these heterospecific neighborhood factors had a greater negative impact on the RGR of individuals (Figure 2b). Both HetAM and HetErM negatively affected the RGR of individuals at the sapling and adult stages. Whereas HetEcM showed a positive effect on saplings, but a non-significant positive effect on adults.

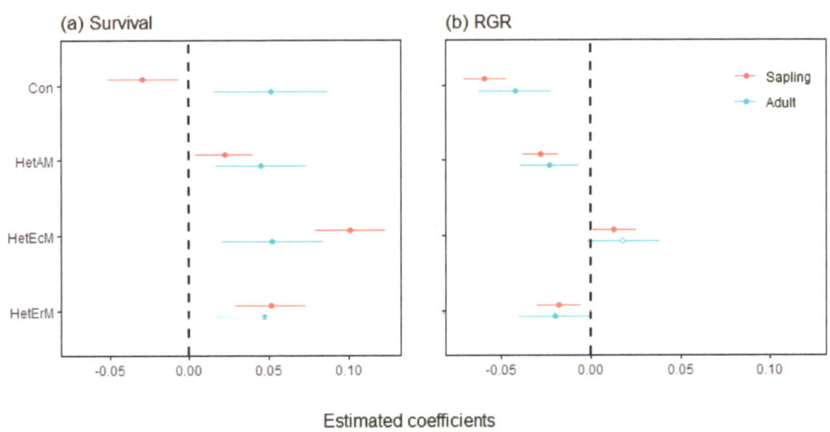

Figure 2. Coefficient estimates (±2SE) of neighborhood factors on survival (**a**) and RGR (**b**) of all individuals at the sapling and adult tree life stages. Solid circles indicate significant effects ($p < 0.05$), while open circles indicate non-significant effects. RGR, relative growth rates; Con, conspecific density; HetAM, heterospecific AM density; HetEcM, heterospecific EcM density; HetErM, heterospecific ErM density.

3.2. Neighbor Effects on Different Tree Mycorrhizal Types

At the sapling stage, the survival of different tree mycorrhizal types experienced strong CNDD (Figure 3a). The CNDD of EcM saplings was the strongest, followed by ErM saplings, and the CNDD of AM saplings was relatively weak. AM and ErM saplings were positively affected by HetEcM. Furthermore, AM and EcM saplings were positively affected by HetErM, while EcM saplings were negatively affected by HetEcM (Figure 3a).

Compared with individual survival, the neighbor effect had a certain difference in the RGR of individuals. AM and ErM saplings were still negatively affected by Con while positively affected by HetEcM. These mycorrhizal saplings were all negatively affected by heterospecific neighbors with the same mycorrhizal type. Furthermore, ErM saplings were also negatively affected by HetAM (Figure 3b).

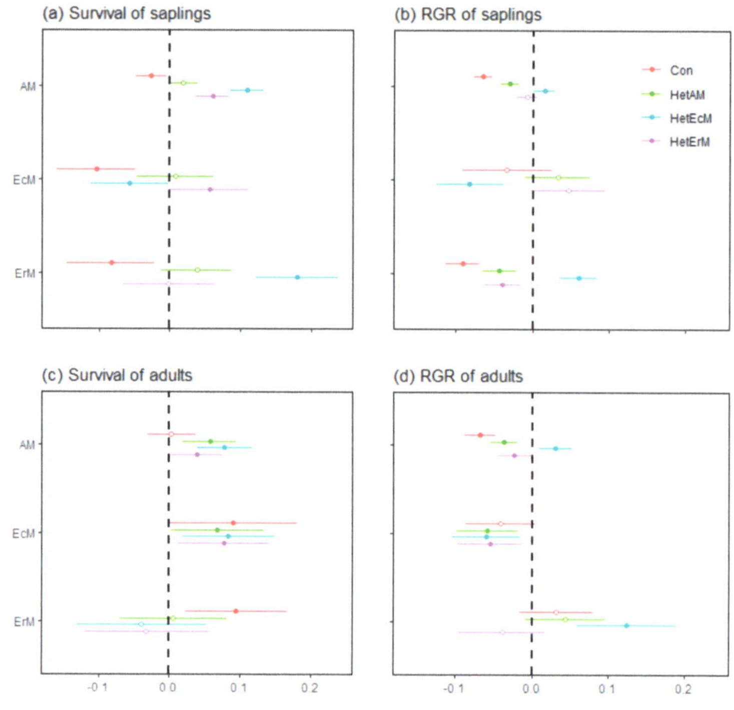

Figure 3. Estimated effects (±2SE) of neighborhood factors on survival and RGR of species mycorrhizal types at the sapling and adult tree life stages. Solid circles indicate significant effects ($p < 0.05$), while open circles indicate non-significant effects. RGR, relative growth rates; Con, conspecific density; HetAM, heterospecific AM density; HetEcM, heterospecific EcM density; HetErM, heterospecific ErM density.

During the adult stage, many neighbor effects showed positive effects on the survival of different tree mycorrhizal types. Specifically, Con shifted to positive effects on EcM and ErM adults. HetAM, HetEcM and HetErM positively affected AM and EcM adults (Figure 3c). The neighbor effect of AM adults RGR was quite similar to that of saplings, with only HetErM having a significant negative effect on AM adults. EcM adults were negatively affected by HetAM, HetEcM and HetErM, whereas ErM adults were only positively affected by HetEcM (Figure 3d).

3.3. *Relationships between R_{EA} and Species Diversity and Average RGR*

There was a nonlinearity along the square root of R_{EA} for species diversity and average RGR (Table S4). Species diversity increased rapidly as the square root of R_{EA} fell below the threshold ($R_{EA} = 0.765^2 = 0.585$; Figure 4a) and turned to decrease above the threshold. The average RGR had the same pattern as species diversity, but it changed relatively slowly compared to species diversity (threshold $R_{EA} = 1.093^2 = 1.195$; Figure 4b), while both total BA and the increment of total BA increased with the square root of R_{EA} (Figure 4c,d).

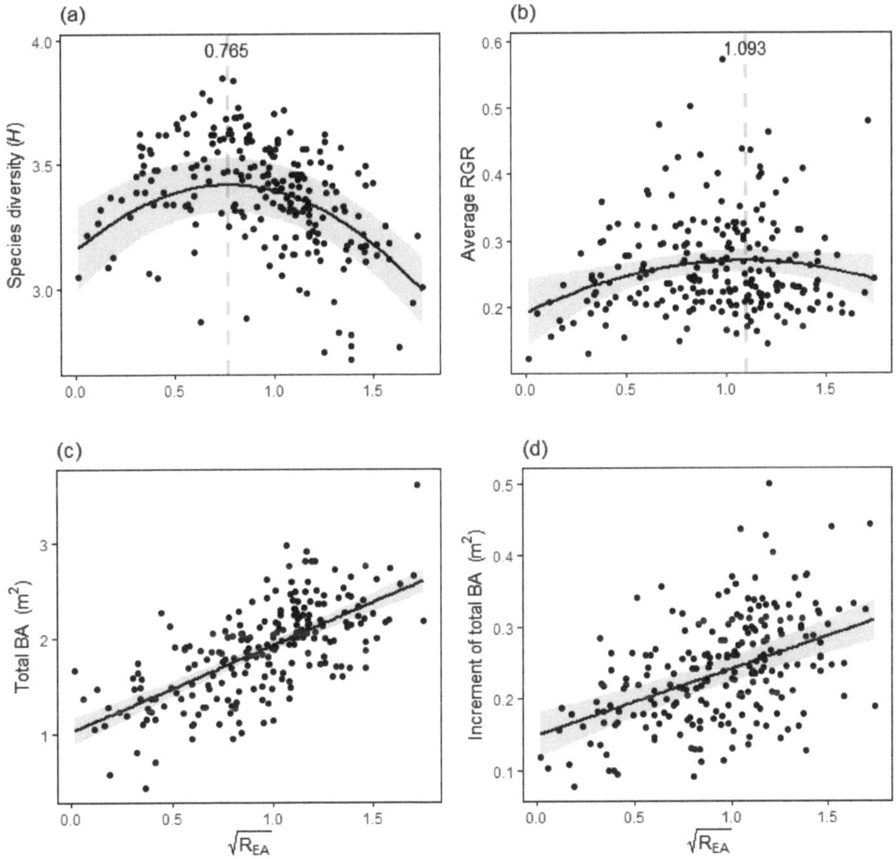

Figure 4. Correlations between the square root of R_{EA} with species diversity (*H*, Shannon–Weiner index) (**a**), average RGR (**b**), total BA (**c**) and increment of total BA (**d**). RGR, relative growth rates; BA, basal area; R_{EA}, the ratio of BA of EcM to BA of AM tree species.

4. Discussion

CNDD varied widely with tree mycorrhizal types and tree life stages. Due to the large variances in ecological characteristics among species, the impact of different neighbors on the survival and growth of focal individuals can be complicated. The results of this study showed that separating heterospecific neighbors into distinct heterospecific mycorrhizal tree neighbors has important implications for further understanding of density-dependent effects on individuals along tree life stages. In addition, compared with other tree mycorrhizal types, the stronger CNDD of EcM tree species at the sapling stage and AM tree species at the adult stage was beneficial to the increase of species diversity. This allowed species diversity to be highest when EcM tree species had a lower proportion of BA and AM tree species had a higher proportion of BA. The EcM tree species significantly improved the survival and growth of AM and ErM tree species, which, in turn, contributed to the increase in the average RGR and total BA of the neighborhood individuals. However, when AM or EcM tree species dominated at the local scale, it reduced species diversity and average RGR. Therefore, manipulating the appropriate BA ratio of EcM to AM tree species will play an important role in maintaining biodiversity and increasing the forest carbon sink rate.

4.1. The Strength of CNDD Varied among Tree Mycorrhizal Types

Similar to previous studies, in the analysis of all individuals or different tree mycorrhizal types, the CNDD of sapling survival was much stronger than that of adults [56,66]. This reflected variations in CNDD along the life history. Previous studies have suggested that AM tree species usually suffer from stronger CNDD than EcM tree species [67–69]. Unexpectedly, the survival of EcM saplings was more negatively affected by conspecific density compared with AM and ErM saplings (Figure 3a). This may be related to the accumulation of soil pathogens [70]. In this study, the abundance of 24 EcM species only accounted for 12.70% of the entire community, but the total BA of EcM species accounted for 44.34% of the entire community, and more than half of the EcM individuals were adult trees (Table 1). This means that, compared with AM and ErM species, EcM species have a higher proportion of large trees, which may accumulate more specific pathogens and lead to higher mortality of EcM saplings. This makes it hard for EcM species to recruit saplings. The death of EcM saplings leaves space for the growth of AM and ErM saplings, thus improving species diversity. However, since surviving EcM saplings may have grown in locations with fewer pathogens, their growth was not significantly affected by CNDD.

In addition, the impacts of conspecifics on the survival and growth of AM adults tended to be more negative compared with EcM and ErM tree species (Figure 3c,d). This may be due to the lower host specificity of AM fungi, which are less able to obtain resources and resist disease than the more host-specific EcM and ErM fungi [71,72]. Since EcM and ErM adults were less likely to die from CPDD (Figure 3c), species diversity decreased with increasing EcM and ErM densities. In contrast, although the survival probability of AM adults was not affected by CNDD, the growth of AM adults decreased with increasing conspecific density. The slow growth rate of AM adults provides opportunities for the growth of EcM and ErM tree species, thereby increasing species diversity. The CNDD of EcM saplings and AM adults was stronger than that of species associated with other mycorrhizal types at the same tree life stage, indicating that tree mycorrhizal types have different contributions in maintaining species diversity along tree life stages.

4.2. The Different Effects of Heterospecific Mycorrhizal Type Neighbors

Consistent with most previous studies, the effect of heterospecific mycorrhizal on species, especially saplings, was weaker than that of conspecific density (Figure 2) [73]. However, different from previous studies, when the heterospecific densities were divided into different tree mycorrhizal types, the effects of these types on individual survival and growth were quite different. The positive effect of EcM density on the survival and growth of individuals of other mycorrhizal types was significantly greater than that of AM and ErM densities (Figure 3). The mantle and Hartig nets formed by EcM fungi in roots and antibiotic compounds produced by EcM fungi protect roots from soil-borne pathogens [34,42,74]. The existence of EcM neighbors hindered the accumulation of soil-borne pathogens in the environment, which, in turn, promoted the growth and survival of AM or ErM tree species. Previous studies have suggested that EcM tree species reduce species diversity with weak CNDD [38,75], but paid little attention to the role of EcM tree species in shaping community composition and increasing forest carbon sink rates by promoting the growth and survival of other species.

In addition, heterospecific mycorrhizal tree neighbors can promote the survival of focal individuals, while heterospecific AM and ErM neighbors hamper individual growth (Figure 2). While heterospecific AM and ErM neighbors attenuate the impact of natural enemies and increase the survival of focal individuals through herd immunity effects [39–41], these individuals also compete for resources with focal individuals, thereby slowing their growth.

Except for ErM adults, the growth of both saplings and adults of other tree mycorrhizal types was significantly inhibited by heterospecific neighbors associated with the same mycorrhizal fungus type (Figure 3b,d). Allsopp and Stock [76] also found that, with the increase of conspecific density, the mass of mycorrhizal plants decreased more rapidly than that of non-mycorrhizal plants. There are three possible reasons. First, the same mycorrhizal tree species obtain resources in a similar way through mycorrhiza [77], and competition for resources slows down its growth. Second, since the proportion of colonized root length increases with density of the same mycorrhizal tree species, more carbohydrates are transporting from the host plant to mycorrhizal fungi, which slows the growth rate of the host plant [78–80]. Third, there are common pathogens that infest species related to the same mycorrhizal type [21,67,81]. When the stem density of the same mycorrhizal species increased, the content of such pathogens in the neighborhood also gradually accumulated, which affected the growth of focal individuals.

4.3. The R_{EA} Affects Species Diversity and Forest Carbon Sink Rates

These tree mycorrhizal types play different roles in the process of community assembly. This study found that species diversity was lower when AM or EcM tree species predominated. Species diversity reached its highest value only when R_{EA} reached 0.585 (this number may vary by region or time) (Figure 4a). Compared with EcM tree species, AM tree species can indeed maintain higher species diversity at larger BA ratios. This is partially consistent with the previous studies [38,69]. This may be related to the stronger CNDD effect on AM adults and more species in AM tree types. However, the diversity decreased when the BA of AM or EcM tree species continued to increase. Carteron et al. [82] also found relatively low species diversity in forests dominated by AM or EcM tree species in the U.S. Since adults occupy more space, an increase in adults that are more likely to survive with weaker CNDD will reduce the total number of individuals in the area, leading to a reduction in species diversity.

Similar to the above results, the average RGR reached the highest value when the R_{EA} was 1.195 (Figure 4b). The results indicated that EcM tree species had a greater ability to promote the growth of surrounding individuals than AM tree species, so EcM tree species were more conducive to the improvement of forest carbon sink rates. Van Der Heijden and Horton [72] found that the EcM mycorrhizal network exchanged resources more efficiently, so EcM fungi were more able to promote seedling growth than AM fungi. However, the average RGR decreased when the BA of EcM tree species was too high. This is consistent with the above results of this study (Figure 3b,d). The RGR of EcM species decreased with increasing heterospecific EcM neighbors. When the BA of EcM tree species is higher, EcM individuals with slower RGR will reduce the average RGR of all individuals. However, we found that total BA and the increment of total BA increased with increasing R_{EA} (Figure 4c,d). This means that EcM species play a relatively important role in the increment of forest carbon sink, especially in the acceleration of forest carbon sink rates. Since there were great differences in the effects of EcM and AM tree species on species diversity or average RGR, finding the optimal proportion of EcM and AM tree species for local scale assemblages will be important for biodiversity conservation and the increment of forest carbon sink rates.

5. Conclusions

The strength of CNDD varied with tree mycorrhizal types, which had different regulatory effects on species diversity at different tree life stages. EcM species significantly increased the survival and growth of AM and ErM tree species, as well as the growth and total BA of surrounding individuals. However, when AM or EcM tree species predominated, species diversity is suppressed and forest carbon sink rates are lowered. Therefore, in the context of global change, manipulating the appropriate BA ratios of AM and EcM tree species will play an important role in maintaining species diversity and increasing forest carbon sink rates. However, due to environmental differences, the optimal BA ratio of AM and EcM tree species will vary with latitude or forest type, and further exploration and research are needed.

Supplementary Materials: The following supporting information can be downloaded at: https://www.mdpi.com/article/10.3390/plants11182340/s1, Figure S1: Log-likelihood as a function of exponent *c* at 5–30 m across all individuals at sapling and adult life stages; Figure S2: Log-likelihood as a function of exponent *c* at 5–30 m in different tree mycorrhizal types at sapling and adult life stages; Figure S3: Estimates (±2SE) of neighborhood factors on survival (a) and RGR (b) of all individuals at sapling and adult life stages at 5–30 m; Figure S4: Estimated effects (±2SE) of neighborhood factors on survival and RGR of tree mycorrhizal types at sapling and adult life stages at 5–30 m; Figure S5: Correlations between BA of EcM species (BA_{EcM}) and BA of AM species (BA_{AM}) with species diversity (*H*, Shannon–Weiner index) (a and e), average RGR (b and f), total BA (c and g) and increment of total BA (d and h); Table S1: 179 focal species used in the analysis of neighborhood effects on survival and growth; Table S2: Optimal scales and *c* for survival and relative growth rate (RGR) of all individuals and tree mycorrhizal types at sapling and adult life stages at 5–30 m; Table S3: Coefficient estimates of the correlation between BA_{EcM} and BA_{AM} with species diversity, average RGR, total BA and increment of total BA; Table S4: Coefficient estimates of the correlation between the square root of R_{EA} ($\sqrt{R_{EA}}$) with species diversity, average RGR, total BA and increment of total BA.

Author Contributions: Conceptualization, B.W. and M.Y.; methodology, B.W.; formal analysis, B.W.; investigation, B.W., L.Z., J.L., F.Z. and Z.L.; writing—original draft preparation, B.W.; writing—review and editing, B.W., L.Z., J.L., Y.J., G.S. and Y.X.; visualization, B.W.; supervision, M.Y.; funding acquisition, M.Y. All authors have read and agreed to the published version of the manuscript.

Funding: This research was funded by the International Collaborative Project of the National Key R&D Plan of China, grant number 2018YFE0112800; and the "Pioneer" and "Leading Goose" R&D Program of Zhejiang, grant number 2022C02053; Scientific Research Project of Baishanzu National Park, grant number 2021KFLY10.

Institutional Review Board Statement: Not applicable.

Informed Consent Statement: Not applicable.

Data Availability Statement: Not applicable.

Acknowledgments: We are grateful to Shenhao Yao, Qianqian Wu and other lab mates of Zhejiang University, Jinwang Wang and other teachers of the Zhejiang Institute of Subtropical Crops, Zhenhong He, Wenfeng Song and other technicians of the Zhejiang Wuyanling National Nature Reserve Management Bureau, Zhenyuan He, Shaobiao Lei and other local farmers for their assistances in the field and lab. We also thank the Zhejiang Wuyanling National Nature Reserve Management Bureau for its strong support for this study.

Conflicts of Interest: The authors declare no conflict of interest.

References

1. Pimm, S.L.; Jenkins, C.N.; Abell, R.; Brooks, T.M.; Gittleman, J.L.; Joppa, L.N.; Raven, P.H.; Roberts, C.M.; Sexton, J.O. The biodiversity of species and their rates of extinction, distribution, and protection. *Science* **2014**, *344*, 1246752. [CrossRef] [PubMed]
2. Ripple, W.J.; Wolf, C.; Newsome, T.M.; Galetti, M.; Alamgir, M.; Crist, E.; Mahmoud, M.I.; Laurance, W.F.; Scientist, S.F.C. World scientists' warning to humanity: A second notice. *Bioscience* **2017**, *67*, 1026–1028. [CrossRef]
3. Thomas, C.D.; Cameron, A.; Green, R.E.; Bakkenes, M.; Beaumont, L.J.; Collingham, Y.C.; Erasmus, B.F.; De Siqueira, M.F.; Grainger, A.; Hannah, L. Extinction risk from climate change. *Nature* **2004**, *427*, 145–148. [CrossRef]
4. Urban, M.C. Accelerating extinction risk from climate change. *Science* **2015**, *348*, 571–573. [CrossRef] [PubMed]
5. Intergovernmental Panel on Climate Change. *Climate Change 2013: The Physical Science Basis: Working Group I Contribution to the Fifth Assessment Report of the Intergovernmental Panel on Climate Change*; Cambridge University Press: Cambridge, UK; New York, NY, USA, 2013.
6. Bunker, D.E.; DeClerck, F.; Bradford, J.C.; Colwell, R.K.; Perfecto, I.; Phillips, O.L.; Sankaran, M.; Naeem, S. Species loss and aboveground carbon storage in a tropical forest. *Science* **2005**, *310*, 1029–1031. [CrossRef] [PubMed]
7. Terrer, C.; Phillips, R.P.; Hungate, B.A.; Rosende, J.; Pett-Ridge, J.; Craig, M.E.; van Groenigen, K.J.; Keenan, T.F.; Sulman, B.N.; Stocker, B.D. A trade-off between plant and soil carbon storage under elevated CO_2. *Nature* **2021**, *591*, 599–603. [CrossRef] [PubMed]
8. Trisos, C.H.; Merow, C.; Pigot, A.L. The projected timing of abrupt ecological disruption from climate change. *Nature* **2020**, *580*, 496–501. [CrossRef]
9. Ammer, C. Diversity and forest productivity in a changing climate. *New Phytol.* **2019**, *221*, 50–66. [CrossRef]

10. Liang, J.; Crowther, T.W.; Picard, N.; Wiser, S.; Zhou, M.; Alberti, G.; Schulze, E.; McGuire, A.D.; Bozzato, F.; Pretzsch, H. Positive biodiversity-productivity relationship predominant in global forests. *Science* **2016**, *354*, f8957. [CrossRef]
11. Liu, X.; Trogisch, S.; He, J.; Niklaus, P.A.; Bruelheide, H.; Tang, Z.; Erfmeier, A.; Scherer-Lorenzen, M.; Pietsch, K.A.; Yang, B. Tree species richness increases ecosystem carbon storage in subtropical forests. *Proc. R. Soc. B* **2018**, *285*, 20181240. [CrossRef]
12. Chesson, P. Mechanisms of maintenance of species diversity. *Annu. Rev. Ecol. Syst.* **2000**, *31*, 343–366. [CrossRef]
13. Chesson, P. Updates on mechanisms of maintenance of species diversity. *J. Ecol.* **2018**, *106*, 1773–1794. [CrossRef]
14. HilleRisLambers, J.; Adler, P.B.; Harpole, W.S.; Levine, J.M.; Mayfield, M.M. Rethinking community assembly through the lens of coexistence theory. *Annu. Rev. Ecol. Evol. Syst.* **2012**, *43*, 2012. [CrossRef]
15. Hubbell, S. *The Unified Neutral Theory of Biodiversity and Biogeography*; Princeton University Press: Princeton, NJ, USA, 2001.
16. Vellend, M. *The Theory of Ecological Communities (MPB-57)*; Princeton University Press: Princeton, NJ, USA, 2016.
17. Comita, L.S.; Queenborough, S.A.; Murphy, S.J.; Eck, J.L.; Xu, K.; Krishnadas, M.; Beckman, N.; Zhu, Y. Testing predictions of the J anzen–C onnell hypothesis: A meta-analysis of experimental evidence for distance-and density-dependent seed and seedling survival. *J. Ecol.* **2014**, *102*, 845–856. [CrossRef] [PubMed]
18. Connell, J.H. On the role of natural enemies in preventing competitive exclusion in some marine animals and in rain forest trees. In *Dynamics of Populations*; Centre for Agricultural Publishing and Documentation: Wageningen, The Netherlands, 1971; pp. 298–312.
19. Janzen, D.H. Herbivores and the number of tree species in tropical forests. *Am. Nat.* **1970**, *104*, 501–528. [CrossRef]
20. LaManna, J.A.; Mangan, S.A.; Myers, J.A. Conspecific negative density dependence and why its study should not be abandoned. *Ecosphere* **2021**, *12*, e3322. [CrossRef]
21. Bagchi, R.; Gallery, R.E.; Gripenberg, S.; Gurr, S.J.; Narayan, L.; Addis, C.E.; Freckleton, R.P.; Lewis, O.T. Pathogens and insect herbivores drive rainforest plant diversity and composition. *Nature* **2014**, *506*, 85–88. [CrossRef]
22. Crawford, K.M.; Bauer, J.T.; Comita, L.S.; Eppinga, M.B.; Johnson, D.J.; Mangan, S.A.; Queenborough, S.A.; Strand, A.E.; Suding, K.N.; Umbanhowar, J. When and where plant-soil feedback may promote plant coexistence: A meta-analysis. *Ecol. Lett.* **2019**, *22*, 1274–1284. [CrossRef]
23. Johnson, D.J.; Clay, K.; Phillips, R.P. Mycorrhizal associations and the spatial structure of an old-growth forest community. *Oecologia* **2018**, *186*, 195–204. [CrossRef]
24. Tedersoo, L.; Bahram, M.; Zobel, M. How mycorrhizal associations drive plant population and community biology. *Science* **2020**, *367*, a1223. [CrossRef]
25. Ferlian, O.; Cesarz, S.; Craven, D.; Hines, J.; Barry, K.; Bruelheide, H.; Buscot, F.; Haider, S.; Heklau, H.; Herrmann, S.; et al. Mycorrhiza in tree diversity-ecosystem function relationships: Conceptual framework and experimental implementation. *Ecosphere* **2018**, *9*, e2226. [CrossRef]
26. Smith, S.E.; Read, D.J. *Mycorrhizal Symbiosis*, 3rd ed.; Academic Press: London, UK, 2008.
27. Van der Heijden, M.G.; Martin, F.M.; Selosse, M.A.; Sanders, I.R. Mycorrhizal ecology and evolution: The past, the present, and the future. *New Phytol.* **2015**, *205*, 1406–1423. [CrossRef] [PubMed]
28. Genre, A.; Lanfranco, L.; Perotto, S.; Bonfante, P. Unique and common traits in mycorrhizal symbioses. *Nat. Rev. Microbiol.* **2020**, *18*, 649–660. [CrossRef] [PubMed]
29. Chen, L.; Swenson, N.G.; Ji, N.; Mi, X.; Ren, H.; Guo, L.; Ma, K. Differential soil fungus accumulation and density dependence of trees in a subtropical forest. *Science* **2019**, *366*, 124–128. [CrossRef]
30. Jiang, F.; Zhu, K.; Cadotte, M.W.; Jin, G. Tree mycorrhizal type mediates the strength of negative density dependence in temperate forests. *J. Ecol.* **2020**, *108*, 2601–2610. [CrossRef]
31. Chapman, S.K.; Langley, J.A.; Hart, S.C.; Koch, G.W. Plants actively control nitrogen cycling: Uncorking the microbial bottleneck. *New Phytol.* **2006**, *169*, 27–34. [CrossRef] [PubMed]
32. Dickie, I.A.; Koele, N.; Blum, J.D.; Gleason, J.D.; McGlone, M.S. Mycorrhizas in changing ecosystems. *Botany* **2014**, *92*, 149–160. [CrossRef]
33. Phillips, R.P.; Brzostek, E.; Midgley, M.G. The mycorrhizal-associated nutrient economy: A new framework for predicting carbon–nutrient couplings in temperate forests. *New Phytol.* **2013**, *199*, 41–51. [CrossRef]
34. Bennett, J.A.; Maherali, H.; Reinhart, K.O.; Lekberg, Y.; Hart, M.M.; Klironomos, J. Plant-soil feedbacks and mycorrhizal type influence temperate forest population dynamics. *Science* **2017**, *355*, 181–184. [CrossRef]
35. Mao, Z.; Corrales, A.; Zhu, K.; Yuan, Z.; Lin, F.; Ye, J.; Hao, Z.; Wang, X. Tree mycorrhizal associations mediate soil fertility effects on forest community structure in a temperate forest. *New Phytol.* **2019**, *223*, 475–486. [CrossRef]
36. Corrales, A.; Mangan, S.A.; Turner, B.L.; Dalling, J.W. An ectomycorrhizal nitrogen economy facilitates monodominance in a neotropical forest. *Ecol. Lett.* **2016**, *19*, 383–392. [CrossRef] [PubMed]
37. McGuire, K.L. Common ectomycorrhizal networks may maintain monodominance in a tropical rain forest. *Ecology* **2007**, *88*, 567–574. [CrossRef]
38. Qin, J.; Geng, Y.; Li, X.; Zhang, C.; Zhao, X.; von Gadow, K. Mycorrhizal type and soil pathogenic fungi mediate tree survival and density dependence in a temperate forest. *For. Ecol. Manag.* **2021**, *496*, 119459. [CrossRef]
39. Peters, H.A. Neighbour-regulated mortality: The influence of positive and negative density dependence on tree populations in species-rich tropical forests. *Ecol. Lett.* **2003**, *6*, 757–765. [CrossRef]

40. Wills, C.; Green, D.R. A Genetic Herd-Immunity Model for the Maintenance of MHC Polymorphism. *Immunol. Rev.* **1995**, *143*, 263–292. [CrossRef] [PubMed]
41. Zhu, Y.; Comita, L.S.; Hubbell, S.P.; Ma, K. Conspecific and phylogenetic density-dependent survival differs across life stages in a tropical forest. *J. Ecol.* **2015**, *103*, 957–966. [CrossRef]
42. Lambers, H.; Albornoz, F.; Kotula, L.; Laliberté, E.; Ranathunge, K.; Teste, F.P.; Zemunik, G. How belowground interactions contribute to the coexistence of mycorrhizal and non-mycorrhizal species in severely phosphorus-impoverished hyperdiverse ecosystems. *Plant Soil* **2018**, *424*, 11–33. [CrossRef]
43. Liang, M.; Johnson, D.; Burslem, D.F.; Yu, S.; Fang, M.; Taylor, J.D.; Taylor, A.F.; Helgason, T.; Liu, X. Soil fungal networks maintain local dominance of ectomycorrhizal trees. *Nat. Commun.* **2020**, *11*, 2636. [CrossRef]
44. Zhong, L.; Chang-Yang, C.; Lu, P.; Gu, X.; Lei, Z.; Cai, Y.; Zheng, F.; Sun, I.; Yu, M. Community structure and species composition of the secondary evergreen broad-leaved forest: The analyses for a 9 ha forest dynamics plot in Wuyanling Nature Reserve, Zhejiang Province, East China (in Chinese). *Biodivers. Sci.* **2015**, *23*, 619. [CrossRef]
45. Anderson Teixeira, K.J.; Davies, S.J.; Bennett, A.C.; Gonzalez Akre, E.B.; Muller Landau, H.C.; Joseph Wright, S.; Abu Salim, K.; Almeyda Zambrano, A.M.; Alonso, A.; Baltzer, J.L. CTFS-Forest GEO: A worldwide network monitoring forests in an era of global change. *Global Chang. Biol.* **2015**, *21*, 528–549. [CrossRef]
46. LaManna, J.A.; Walton, M.L.; Turner, B.L.; Myers, J.A. Negative density dependence is stronger in resource-rich environments and diversifies communities when stronger for common but not rare species. *Ecol. Lett.* **2016**, *19*, 657–667. [CrossRef] [PubMed]
47. Liu, J.; Zhong, Y.; Zhong, L.; Wei, B.; Zheng, S.; Xie, Y.; Jin, Y.; Yu, M. The asymmetric relationships of the distribution of conspecific saplings and adults in forest fragments. *J. Plant Ecol.* **2020**, *13*, 398–404. [CrossRef]
48. Pu, X.; Jin, G. Conspecific and phylogenetic density-dependent survival differs across life stages in two temperate old-growth forests in Northeast China. *For. Ecol. Manag.* **2018**, *424*, 95–104. [CrossRef]
49. Editorial Committee of the Flora of China. *Flora of China*; Beijing Science Press: Beijing, China, 2018. (In Chinese)
50. Editorial Committee of Flora of Zhejiang. *Flora of Zhejiang*; Zhejiang Science Technology Press: Hangzhou, China, 1993. (In Chinese)
51. Soudzilovskaia, N.A.; Vaessen, S.; Barcelo, M.; He, J.; Rahimlou, S.; Abarenkov, K.; Brundrett, M.C.; Gomes, S.I.; Merckx, V.; Tedersoo, L. FungalRoot: Global online database of plant mycorrhizal associations. *New Phytol.* **2020**, *227*, 955–966. [CrossRef]
52. Keller, A.B.; Phillips, R.P. Leaf litter decay rates differ between mycorrhizal groups in temperate, but not tropical, forests. *New Phytol.* **2019**, *222*, 556–564. [CrossRef]
53. Detto, M.; Visser, M.D.; Wright, S.J.; Pacala, S.W. Bias in the detection of negative density dependence in plant communities. *Ecol. Lett.* **2019**, *22*, 1923–1939. [CrossRef]
54. Bolker, B.M.; Brooks, M.E.; Clark, C.J.; Geange, S.W.; Poulsen, J.R.; Stevens, M.H.H.; White, J.S. Generalized linear mixed models: A practical guide for ecology and evolution. *Trends Ecol. Evol.* **2009**, *24*, 127–135. [CrossRef]
55. Zuur, A.F.; Ieno, E.N.; Walker, N.J.; Saveliev, A.A.; Smith, G.M. *Mixed Effects Models and Extensions in Ecology with R*; Springer: New York, NY, USA, 2009; Volume 574.
56. Yao, J.; Bachelot, B.; Meng, L.; Qin, J.; Zhao, X.; Zhang, C. Abiotic niche partitioning and negative density dependence across multiple life stages in a temperate forest in northeastern China. *J. Ecol.* **2020**, *108*, 1299–1310. [CrossRef]
57. Chen, L.; Comita, L.S.; Wright, S.J.; Swenson, N.G.; Zimmerman, J.K.; Mi, X.; Hao, Z.; Ye, W.; Hubbell, S.P.; Kress, W.J. Forest tree neighborhoods are structured more by negative conspecific density dependence than by interactions among closely related species. *Ecography* **2018**, *41*, 1114–1123. [CrossRef]
58. Lu, M.; Du, H.; Song, T.; Peng, W.; Su, L.; Zhang, H.; Zeng, Z.; Wang, K.; Zeng, F. Effects of density dependence in an evergreen-deciduous broadleaf karst forest in southwest China. *For. Ecol. Manag.* **2021**, *490*, 119142. [CrossRef]
59. May, F.; Wiegand, T.; Huth, A.; Chase, J.M. Scale-dependent effects of conspecific negative density dependence and immigration on biodiversity maintenance. *Oikos* **2020**, *129*, 1072–1083. [CrossRef]
60. Burnham, K.P.; Anderson, D.R. *Model Selection and Inference: A Practical Information Theoretic Approach*; Springer Science & Business Media: New York, NY, USA, 2002.
61. Crawley, M.J. *The R Book*, 2nd ed.; John Wiley & Sons: Chichester, UK, 2012.
62. R Core Team. R: A Language and Environment for Statistical Computing. Available online: https://www.R-project.org/ (accessed on 10 March 2022).
63. Bates, D.; Mächler, M.; Bolker, B.; Walker, S. Fitting Linear Mixed-Effects Models Using lme4. *J. Stat. Softw.* **2015**, *67*, 1–48. [CrossRef]
64. Kuznetsova, A.; Brockhoff, P.B.; Christensen, R.H. LmerTest package: Tests in linear mixed effects models. *J. Stat. Softw.* **2017**, *82*, 1–26. [CrossRef]
65. Pinheiro, J.; Bates, D.; R Core Team. Nlme: Linear and Nonlinear Mixed Effects Models. R package Version 3.1-157. Available online: https://CRAN.R-project.org/package=nlme (accessed on 15 June 2022).
66. Zhu, Y.; Queenborough, S.A.; Condit, R.; Hubbell, S.P.; Ma, K.P.; Comita, L.S. Density-dependent survival varies with species life-history strategy in a tropical forest. *Ecol. Lett.* **2018**, *21*, 506–515. [CrossRef] [PubMed]
67. Bennett, J.A.; Klironomos, J. Mechanisms of plant–soil feedback: Interactions among biotic and abiotic drivers. *New Phytol.* **2019**, *222*, 91–96. [CrossRef] [PubMed]

68. Jia, S.; Wang, X.; Yuan, Z.; Lin, F.; Ye, J.; Lin, G.; Hao, Z.; Bagchi, R. Tree species traits affect which natural enemies drive the Janzen-Connell effect in a temperate forest. *Nat. Commun.* **2020**, *11*, 286. [CrossRef]
69. Jiang, F.; Lutz, J.A.; Guo, Q.; Hao, Z.; Wang, X.; Gilbert, G.S.; Mao, Z.; Orwig, D.A.; Parker, G.G.; Sang, W. Mycorrhizal type influences plant density dependence and species richness across 15 temperate forests. *Ecology* **2021**, *102*, e3259. [CrossRef]
70. Liu, Y.; He, F. Incorporating the disease triangle framework for testing the effect of soil-borne pathogens on tree species diversity. *Funct. Ecol.* **2019**, *33*, 1211–1222. [CrossRef]
71. Schroeder, J.W.; Dobson, A.; Mangan, S.A.; Petticord, D.F.; Herre, E.A. Mutualist and pathogen traits interact to affect plant community structure in a spatially explicit model. *Nat. Commun.* **2020**, *11*, 2204. [CrossRef]
72. Van der Heijden, M.G.; Horton, T.R. Socialism in soil? The importance of mycorrhizal fungal networks for facilitation in natural ecosystems. *J. Ecol.* **2009**, *97*, 1139–1150. [CrossRef]
73. Brown, A.J.; Payne, C.J.; White, P.S.; Peet, R.K. Shade tolerance and mycorrhizal type may influence sapling susceptibility to conspecific negative density dependence. *J. Ecol.* **2020**, *108*, 325–336. [CrossRef]
74. Marx, D. Ectomycorrhizae as Biological Deterrents to Pathogenic Root Infections. *Annu. Rev. Phytopathol.* **1972**, *10*, 429–454. [CrossRef] [PubMed]
75. Sasaki, T.; Konno, M.; Hasegawa, Y.; Imaji, A.; Terabaru, M.; Nakamura, R.; Ohira, N.; Matsukura, K.; Seiwa, K. Role of mycorrhizal associations in tree spatial distribution patterns based on size class in an old-growth forest. *Oecologia* **2019**, *189*, 971–980. [CrossRef] [PubMed]
76. Allsopp, N.; Stock, W.D. Density dependent interactions between VA mycorrhizal fungi and even-aged seedlings of two perennial Fabaceae species. *Oecologia* **1992**, *91*, 281–287. [CrossRef] [PubMed]
77. Zhang, H.Y.; Lü, X.T.; Hartmann, H.; Keller, A.; Han, X.G.; Trumbore, S.; Phillips, R.P. Foliar nutrient resorption differs between arbuscular mycorrhizal and ectomycorrhizal trees at local and global scales. *Global Ecol. Biogeogr.* **2018**, *27*, 875–885. [CrossRef]
78. Bennett, A.E.; Groten, K. The Costs and Benefits of Plant–Arbuscular Mycorrhizal Fungal Interactions. *Annu. Rev. Plant Biol.* **2022**, *73*, 649–672. [CrossRef]
79. Fitter, A.H. Costs and benefits of mycorrhizas: Implications for functioning under natural conditions. *Experientia* **1991**, *47*, 350–355. [CrossRef]
80. Genney, D.R.; Hartley, S.H.; Alexander, I.J. Arbuscular mycorrhizal colonization increases with host density in a heathland community. *New Phytol.* **2001**, *152*, 355–363. [CrossRef]
81. Mangan, S.A.; Schnitzer, S.A.; Herre, E.A.; Mack, K.M.; Valencia, M.C.; Sanchez, E.I.; Bever, J.D. Negative plant–soil feedback predicts tree-species relative abundance in a tropical forest. *Nature* **2010**, *466*, 752–755. [CrossRef]
82. Carteron, A.; Vellend, M.; Laliberté, E. Mycorrhizal dominance reduces local tree species diversity across US forests. *Nat. Ecol. Evol.* **2022**, *6*, 370–374. [CrossRef]